ETHNOMATHEMATICS

SUNY Series, Reform in Mathematics Education
Judith Sowder, Editor

ETHNOMATHEMATICS

Challenging Eurocentrism in Mathematics Education

EDITED BY

Arthur B. Powell
and
Marilyn Frankenstein

State University of New York Press

Published by
State University of New York Press, Albany

Printed in the United States of America

For information, address State University of New York Press, State University Plaza, Albany, N.Y., 12246

Production by Diane Ganeles
Marketing by Dana Yanulavich

Library of Congress Cataloging-in-Publication Data

Ethnomathematics : challenging eurocentrism in mathematics education / edited by Arthur B. Powell and Marilyn Frankenstein.
p. cm. — (SUNY series, reform in mathematics education)
Includes bibliographical references and index.
ISBN 0-7914-3351-X (ch : alk. paper). — ISBN 0-7914-3352-8 (pb : alk. paper)
1. Ethnomathematics. 2. Mathematics—Study and teaching. 3. Eurocentrism. I. Powell, Arthur B. II. Frankenstein, Marilyn. III. Series.
GN476.15.E85 1997
510′.7—dc20 96-24925
CIP

10 9 8 7 6 5 4 3 2 1

I, Arthur B. Powell, dedicate this book to my parents Bernice J. Powell and Arthur B. Powell, Sr. for their support of my intellectual development and to my wife, Carolyn M. Somerville, and our children, Samir René and Karma Sherie, for their support, inspiration, and love. Each has taught me invaluable lessons concerning the negative effects of Eurocentrism.

I, Marilyn Frankenstein, dedicate this book to two people who have greatly inspired my work: VANGIE DUPIGNY, a dear friend, a brilliant student who initially thought my curriculum was crazy, who didn't need the "math" but stayed, and understood everything in addition to the math I was trying to teach, now a brilliant teacher, in the schools and in the streets, from whom I continue to learn about what education can mean; LEE LORCH, a dear friend, an internationally renowned mathematician, who has acted consistently, commitedly, against racism, against sexism, against all other inhumane institutional structures, and who has used his professional standing to insist that voices such as those in this book are heard by the mathematical community.

The map on the cover shows the relative area of Africa which is larger than China, the U.S.A., India, what is commonly referred to as "Europe," Argentina, and New Zealand combined. This picture generally surprises people because the map most of us are familiar with greatly distorts land areas, enlarging "Europe" and North America and shrinking South America and Africa. Since any two-dimensional map of our three-dimensional earth must contain distortions, the choice of a map cannot be a "neutral" decision. Instead, map choice involves political struggle about which of these distortions is acceptable to us and what other understandings of ours are distorted by these false pictures.

Contents

Acknowledgments

What is particularly significant about the academic discipline of ethnomathematics is that it emerged from intellectual influences of emancipatory struggles worldwide and that interest and work in the discipline has not been limited to the academy. Many people have contributed to the emergence of ethnomathematics through their organizational and curricular efforts, thereby changing ideas and practices within the mathematics education community. Among the many, we want to acknowledge the central role of the authors in this volume. They have acted in the spheres of mathematics and mathematics education as a renowned nineteenth-century intellectual urged at the end of his critique of philosophy: The philosophers have only *interpreted* the world in various ways; the point is to *change* it.[1]

There have been many people that have contributed to the idea of this book and to its production. Jointly, we want to thank a number of people at Rutgers University-Newark. We are grateful to Eleanore Harris for, while we were both in southern Africa, she ensured that our initial proposal reached our publisher and authors, and kept us abreast of related correspondence. Since the beginning, she has provided us with tremendous clerical support. We also thank Mahendra Ramnauth, Kevin Powell, and Michael Bramlett, each of whom assisted with research and computer-related tasks; and to Daniel Ness, Teachers College, for typographical corrections.

For me, Arthur, it is a pleasure to acknowledge the incalculable support of Rutgers University (Newark College of Arts and Sciences) and inspiration of colleagues. Most research and communication costs

1. Marx's "Theses on Feuerbach," written in the spring of 1845, contains this as its eleventh and final thesis (Karl Marx and Frederick Engels, Collect Works, Volume 5, New York: International Publishers, 1976, pp. 5 and 8). Lorch (1996) writes that "In the main entrance hall to Berlin's Humboldt University [Germany's most prestigious], this quotation is engraved in marble in huge letters."

have been graciously donated by my department and college. I am indebted to my colleague and friend Professor William Jones who was the first to insist that, besides class, I examine carefully and recognize the role of culture in the African Diaspora and understand how it shapes educational practices in African American communities. Consequently, in 1986, I participated in campus seminars on the incorporation of new scholarship on race, class, gender, and culture into our college's curriculum and, as a result, developed an extensive bibliography on mathematics and culture, which led me to ethnomathematics. Finally, I owe a special and warm word of appreciation to my wife, Carolyn M. Somerville, who always supported me in this work and who debated with me the ethnomathematics of a particular chapter and eventually assisted me in preparing its inclusion.

I, Marilyn, acknowledge the vital intellectual and political support that the students, faculty, and staff at the College of Public and Community Service (CPCS) provide. The institutional philosophy of CPCS includes a vision of the activist role that education can play in creating a just, humane world, and a vision that intellectual and political activism dialectically support each other. Within that context my work as an anti-racist, anti-capitalist, anti-sexist (and so on . . .) activist academic has not merely been "tolerated," but more significantly, has been valued.

We also want to acknowledge the support of SUNY Press. We greatly appreciated Priscilla Ross, our acquisition editor, who treated our work as a serious intellectual project, not merely a marketing item. She listened to all our arguments, understood our choices, and was instrumental in insuring that our ideas were intact in the final manuscript. Also, we are grateful to Diane Ganeles, our production editor, who promptly responded to all our queries and helped us with all the details of moving the book from manuscript to publication.

For permission to reprint, we wish to thank the following:

Routledge, London, for the use with alterations in the introduction to sections 1, 2, 3, 4, and 6 of our chapter, "Toward Liberatory Mathematics: Paulo Freire's Epistemology and Ethnomathematics," which appeared in *Politics of Liberation: Paths from Freire*, edited by Peter L. McLaren and Colin Lankshear, 1994.

A special note of thanks to David Wheeler, founder and editor of *For the Learning of Mathematics*, for use of the following articles:

D'Ambrosio, U. (1985). Ethnomathematics and Its Place in the History and Pedagogy of Mathematics. *For the Learning of Mathematics* 5(1): 41–48; reprinted here as chapter 1.

Walkerdine, V. (1990). Difference, Cognition, and Mathematics

Education. *For the Learning of Mathematics* 10(3):51–5; reprinted here as chapter 9.

Harris, M. (1987). An Example of Traditional Women's Work as a Mathematics Resource. *For the Learning of Mathematics* 7(3): 26–28; reprinted here as chapter 10.

Borba, M. (1990). Ethnomathematics and Education. *For the Learning of Mathematics* 10(1): 39–43; reprinted here as chapter 12.

Fasheh, M. (1982). Mathematics, Culture and Authority. *For the Learning of Mathematics* 3(2): 2–8; reprinted here as chapter 13.

Zaslavsky, C. (1991). World Cultures in the Mathematics Class. *For the Learning of Mathematics* 11(2): 32–36; reprinted here as chapter 15.

Knijnik, G. (1992). An Ethnomathematical Approach to Mathematics Education: Culture, Mathematics, Education, and the Landless of Southern Brazil. *For the Learning of Mathematics* 13(3): 23–26; reprinted here as chapter 18 as An Ethnomathematical Approach in Mathematical Education: A Matter of Political Power.

Science History Publications, Ltd. for the use of Ascher, M. and Ascher, R. (1986). Ethnomathematics. *History of Science,* 14: 125–144; reprinted here as chapter 2.

Institute of Race Relations for the use of Joseph, G. G. (1987). Foundations of Eurocentrism in Mathematics. *Race & Class* 28(3): 13–28; reprinted here as chapter 3.

The University of Chicago Press for the use of Bernal, M. (1992). Animadversions on the Origins of Western Science. *ISIS,* 83: 596–607; reprinted here as chapter 4.

Transaction Books, New Brunswick, NJ, for the use of Lumpkin, B. (1983). Africa in the Mainstream of Mathematics History. In I. van Sertima (Ed.), *Black in Science: Ancient and Modern* (pp. 100–109); reprinted here as chapter 5.

Lawrence Erlbaum Associates, Hillsdale, NJ, for the use of Ginsburg, H. P. (1986). The Myth of the Deprived Child: New Thoughts on Poor Children. In U. Neisser (Ed.), *The School Achievement of Minority Children: New Perspectives* (pp. 169–189); reprinted here as chapter 6.

Control Publications, Melbourne, Australia, for use of Martin, B. (1988). Mathematics and Social Interests. *Search: Science and Technology in Australia and New Zealand* 19(4): 209–214; reprinted here as chapter 7.

Science & Society Quarterly, New York, NY, for use of Struik, D. J. (1948). Marx and Mathematics. *Science & Society* 12(1): 181–196; reprinted here as chapter 8.

Kluwer Academic Publishers, Dordrecht, The Netherlands, for the use of Gerdes, P. (1988). On Culture, Geometrical Thinking and Mathematics Education. *Educational Studies in Mathematics* 19(3): 137–162; reprinted here as chapter 11.

The Journal of Negro Education, Howard University, Washington, DC, for the use of Anderson, S. E. (1990). Worldmath Curriculum: Fighting Eurocentrism in Mathematics. *The Journal of Negro Education* 59(3): 348–359; reprinted here as chapter 14.

The University of Pennsylvania Press, Philadelphia, PA, for the use of a modified version of chapter 5 of Pinxten, R., van Dooren, I., and Harvey, F. (1983). *The Anthropology of Space*; reprinted here as chapter 17.

Foreword

It is an honor to be invited to write this foreword and it gives me much pleasure to do so.

Writing a foreword always gives one the opportunity to be personal. I am writing this foreword twelve years after the initial proposal for creating the International Study Group on Ethnomathematics (ISGEm), which followed the Fifth International Congress on Mathematics Education in Adelaide, Australia. I think my opening plenary talk in that meeting was influential; it marked the beginning of a new drive towards ethnomathematics. Although much was already going on, the intensity of what happened since then is impressive.

We cannot mistake recognizing the importance of this when we see the growth of ethnomathematics and its challenge to Eurocentrism, the spelling out of criticalmathematics education,[1] and the central position of mathematics in the trend towards multicultural education. All these fields of inquiry are essentially a new reflection on education, in particular, mathematics education and its social and cognitive implications; on history, in particular, the history of mathematics; on epistemology, in particular, epistemology of mathematics; and on mathematics itself.

The word *ethnomathematics* has been used for some time, mostly concerning mathematics practiced in "cultures without written expression," identified with cultures called "primitive." However, there is now a broader use of the word "ethnomathematics," which is given expression in this book. We see different attempts at defining and even proposals for alternative names, all of which is normal and indeed reaffirms the emergence of a new area of studies. The undeni-

1. *Editor's note:* This refers to another emerging area in mathematics education. An attempt to coordinate the development of these ideas is the Criticalmathematics Educators Group (CmEG). For a copy of this group's newsletter, contact one of the editors of this book.

able fact is that nowadays we search for a new, broader view of mathematics and its social implications. This collection represents an expression from a criticalmathematics education perspective of such a search.

Ethnomathematics was forged in the cauldron of experiences, reflections, delusions, and hopes of the uses of modern science, particularly mathematics, for a better quality of life for the entire human species. We all share the dream of equity and dignity in the relation of every human being to the other, of understanding our place in the cosmic reality, of achieving inner peace, and of finding a relation of equilibrium with other species and with nature as a whole. Some of our colleagues may still come with the question: "But what do mathematics and mathematics education have to do with all this." And they may go even further: "These are the domains of the social and political sciences, of philosophy and religion, of psychology and psychoanalysis, of environmental sciences and ecology, not of mathematics and mathematics education."

I see ethnomathematics as a way of going back to basics. Of course, basics in the broad sense mentioned above, with the global objectives that constitute our common dream. Some people still may not see what this has to do with such a specific mode of thought as mathematics, which has its own codes, norms, rules, and values—including rigor, precision, non-contradiction—identified with what some call "positivistic rationality." However, it is indeed conceivable to ask about other codes, norms, rules, and values belonging to alternative rationalities.

The complexity of every society, so different one from another, is responsible for the generation of codes, norms, rules, and values in the direction of organizing, classifying, comparing, and ordering the action of its individuals. Instances of these codes, norms, rules, and values are instruments of analyses, of explanations, and of actions, such as more or less, small and big, few or many, near and far, and in and out. These codes, norms, rules, and values—for instance, cardinality and ordinality, counting and measuring, and sorting and comparing—take different forms according to the cultures in which they were generated, organized, and accepted. To recover these forms and behaviors in different cultural environments has been the main thrust of ethnomathematics, which has found a common ground with the objectives of intellectual movements in psychology and anthropology.

The discovery of the other in the human species has been a major step towards social life. How does the recognition of the other manifest in our everyday behavior? What kind of relations do we establish with the other, present or remote in time and space? Communications is seen as the main instrument in these relations. Language and music, drawings and paintings, artifacts and constructions, arts and religions, divinations and sciences are all efforts to relate to the immediate environment or to transcend in space and in time—going back to the past or foreseeing and foretelling the future.

In the last three centuries, we have witnessed the development of new modes of property and of production, new ways of organizing daily life in urban communities, new relations in family and in society at large. Each of these developments carry their own specificities that connect to the dominating Eurocentric conception of mathematics. Since colonial times, this conception was imposed globally and later accepted as the pattern of "rational" human behavior. The results of this intended globalization under the control of the imperialist powers are far from being acceptable.

The concept of ethnomathematics comes, consciously or unconsciously, from these reflections. More than just learning about the styles of knowing and doing of marginalized cultures, both in nations of the periphery as well as in so-called developed countries, we are also concerned with the reasons why the behavior of human beings towards fellow human beings became so despicable: people burnt and gassed in concentration camps, cities bombed, human dignity reduced to mere rhetorics, human beings tortured by other human beings, and cultural acts censored and repressed. All of these acts have been performed officially by countries that fete a large number of Nobel laureates and boast the highest degree of scientific development—hence rationality—validated by good mathematics. For what reasons has humanity gone so low? As a mathematician, I was always in search of the humanitarian values implicit in my specialty. Couldn't we, as mathematicians, see ourselves as heralds of a new era for humankind. Why not? But for this we need to recognize the ethical components in science and mathematics. Of course, the values of mathematics are always spelled out. Even a humanist like Bertrand Russell would say that "Mathematics possesses . . . supreme beauty—a beauty cold and austere, like that of sculpture." Naturally, everyone learns in school the fundamentals of mathematics—the basics!—that have, in such a cold and austerely precise way, produced bombs and destructive technology. Ethnomathematics may help us in our quest for the affection and love in this sculpture.

I felt the need for a broader view of the history of science and mathematics, as part of the history of ideas, to help us understand why mathematics became so central in building up modern thought and modern society. It is thus natural to look into the history of mathematics to bridge our understanding of the natural, the social, the political, the economical, the religious, and many other influences on the styles and motivations in the development of mathematics.

It is not enough to say that the Romans refused Greek philosophy because they had a practical sense, that the European Middle Ages were the ages of the darkness, that modern mathematics is a result of the thoughts of a few academicians of the Renaissance and Modern Age who captured the Greek message that was preserved and conveyed by the Arabs. It is so naive to accept these as explanations.

We are naturally led to ask much broader questions: Is it possible to understand Greek mathematics without reading and interpreting Quintus Curtius Rufus' History of Alexander. And how can we talk about Medieval mathematics without an analysis of Vitruvius and of the urbanization of Europe? But how can we talk about urbanization without referring to Machu Pichu? How can we appreciate the construction of cathedrals without seeing them as intimately related to the developments of Medieval geometry. But how can we talk about cathedrals and ignore the big mosques of Mopti and Djenne? Of course, an analysis of the social and political construction of Europe is needed to understand the steps towards mathematical knowledge in the Middle Ages. This is why it is indeed relevant for us to learn of the Arthurian epics, which is closely related to the atmosphere created in the English monasteries that paved the way for Roger Bacon, Ockham, and others. Of course, Cambridge and Oxford are related to this. But the epic of Sundyatha is at least as impressive as that of Arthur, and this has to do with the emergence of the Masa. Naturally, the importance of the medieval university at Tombouctou is related to this.

A string of questions are fundamentally important to understand the cultural dynamics of the encounters. Who were the intellectuals who met the conquerors in Africa—and can we use a similar line of questioning concerning Asia and Latin America? How did these intellectuals, all having the search for explanations as their leitmotiv, succeed in organizing a colloquium? Of course, a colloquium is the outcome of every authentic intercultural encounter. If these colloquia did not happen—which is inconceivable—why was it so?

Let us consider the "heroes" of modern times. Napoleon, Sir Cecil Rhodes, and many others who are in just about every encyclopedia. But only a few encyclopedias list Samory Toure and Pancho Villa, and when they do these heroes appear as bandits—or at best rebels.

The way history has been manipulated is disgusting. This is no less true with the history of mathematics.

Broadening our view of history has everything to do with understanding the role of mathematics and mathematics education in our society. This broadened view contributes to my views of ethnomathematics as a program in the history, epistemology, and pedagogy, in particular, of mathematics.

We are now looking at ethnomathematics as focusing these issues from within. Of course, many still try to ignore the questions mentioned above that shape this approach. About twenty years ago I was asked by a colleague: "But what can Nietzsche say about the objectives of mathematics education? And who is this Paulo Freire?" Questions of this kind are recurrent. People frequently ask what do Sartre and Foucault, and Flaubert and Musil, and Dr. Kevorkian, and O. J. Simpson have to do with mathematics education?

These questions may come because these people are afraid, not of ethnomathematics, but of themselves since ethnomathematics questions what is ingrained in most scientists, particularly science educators, and in a very special way, mathematicians and mathematics educators: their absolute belief in what they practice and teach. Many exercise their profession without questioning the why and, even worse, without sharing the knowledge and views of their students, captives in the process.

There was a great transformation in the world about the same time when the word mathematics began to be used in a sense similar to what we have today, which goes back to the fourteenth and fifteenth centuries. This is no coincidence. Since then, new modes of explanation, of production and of property were imposed on all of humanity. New concepts of space and of time and new perceptions of big and small emerged.

It is impossible to understand the history of ideas without understanding the complexity of concepts and perceptions, of possibilities and difficulties, of desires and ambitions, of costs and rewards of new thinking. How can we talk about the history of mathematics without analyzing these general and broad categories?

From the conceptualization of a global world through the development of means for effective globalization, we see the major steps in the development of modern science and mathematics. The concept of a global world is intimately related to the emergence of religions of conversion, essentially the Christian and Islamic faiths. But indeed globalization is made possible with the intensification of the means of mass transportation—ships, trains, automobiles, and airplanes—and of communication—telephone, radio, and television. Indeed modern science and mathematics are intrinsic to the world as it is today. The same as Greek geometry was intrinsic to Greek thought. The same as each culture's mathematics is intrinsic to the corresponding cultural environment. If we question the current world order, there is no way of avoiding questioning the mode of thought par excellence of modern times, science and mathematics. Thus, the need to look into the political dimensions of science and mathematics.

We all are concerned with the future. But not only the more immediate times. Our concern is the future that we will not see; the future that induces us to write, to produce art, books, and knowledge; the future in which we want our name to be remembered with respect and affection and our person to be referred to as a good example. The same future that induces others to amass fortunes and build power. Among these concerns we find reasons for the multiple driving forces in the life of every human being. As educators we have to go in another direction. We walk into the future led by the hands of the new generations. Thus, our concern is the future of our children, of our students and their children, of our grandchildren and their children. Hence, education is an act of love. Mathematics education is no different. Why should it be odd to discuss these items, to bring love into our reflections, when we are talking about mathematics?

A look at history will show that these issues have been relevant to science and mathematics. However, they were gradually removed in the name of an ethics based on a set of values intrinsic to mathematics, such as rigor, precision, resilience, and others of the same kind. Yet, I am also concerned with an ethics of respect, solidarity, and cooperation. In fact, we know that so much is involved in the acquisition of knowledge. Knowledge results from the complexity of sensorial, intuitive, emotional, and rational components. Is this incompatible with mathematics? If not, how do they relate?

Ethnomathematics has everything to do with all this. As a research program, ethnomathematics invites us to look into how knowledge was built throughout history in different cultural environments. It is a comparative study of the techniques, modes, arts, and styles of explaining, understanding, learning about, and coping with the reality in different natural and cultural environments. An etymological abuse leads me to use the words, respectively, *ethno* and *mathema* for their categories of analysis and *tics* (from techne).

In historical terms, this is an analysis of the generation of knowledge, of its intellectual and social organization, and of its diffusion.

These phases in analyzing knowledge are usually studied in isolation one from another and are identified with disciplines labeled cognition, epistemology, history, sociology, and education. The holistic approach looks primarily into the interrelations of all these phases in the analysis of knowledge.

Where does ethnomathematics stand at this moment? Much work has been done. Some doctoral theses have been submitted, several books have been published, and new research is going on in several parts of the world. This book is a particularly important contribution because it brings together in one collection a diverse group of articles relating to ethnomathematics. Powell and Frankenstein's organization and analysis in extensive section introductions show how these classic articles have made significant contributions to the field. As a result, the field is moved forward, the emerging discipline of ethnomathematics is strengthened. Ethnomathematics still has for many the connotations of a non-academic practice and of an anti-science theory. Contributions, such as this book, show ethnomathematics in its true dimension as an holistic and transdisciplinary view of knowledge.

Ubiratan D'Ambrosio

São Paulo

Introduction

Arthur B. Powell and Marilyn Frankenstein

Perhaps the most telling point to mention in discussing an educational challenge to Eurocentrism is that

> Geographically, Europe does not exist, since it is only a peninsula on the vast Eurasian continent . . . Europe has always been a political and cultural definition. . . . Before the 19th century, geographers generally referred to it as "Christendom." When colonialism began to spread Western culture and religion to all corners of the globe, some British and German geographers began to delineate the eastern boundaries of a European continent. What they were actually doing was trying to draw the eastern limits of "western civilization" and the white race (Grossman, 1994, p. 39).

This is an important illustration of how false "facts" become part of our taken-for-granted knowledge of the world. That assumed "knowledge" extends beyond the mere creation of this fictitious geographic entity to proclaiming Europe's centrality in the creation of knowledge and the development of "civilization." In the Eurocentric account, Europe (and "Europeanized" areas like the U.S.A.) has always been and currently is the superior Center from which knowledge, creativity, technology, culture, and so forth flow forth to the inferior Periphery, the so-called underdeveloped countries.

Of course, there are significant intellectual challenges to Eurocentrism. Amin (1989) argues against this account by showing the central contributions of the Arab-Islamic cultures to world knowledge, and by showing how the Eurocentric version of "humanist universalism . . . negates any such universalism. For Eurocentrism has brought with it the destruction of peoples and civilizations who have resisted its spread" (p. 114). Diop (1991) demonstrates that the Greek foundations of European knowledge are themselves founded upon Black Egyptian civilization. Bernal (1987) illustrates how Eurocentrism developed in eighteenth-century Europe as the rationale for various forms of European slavery and imperialism. Blaut (1993) further shows that the successful conquest of the Americas and the spread of European colonialism, actions which were responsible for the selective development of Europe and underdevelopment of Asia, Africa, and Latin America, "is not to be explained in terms of any internal characteristics of Europe, but instead reflects the mundane realities of location" (p.2).

In spite of this scholarship, the Eurocentric myth persists and influences school curricula, even in a supposedly neutral discipline like mathematics. This book challenges the particular ways in which Eurocentrism permeates mathematics education: that the "academic" mathematics taught in schools worldwide was created solely by European males and diffused to the Periphery; that mathematical knowledge exists outside of and unaffected by culture; and that only a narrow part of human activity is mathematical and, moreover, worthy of serious contemplation as "legitimate" mathematics. This challenge has brought together knowledge from mathematics, mathematics education, history, anthropology, cognitive psychology, feminist studies, and studies of the Americas, Asia, Africa, White America, Native America, and African America to create a new discipline: ethnomathematics. This book also attempts to organize the various intellectual currents in ethnomathematics, from an anti-Eurocentric, liberatory perspective. We are critically selective, not just interested, for example, in the mathematics of Angolan sand drawings, but also in the politics of imperialism that arrested the development of this cultural tradition, and in the politics of cultural imperialism that discounts the mathematical activity involved in creating Angolan sand drawings.

This book is organized into sections that focus on specific challenges to Eurocentrism in mathematics education. Each section begins with an extensive introduction, followed by contributions we judge to be path-breaking to the development of that area of ethnomathematics. The first section, "Ethnomathematical knowledge," defines the field and points to other challenges to Eurocentrism. The second section, "Uncovering distorted and hidden history of mathematical knowledge," challenges the historiographic project of Eurocentrism. The third section, "Considering interactions between culture and mathematical knowledge," inquires into who does mathematics and how various practices influence mathematical activity. The fourth section, "Reconsidering what counts as mathematical knowledge," examines non-academic sources of mathematical knowledge. The fifth section, "Ethnomathematical praxis in the curriculum," discusses possibilities for incorporating broader notions of mathematics into traditional and nontraditional educational settings. Finally, section six, "Ethnomathematical research," analyzes research activity in the field and provides an example of a methodological approach that enables political challenges to the politics of silence and poverty.

A theme that emerges throughout these various directions of ethnomathematical thought concerns the need to reconsider the discrete categories common in academic thought. Asante (1987) argues that an underlying theoretical tenet of an Afrocentric perspective is that "oppositional dichotomies in real, every day experience do not exist" (p.14). For Freire (1970, 1982) this means breaking down the dichotomy between subjectivity and objectivity, between action and reflection, between teaching and learning, and between knowledge and its applications. For Fasheh (1989) and Adams (1983) this means that thought which is labeled "logic" and thought which is labeled "intuition" continuously and dialectically interact with each other. For D'Ambrosio (1987) this means that the notion that "there is only one underlying logic governing all thought" is too static. For Diop (1991) this means that the interactions between "logic" and "experience" change our definition of "logic" over time (p.363). For Lave (1988) this means understanding how "activity-in-setting" is seamlessly stretched across persons acting." For Diop (1991) this means that the distinctions between "Western," "Eastern," and "African" knowledge distort the human process of creating knowledge which result from interactions among humans and with the world. Throughout this book, we emphasize that underlying all these false dichotomies is the split between practical, everyday knowledge and abstract, theoretical knowledge. Understanding these dialectical interconnections, we believe, leads us to connect mathematics to all other disciplines, and to view mathematical knowledge as one aspect of humans trying to understand and act in the world. We see ethnomathematics as a powerful and insightful vehicle for conceptualizing these connections.

References

Adams III, H. H. (1983). African observers of the universe: The Sirius question. In I. Van Sertima (Ed.). *Blacks in science: Ancient and modern* (pp. 27–46). New Brunswick, NJ: Transaction.

Amin, S. (1989). *Eurocentrism.* New York: Monthly Review.

Asante, M. K. (1987). *The Afrocentric idea.* Philadelphia: Temple University.

Bernal, M. (1987). *Black Athena: The Afro-asiatic roots of classical civilization.* Vol. 1. London: Free Association.

Blaut, J. M. (1993). *The colonizer's model of the world: Geographical diffusionism and Eurocentric history.* New York: Guilford.

D'Ambrosio, U. (1987). Reflections on ethnomathematics. *International Study Group on Ethnomathematics Newsletter* 3(1):3–5.

Diop, C. A. (1991). *Civilization or barbarism: An authentic anthropology.* New York: Lawrence Hill Books.

Fasheh, M. (1989). Mathematics in a social context: Math within education as praxis versus within education as hegemony. In C. Keitel, P. Damerow, A. Bishop, and P. Gerdes (Eds.). *Mathematics, education, and society* (pp. 84–86). Paris: UNESCO.

Freire, P. (1970). *Pedagogy of the oppressed.* New York: Seabury.

———. (1982). "Education for critical consciousness." Boston College course notes taken by M. Frankenstein.

Grossman, Z. (1994). Erecting the new wall: Geopolitics and the restructuring of Europe. *Z Magazine* (March):39–45.

Lave, J. (1988). *Cognition in practice.* Cambridge, England: Cambridge.

Section I

Ethnomathematical Knowledge

Arthur B. Powell and Marilyn Frankenstein

Ethnomathematics emerged as a new conceptual category from the discourse on the interplay among mathematics, education, culture, and politics. Naturally, it has various definitions and associated perspectives; each definition and perspective and the term itself, has been debated and then rejected or embraced in scholarly journals and in other academic forums.[1] Among recent, written efforts to define and describe the terrain of ethnomathematics, two dominant positions are represented by the ideas of Ascher and Ascher (1986; Ascher 1991), and D'Ambrosio (1985/reprinted here as chapter 1, 1987, 1988, 1990).

Ascher and Ascher (1986/reprinted here as chapter 2) define ethnomathematics as "the study of the mathematical ideas of nonliterate peoples" (p. 125). While acknowledging that mathematical ideas exist in all cultures, Ascher (1991) points out that this does not imply that, across cultures, mathematical ideas are the same.

> In Western culture and among the Tshokwe of Africa, the cultural surroundings of the graph theoretical ideas are not the same, nor should we expect that they would be. The strip patterns that we see around us, those of the Incas, and those of the Maori are quite different in style, in usage, and in their other cultural linkages. Shared is the creation of strip patterns and an interest in them, but not necessarily shared is the motivation for their creation, nor the world view or aesthetic that leads to the particular strip that results (p. 186). . . .
>
> . . . [Mathematical] ideas exist in all cultures, but which ones are emphasized, how they are expressed, and their particular contexts will vary from culture to culture (p. 187). . .

> . . . The differences, however, are *not* the ability to think abstractly or logically. They are in the subjects of thought, the cultural premises, and what situations call forth which thought processes (p. 190).

These statements reveal the anthropological and mathematical roots and concerns of the Aschers' project. Their project also has ideological concerns: They intend to challenge Eurocentric historical and anthropological notions about the locus of mathematical ideas, including pernicious statements in the mathematical literature concerning the value of the mathematical ideas of nonliterate, non-Western peoples.[2] As they point out, most statements about nonliterate peoples are usually (1) in preliminary chapters in histories of mathematics or in texts on the spirit of the subject, and (2) theoretically and factually flawed (1986, p. 125). Nonliterate peoples are thought of as primitive or existing earlier along a linear evolutionary path. As such, their ideas are placed at the beginning of discussions of mathematics. In contrast, Ascher and Ascher (1975, 1981, 1986) and Ascher (1983, 1987, 1988a, 1988b, 1990, 1991), have demonstrated that certain notions of nonliterate peoples are akin to and as complex as those of modern, "Western" mathematics; they have broadened the history of mathematics by imbuing it with a multicultural, global perspective. However, circumscribing the terrain of ethnomathematics to the mathematical ideas of nonliterate, non-Western peoples, we insist, is too small a circle. The radius should be longer since much lies in the complement of the circle.

To discover the complement requires a broader perspective of ethnomathematics. Insightfully, D'Ambrosio (1985/reprinted here as chapter 1), the founder and most significant theoretician of the ethnomathematics program, points out that belief in the universality of mathematics can limit one from considering and recognizing that different modes of thought or culture may lead to different forms of mathematics, radically different ways of counting, ordering, sorting, measuring, inferring, classifying, and modeling. That is, once we abandon notions of general universality, which often cover for Eurocentric particularities, we can acquire an anthropological awareness: different cultures can produce different mathematics and the mathematics of one culture can change over time, reflecting changes in the culture. For D'Ambrosio ethnomathematics, existing at the crossroads of the history of mathematics and cultural anthropology, overcomes the Egyptian and Greek distinction between scholarly and practical mathematics, a distinction rooted in socioeconomic class differentiation (pp. 44–45). Now in the twentieth century, this distinction is manifested in the contrast between the "academic" mathematics that is taught in schools, which allows an elite to assume management of a society's productive forces, and the "everyday" mathematics, which allows individuals to function effectively in the world. On the other hand, ethnomathematics is

> the mathematics which is practised among identifiable cultural groups, such as national-tribal societies, labor groups, children of a certain age bracket, professional classes, and so on. Its identity depends largely on focuses of interest, on motivation, and on certain codes and jargons which do not belong to the realm of academic mathematics. We may go even further in this concept of ethnomathematics to include for example much of the mathematics which is currently practised by engineers, mainly calculus, which does not respond to the concept of rigor and formalism developed in academic courses of calculus (D'Ambrosio 1985, p. 45).

Here we have a conception of ethnomathematics which embraces a broader spectrum of humanity than the previous one. Within this conception, cultural groups within Western societies also have an ethnomathematics. Moreover, D'Ambrosio (1987) argues that we should neither minimize nor ignore the influence of cultural atmosphere and motivation. As with the production of other cultural products, music, for example, mathematical ideas take shape within particular contexts and which ideas are produced is connected to contextual content.

> This calls for a somewhat different way of looking into the History of Science and the epistemological foundations of scientific knowledge. It calls for an ethnological interpretation of mental processes and the recognition of different modes of thought, as well as different logics of explanation, which depend upon experiential background of the cultural group being considered. Thus we are led to disclaim the assertion that there is only one underlying logic governing all thought (p. 3).

Here, then, different cultural groups—industrial engineers, children, peasants, computer scientists, for example—have distinct ways of reasoning, of measuring, of coding, of classifying, and so on. Consequently each group has their own ethnomathematics, *including* academic mathematicians. Further, it is the informal and ad hoc aspects of ethnomathematics that broaden it to include more than academic mathematics. This point has been aptly elaborated by both Borba (1990/reprinted here as chapter 12) and Mtetwa (1992). For instance, stating that ethnomathematics is "[m]athematical knowledge expressed in the language code of a given sociocultural group," Borba points out that this implies that "[e]ven the mathematics produced by professional mathematicians can be seen as a form of ethnomathematics . . ." (p. 40). Further, he echoes the critique of universality:

> Although academic mathematics may be international in that it is currently in use in many parts of the world, it is not international in that only a small percentage of the population of the world is likely to use academic mathematics (p. 40). . . .
>
> Hence ethnomathematics should not be misunderstood as "vulgar" or "second class" mathematics, but as *different* cultural expressions of mathematical ideas (p. 41).

Beyond critiquing the imperialism of academic mathematics, Borba argues for a recognition of diverse expressions of mathematical ideas instead of one ethnomathematics dominating another. The genesis of ethnomathematical ideas depends on the cognitive practices of a culturally differentiated group, and those ideas maintain, evolve, or disappear according to the dynamics of the group and its relation to other cultural groups. At some stage, a professional class of mathematicians may decide to theorize an aspect of ethnomathematical knowledge; they appropriate it and later return it in a codified version. In this context, D'Ambrosio writes:

> We may look for examples in mathematics of the parallel development of the scientific discipline outside the established and accepted model of the profession. One such example is Dirac's delta function which, about 20 years after being in full use among physicists, was expropriated and became a mathematical object, structured by the theory of distributions (1985, p. 47).

D'Ambrosio's broader view of ethnomathematics accounts for the dialectical transformation of knowledge within and among societies. Moreover, his epistemology is consistent with Freire's (1970, 1973) in that D'Ambrosio views mathematical knowledge as dynamic and the result of human activity, not as static and ordained. Necessarily, this conception of ethnomathematics admits a critique of the historiography of mathematics (D'Ambrosio, 1988). That is, there are mathematical notions of peoples that written history has hidden, frozen, or stolen. Including these ideas makes it clear that what is labeled "Western" mathematics is more accurately called "world mathematics" (Anderson, 1990/reprinted here as chapter 14). We argue that ethnomathematics includes the mathematical ideas of peoples, manifested in written or non-written, oral or non-oral forms, many of which have been either ignored or otherwise distorted by conventional histories of mathematics. We and other mathematics educators are pushing the boundaries of both ethnomathematics and academic mathematics so that the two fields merge to encompass all of the intellectual enterprises and other actions of everyday life having to do with mathematics.[3] Fasheh goes so far as to define the underlying project of ethnomathematics as "working hard to understand the logic of other peoples, of other ways of thinking."[4]

Notes

1. One such forum is the newsletter of the International Study Group on Ethnomathematics (ISGEm). To subscribe or receive further information, contact Gloria Gilmer, 9155 North 70th Street, Milwaukee, WI 53223, U.S.A.

2. Joseph's work (1987/reprinted here as chapter 3, 1991), particularly the latter one, *The crest of the peacock: The non-European roots of mathematics,* represent other recent, significant challenges to Eurocentric historiography.

3. For example, see Ascher (1983), Crowe (1971, 1975), Gattegno (1988), Gerdes (1986, 1988a/reprinted here as chapter 11, 1988b), Harris (1987), Ginsburg (1986/reprinted here as chapter 6), Joseph (1991, 1993), and Zaslavsky (1973, 1990, 1991, 1992, 1996).

4. Fasheh made this remark during his panel presentation—"Mathematics Education in the Global Village: What can we expect from ethnomathematics?"—at the Sixth International Congress on Mathematics Education in Budapest, Hungary, July 1988.

References

Anderson, S. E. (1990). Worldmath curriculum: Fighting Eurocentrism in mathematics. *Journal of Negro Education* 59(3): 348–359.

Ascher, M. (1983). The logical-numerical system of Inca quipus. *Annals of the History of Computing* 5(3): 268–278.

———. (1987). Mu Torere: An analysis of a Maori game. *Mathematics Magazine* 60(2): 90–100.

———. (1988a). Graphs in cultures (II): A study in ethnomathematics. *Archives for History of Exact Sciences* 39(1): 75–95.

———. (1988b). Graphs in cultures: A study in ethnomathematics. *Historia Mathematica* 15(3): 201–227.

———. (1990). A river-crossing problem in cross-cultural perspective. *Mathematics Magazine* 63 (1): 26–29.

———. (1991). *Ethnomathematics: A multicultural view of mathematical ideas*. Belmont, CA: Brooks/Cole.

Ascher, M., and Ascher, R. (1975). The quipu as a visible language. *Visible Language* 9(4): 329–356.

———. (1981). *Code of the quipu: A study in media, mathematics, and culture*. Ann Arbor: University of Michigan Press.

———. (1986). Ethnomathematics. *History of Science* 24: 125–144.

Borba, M. C. (1990). Ethnomathematics and education. *For the Learning of Mathematics* 10(1): 39–43.

Crowe, D. W. (1971). The geometry of African art I: Bakuba art. *Journal of Geometry* 1: 169–182.

———. (1975). The geometry of African art II: A catalog of Benin patterns. *Historia Mathematica* 2: 253–271.

D'Ambrosio, U. (1985). Ethnomathematics and its place in the history and pedagogy of mathematics. *For the Learning of Mathematics* 5(1): 44–48.

———. (1987). Reflections on ethnomathematics. *International Study Group on Ethnomathematics Newsletter* 3(1): 3–5.

———. (1988). Ethnomathematics: A research program in the history of ideas and in cognition. *International Study Group on Ethnomathematics Newsletter* 4(1): 5–8.

———. (1990). *Etnomatemática: Arte ou técnica de explicar e conhecer* [Ethnomathematics: Art or technique of explaining and knowing]. São Paulo: Editora Atica.

Freire, P. (1970). *Pedagogy of the oppressed*. New York: Seabury.

———. (1973). *Education for critical consciousness*. New York: Seabury.

Gattegno, C. (1988). *The science of education: Part 2B: The awareness of mathematization*. New York: Educational Solutions.

Gerdes, P. (1986). How to recognize hidden geometrical thinking: A contribution to the development of anthropological mathematics. *For the Learning of Mathematics* 6(2): 10–12,17.

———. (1988a). On culture, geometrical thinking and mathematics education. *Educational Studies in Mathematics* 19: 137–162.

———. (1988b). On some possible uses of traditional Angolan sand drawing in the mathematics classroom. *Journal of the Mathematical Association of Nigeria* 18(1): 107–125.

Ginsburg, H. P. (1986). The myth of the deprived child: New thoughts on poor children. In U. Neisser (Ed.). *The school achievement of minority children: New perspectives* (pp. 169–189). Hillsdale, New Jersey: Lawrence Erlbaum Associates.

Harris, M. (1987). Mathematics and fabrics. *Mathematics Teaching* 120: 43–45.

Joseph, G. G. (1987). Foundations of eurocentrism in mathematics. *Race & Class* 28(3): 13–28.

———. (1991). *Crest of the peacock: The non-European roots of mathematics.* London: I. B. Tauris.

———. (1993). *Multicultural Mathematics.* Oxford: Oxford University.

Mtetwa, D. (1992). "Mathematics" and ethnomathematics: Zimbabwean students' view. *International Study Group on Ethnomathematics Newsletter* 7(1): 1–3.

Zaslavsky, C. (1973). *Africa counts: Number and patterns in African culture.* Boston: Prindle, Weber and Schmidt.

———. (1990). Symmetry in American folk art. *Arithmetic Teacher* 38(1): 6–12.

———. (1991). Multicultural mathematics education for the middle grades. *Arithmetic Teacher* 38(6): 8–13.

———. (1992). Multicultural mathematics: Interdisciplinary cooperative-learning activities. Portland, ME: Walch.

———. (1996). *The multicultural math classroom: Bringing in the world.* Portsmouth, NH: Heinemann.

Chapter 1

Ethnomathematics and its Place in the History and Pedagogy of Mathematics

Ubiratan D'Ambrosio

Editors's comment: In agreement with Gerdes (see chapter 16), we consider Ubiratan D'Ambrosio, a Brazilian mathematician and philosopher of mathematics education, "the intellectual father of the ethnomathematics program." Since the mid-1970s, he has presented his ethnomathematics program in both English and Portuguese in a variety of forums throughout the world. This chapter represents the first comprehensive, theoretical treatment in English of ethnomathematics. These ideas have stimulated the development of the field. This chapter first appeared in *For the Learning of Mathematics* 5(1): 41–48, in 1985. He has a more updated statement in Portuguese in D'Ambrosio (1990). At the end of the present version, the author includes a brief update.

Introductory Remarks

In this paper, we will discuss some basic issues which may lay the ground for an historical approach to the teaching of mathematics in a novel way. Our project relies primarily on developing the concept of *ethnomathematics.*

Our subject lies on the borderline between the history of mathematics and cultural anthropology. We may conceptualize ethnoscience as the study of scientific and, by extension, technological phenomena in direct relation to their social, economic, and cultural backgrounds.[1] There has been much research already on ethnoastronomy, ethnobotany, ethnochemistry, and so on. Not much has been done in ethnomathematics, perhaps because people believe in the universality of mathematics. This seems to be harder to sustain, for recent research, mainly carried on by anthropologists, shows evidence of practices which are typically mathematical, such as counting, ordering, sorting, measuring and weighing, done in radically different ways than those which are commonly taught in the school system. This has encouraged a few studies on the evolution of the concepts of mathematics in a cultural and anthropological framework. But we consider this direction to have been pursued only to a very limited and—we might say—timid extent. A basic book by R. L. Wilder which takes this approach and a recent comment on Wilder's approach by C. Smorinski[2] seem to be the most important attempts by mathematicians. On the other hand, there is a reasonable amount of literature on this by anthropologists. Making a bridge between anthropologists and historians of culture and mathematicians is an important step towards recognizing that different modes of thoughts may lead to different forms of mathematics; this is the field which we may call "ethnomathematics."

Anton Dimitriu's extensive history of logic[3] briefly describes Indian and Chinese logics merely as background for his general historical study of the logics that originated from Greek thought. We know from other sources that, for example, the concept of "the number one" is a quite different concept in the Nyaya-Vaisesika epistemology: "the number one is eternal in eternal substances, whereas two, etc., are always non-eternal," and from this proceeds an arithmetic (p. 119).[4] Practically nothing is known about the logic underlying the Inca treatment of numbers, though what is known through the study of the "quipus" suggests that they used a mixed qualitative-quantitative language.[5]

These remarks invite us to look at the history of mathematics in a broader context so as to incorporate in it other possible forms of mathematics. But we will go further than these considerations in saying that this is not a mere academic exercise, since its implications for the pedagogy of mathematics are clear. We refer to recent advances in theories of cognition which show how strongly culture and cognition are related. Although for a long time there have been indications of a close connection between cognitive mechanisms and cultural environment, a reductionist tendency, which goes back to Descartes and has to a certain extent grown in parallel with the development of mathematics, tended to dominate education until recently, implying a culture-free cognition. Recently a holistic recognition of the interpenetration of biology and culture has opened up a fertile ground of research on culture and mathematical cognition.[6] This has clear implications for mathematics education, as has been amply discussed in my books *Socio-cultural Bases for Mathematics Education,*[7] and *Several Dimensions* of Science Education.[8]

A Historical Overview of Mathematics Education

Let us look very briefly into some aspects of mathematics education throughout history. We need some sort of periodization for this overview which corresponds, to a certain extent, to major turns in the sociocultural composition of Western history. (We disregard for this purpose other cultures and civilizations.)

Up to the time of Plato, our reference is the beginning and growth of mathematics in two clearly distinct branches: what we might call "scholarly" mathematics, which was incorporated in the ideal education of Greeks, and another, which we may call "practical" mathematics, reserved to manual workers mainly. In the Egyptian origins of mathematical practice, there was the space reserved for "practical" mathematics behind it, which was taught to workers. This distinction was carried on into Greek times and Plato clearly says that "all these studies [ciphering and arithmetic, mensurations, relations of planetary orbits] into their minute details is not for the masses but for a selected few," (Laws VII, 818)[9] and "we should induce those who are to share the highest functions of State to enter upon that study of calculation and take hold of it, . . . not for the purpose of buying and selling, as if they were preparing to be merchants or hucksters" (*Republic* VII 525b).[10] This distinction between scholarly and practical mathematics, reserved for different social classes, is carried on by the Romans with the "trivium" and "quadrivium" and a practical training for laborers. In the Middle Ages, we begin to see a convergence of both in one direction: that is, practical mathematics begins to use some ideas from scholarly mathematics in the field of geometry. Practical geometry is a subject in its own right in the Middle Ages. This approximation of practical to theoretical geometry follows the translation from the Arabic of Euclid's *Elements* by Adelard of Bath, (early twelfth century). Dominicus Gomdissalinus, in his classification of sciences, says that "it would be disgraceful for someone to exercise any art and not know what it is, and what subject matter it has, and the other things that are premised of it," as cited in (p. 8).[11] With respect to ciphering and counting, changes start to take place with the introduction of Arabic numerals; the treatise of Fibonnaci (p. 481)[12] is probably the first to begin this mixing of the practical and theoretical aspects of arithmetic.

The next step in our periodization is the Renaissance when a new labor structure emerges: changes take place in the domain of architecture since drawing makes plans accessible to bricklayers, and machinery can be drawn and reproduced by others than the inventors. In painting, schools are found to be more efficient and treatises become available. The approximation is felt by scholars who start to use the vernacular for their scholarly works, sometimes writing in a nontechnical language and in a style accessible to non-scholars. The best known examples may be Galileo, and Newton, with his "Optiks."

The approximation of practical mathematics to scholarly mathematics increases in pace in the industrial era, not only for reasons of necessity in dealing with increasingly complex machinery and instruction manuals, but also for social reasons. Exclusively scholarly training would not suffice for the children of an aristocracy which had to be prepared to keep its social and economical predominance in a new order in note 12 (p. 482). The approximation of scholarly mathematics and practical mathematics begins to enter the school system, if we may so call education in these ages.

Finally, we reach a last step in this rough periodization in attaining the twentieth century and the widespread concept of mass education. More urgently than for Plato the question of *what* mathematics should be taught in mass educational systems is posed. The answer has been that it should be a mathematics that maintains the economic and social structure, reminiscent of that given to the aristocracy when a good training in mathematics was essential for preparing the elite (as advocated by Plato), and at the same time allows this elite to assume effective management of the productive sector. Mathematics is adapted and given a place as "scholarly practical" mathematics which we will call, from now on, "academic mathematics," that is, the mathematics which is taught and learned in the schools. In contrast to this, we will call *ethnomathematics* the mathematics which is practiced among identifiable cultural groups, such as national-tribal societies, labor groups, children of a certain age bracket, professional classes, and so on. Its identity depends largely on focuses of interest, on motivation, and on certain codes and jargons which do not belong to the realm of academic mathematics. We may go even further in this concept of ethnomathematics to include much of the mathematics which is currently practised by engineers, mainly calculus, which does not respond to the concept of rigor and formalism developed in academic courses of calculus. As an example, the Sylvanus Thompson approach to calculus may fit better into this category of ethnomathematics. And builders and well-diggers and shack-raisers in the slums also use examples of ethnomathematics.

Of course this concept asks for a broader interpretation of what mathematics is. Now we include as mathematics, apart from the Platonic ciphering and arithmetic, mensuration and relations of planetary orbits, the capabilities of classifying, ordering, inferring and modeling. This is a very broad range of human activities which, throughout history, have been expropriated by the scholarly establishment, formalized and codified and incorporated into what we call "academic mathematics," but which remain alive in culturally identified groups and constitute routines in their practices.

Ethnomathematics in History and Pedagogy and the Relations between Them

We would like to insist on the broad conceptualization of mathematics which allows us to identify several practices which are essentially mathematical in their nature. And we also presuppose a broad concept of *ethno-*, to include all culturally identifiable groups with their jargons, codes, symbols, myths, and even specific ways of reasoning and inferring. Of course, this comes from a concept of culture as the result of an hierarchization of behavior, from individual behavior through social behavior to cultural behavior.

The concept relies on a model of individual behavior based on the cycle . . . reality—individual—action—reality . . . , schematically shown as figure 1–1. In this holistic model we will not enter into a discussion of what is reality, or what is an individual, or what is action. We refer to.[13] We simply assume reality in a broad sense, both natural, material, social and psycho-emotional. Now, we observe that links are possible through the mechanism of information (which includes both sensorial and memory, genetic and acquired systems) which produces stimuli in the individual. Through a mechanism of reification these stimuli give rise to strategies (based on codes and models) which allow for action. Action impacts upon reality by introducing facti into this reality, both artifacts and "mentifacts." (We have

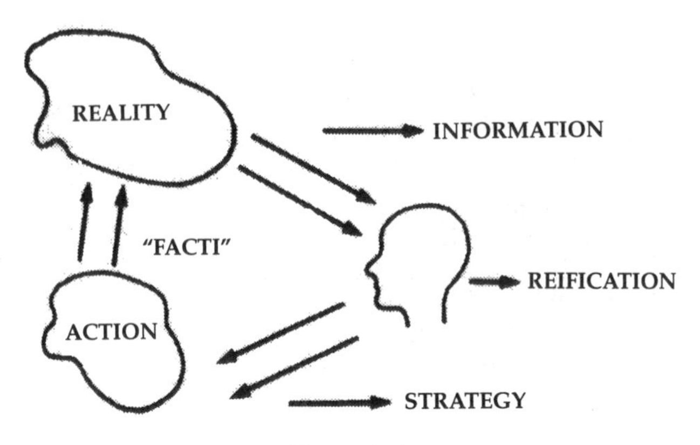

Figure 1–1.

introduced this neologism to mean all the results of intellectual action which do not materialize, such as ideas, concepts, theories, reflections, and thoughts.) These are added to reality, in the broad sense mentioned above, and clearly modify it. The concept of reification has been used by sociobiologists as "the mental activity in which hazily perceived and relatively intangible phenomena, such as complex arrays of objects or activities, are given a factitiously concrete form, simplified and labeled with words or other symbols" (p. 380).[14] We assume this to be the basic mechanism through which strategies for action are defined. This action, be it through artifacts or through mentifacts, modifies reality, which in turn produces additional information which, through this reificative process, modifies or generates new strategies for action, and so on. This ceaseless cycle is the basis for the theoretical framework upon which we base our ethnomathematics concept.

Individual behavior is homogenized in certain ways through mechanisms such as education to build up societal behavior, which in turn generates what we call "culture." Again a scheme such as figure 1–2 allows for the concept of culture as a strategy for societal action. Now, the mechanism of reification, which is characteristic of individual behavior, is replaced by communication, while information, which impacts upon an individual, is replaced by history, which has its effect on society as a whole. (We will not go deeper here into this theoretical framework; this will appear somewhere else.)

As we have mentioned above, culture manifests itself through jargons, codes, myths, symbols, utopias, and ways of reasoning and inferring. Associated with these we have practices such as ciphering

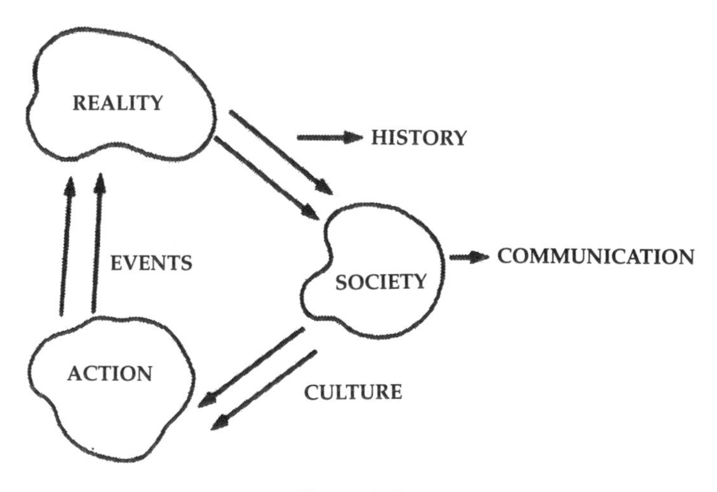

Figure 1–2.

and counting, measuring, classifying, ordering, inferring, modeling, and so on, which constitute ethnomathematics.

The basic question we are then posed is the following: How "theoretical" can ethnomathematics be? It has long been recognized that mathematical practices, such as those mentioned in the end of the previous paragraph, are known to several culturally differentiated groups; and when we say "known" we mean in a way which is substantially different from the Western or academic way of knowing them. This is often seen in the research of anthropologists and, even before ethnography became recognized as a science, in the reports of travelers all over the world. Interest in these accounts has been mainly curiosity or the source of anthropological concern about learning how natives think. We go a step further in trying to find an underlying structure of inquiry in these ad hoc practices. In other terms, we have to pose the following questions:

1. How are ad hoc practices and solution of problems developed into methods?
2. How are methods developed into theories?
3. How are theories developed into scientific invention?

It seems, from a study of the history of science, that these are the steps in the building-up of scientific theories. In particular, the history of mathematics gives quite good illustrations of steps 1, 2, and 3, and research programs in the history of science are in essence based on these three questions.

The main issue is then a methodological one, and it lies in the concept of history itself, in particular of the history of science. We have to agree with the initial sentence in Bellone's excellent book on the second scientific revolution: "There is a temptation hidden in the pages of the history of science—the temptation to derive the birth and death of theories, the formalization and growth of concepts, from a scheme (either logical or philosophical) always valid and everywhere applicable. . . . Instead of dealing with real problems, history would then become a learned review of edifying tales for the benefit of one philosophical school or another" (p. 1).[15] This tendency permeates the analysis of popular practices such as ethnoscience, and in particular ethnomathematics, depriving it of any history. As a consequence, it deprives it of the status of knowledge.

It is appropriate at this moment to make a few remarks about the nature of science nowadays, which is regarded as a large scale professional activity. As we have already mentioned, it developed into this position only since the early nineteenth century. Although scientists communicated among themselves, and scientific periodicals, meetings, and associations were known, the activity of scientists in earlier centuries did not receive any reward as such. What reward there was came more as the result of patronage. Universities were little concerned with preparing scientists or training individuals for scientific work. Only in the nineteenth century did becoming a scientist start to be regarded as a professional activity. And out of this change, the differentiation of science into scientific fields became almost unavoidable. The training of a scientist, now a professional with specific qualifications, was done in his subject, in universities or similar institutions, and mechanisms to qualify him for professional activity were developed. And standards of evaluation of his credentials were developed. Knowledge, particularly scientific knowledge, was granted a status which allowed it to bestow upon individuals the required credentials for their professional activity. This same knowledge, practiced in many strata of society at different levels of sophistication and depth, was expropriated by those who had the responsibility and power to provide professional accreditation.

We may look for examples in mathematics of the parallel development of the scientific discipline outside the established and accepted model of the profession. One such example is Dirac's delta function which, about twenty years after being in full use among physicists, was expropriated and became a mathematical object, structured by the theory of distributions. This process is an aspect of the internal dynamics of knowledge vis-à-vis society.

There is unquestionably a timelag between the appearance of new ideas in mathematics outside the circle of its practitioners and the recognition of these ideas as "theorizable" into mathematics, endowed with the appropriate codes of the discipline, until the expropriation of the idea and its formalization as mathematics. During this period of time the idea is put to use and practiced: it is an example of what we call "ethnomathematics" in its broad sense. Eventually it may become mathematics in the style or mode of thought recognized as such. In many cases it never gets formalized, and the practice continues restricted to the culturally differentiated group which originated it. The mechanism of schooling replaces these practices by other equivalent practices which have acquired the status of mathematics, which have been expropriated in their original forms and returned in a codified version.

We claim a status for these practices, ethnomathematics, which do not reach the level of mathematization in the usual, traditional sense. Paraphrasing the terminology of T. S. Kuhn, we say they are not "normal mathematics" and it is very unlikely they will generate "revolutionary mathematics." Ethnomathematics keeps its own life, evolving as a result of societal change, but the new forms simply replace the former ones, which go into oblivion. The cumulative character of this form of knowledge cannot be recognized, and its status as a scientific discipline becomes questionable. The internal revolutions in ethnomathematics, which result from societal changes as a whole, are not sufficiently linked to "normal ethnomathematics." The chain of historical development, which is the spine of a body of knowledge structured as a discipline, is not recognizable. Consequently ethnomathematics is not recognized as a structured body of knowledge, but rather as a set of ad hoc practices.

It is the purpose of our research program to identify within ethnomathematics a structured body of knowledge. To achieve this it is essential to follow steps 1, 2, and 3 above.

As things stand now, we are collecting examples and data on the practices of culturally differentiated groups which are identifiable as mathematical practices, hence ethnomathematics, and trying to link these practices into a pattern of reasoning, a mode of thought. Using both cognitive theory and cultural anthropology we hope to trace the origin of these practices. In this way, a systematic organization of these practices into a body of knowledge may follow.

Conclusion

For effective educational action not only an intense experience in curriculum development is required, but also investigative and research methods that can absorb and understand ethnomathematics. And this clearly requires the development of quite difficult anthropological research methods relating to mathematics, a field of study as yet poorly cultivated. Together with the social history of mathematics, which aims at understanding the mutual influence of sociocultural, economic, and political factors in the development of mathematics, anthropological mathematics, if we may coin a name for this specialty, is a topic which we believe constitutes an essential research theme in Third World countries, not as a mere academic exercise, as it now draws interest in the developed countries, but as the underlying ground upon which we can develop curriculum in a relevant way.

Curriculum development in Third World countries requires a more global, clearly holistic approach, not only by considering methods, objectives, and contents in solidarity, but mainly by incorporating the results of anthropological findings into the three-dimensional space which we have used to characterize curriculum. This is quite different than what has frequently and mistakenly been done, which is to incorporate these findings individually in each coordinate or component of the curriculum.

This approach has many implications for research priorities in mathematics education for Third World countries and has an obvious counterpart in the development of mathematics as a science. Clearly the distinction between Pure and Applied Mathematics has to be interpreted in a different way. What has been labeled Pure Mathematics, and continues to be called such, is the natural result of the evolution of the discipline within a social, economic, and cultural atmosphere which cannot be disengaged from the main expectations of a certain historical moment. It cannot be disregarded that L. Kronecker ("God created the integers—the rest is the work of men"), Karl Marx, and Charles Darwin were contemporaries. Pure Mathematics, as opposed to Mathematics, came into consideration at about the same time, with obvious political and philosophical undertones. For Third World countries this distinction is highly artificial and ideologically dangerous. Clearly, to revise curriculum and research priorities in such a way as to incorporate national development priorities into the scholarly practices which characterizes university research is a most difficult thing to do. But all the difficulties should not disguise the increasing necessity of pooling human resources for the more urgent and immediate goals of our countries.

This poses a practical problem for the development of mathematics and science in Third World countries. The problem leads naturally to a close for the theme of this paper: that is, the relation between science and ideology.

Ideology, implicit in dress, housing, titles, so superbly denounced by Aimé Césaire in *La Tragédie du Roi Christophe*, takes a more subtle and damaging turn, with even longer and more disrupting effects, when built into the formation of the cadres and intellectual classes of former colonies, which constitute the majority of so-called Third World countries. We should not forget that colonialism grew together in a symbiotic relationship with modern science, in particular with mathematics and technology.

Notes

1. Ubiratan D'Ambrosio, "Science and technology in Latin America during discovery," *Impact of Science on Society* 27, no. 3 (1977):267–274.

2. R. L. Wilder, *Mathematics as a cultural system* Oxford: Pergamon, 1981; see also C. Smorynski, "Mathematics as a cultural system," *Mathematical Intelligencer* 5, no. 1 (1983):9–15.

3. Anton Dimitriu, *History of logic* (4 vols) (Kent, England: Abacus Press, 1977).

4. Karls H. Potter ed., Indian metaphysics and epistemology. *Encyclopedia of Indian Philosophies* (Princeton, N.J.: Princeton University Press, 1977).

5. Marcia Ascher, and Robert Ascher, *Code of the Quipu* (Ann Arbor, MI: The University of Michigan Press, 1981).

6. David F. Lancy, *Cross-cultural studies in cognition and mathematics* (New York: Academic Press, 1983).

7. Ubiratan D'Ambrosio, *Socio-cultural bases for mathematics education* (São Paulo: UNICAMP, Campinas 1985).

8. ———. *Several dimensions of science education* (Santiago: CIDE/REDUC, 1990).

9. Plato, *Dialogues*, ed., E. Hamilton, and H. Cairns (New York: Pantheon Books, 1963).

10. Plato, *Dialogues (Republic)*, ed., E. Hamilton, and H. Cairus (New York: Panthem Books, 1963).

11. Stephen K. Victor ed., *Practical geometry in the high Middle Ages* (Philadelphia: The American Philosophical Society, 1979).

12. Ubiratan D'Ambrosio, "Mathematics and society: some historical considerations and pedagogical implications," *Int. J. Math. Educ. Sci. Technol.*, 2, no. 4, (1980): 479–488.

13. ———. Uniting reality and action: A holistic approach to mathematics education, in *Teaching Teachers Teaching Students*, L. A. Steen and D. J. Albers eds. (Boston: Birkhauser, 1980), 33–42.

14. Charles J. Lumsden, and Edward D. Wilson, *Genes mind and culture* (Cambridge, MA: Harvard University Press, 1981).

15. Enrico Bellone, *A world on paper* (Cambridge, MA: The MIT Press, 1980; orig. ed. 1976).

Chapter 2

Ethnomathematics

Marcia Ascher and Robert Ascher

Editors's comment: Marcia Ascher, a mathematician, and Robert Ascher, an anthropologist, challenge the idea that nonliterate peoples have only "primitive" mathematical ideas. Their work, directed toward professional mathematicians, has provided detailed evidence of sophisticated mathematical ideas among nonliterate peoples, ideas akin to and as complex as those of modern, "Western" mathematics. As such, this chapter, which first appeared in *History of Science*, 14: 125–144, in 1986, played a significant role in introducing ethnomathematics to the English-reading mathematical community.

Authors' Note 1995: This article was written in the early 1980s but did not find journal acceptance for several years. Hence, it reflects our earliest thinking on ethnomathematics. Since then we have clarified our thoughts on two most important points. First of all, we subsequently articulated what we meant by the term "mathematical ideas" which occurs in the first sentence but underlies the entire discussion. To us, mathematical ideas include those involving number, logic, spatial configuration and, more significant, the combination or organization of these into systems and structures. Second, we used the term "nonliterate" to counter the outmoded usage of "primitive" in the mathematics literature. But, in defining the scope of ethnomathematics, that term is too limited. The term "traditional" is better although no single word is sufficient. An entire chapter of *Ethnomathematics: A Multicultural View of Mathematical Ideas* (Belmont, CA: Brooks/Cole, 1991) is devoted to a detailed elaboration of what we see as the scope and implications of ethnomathematics as well as its relationship to other fields of endeavor.

I

Ethnomathematics is the study of mathematical ideas of nonliterate peoples. We recognize as mathematical thought notions that in some way correspond to that label in our culture. For example, all humans, literate or not, impose arbitrary orders on space. Particular orders develop within cultural contexts and their form and content will necessarily be expressive of the culture in which they arise.

A fair number of statements about nonliterate peoples are found in mathematical literature, usually in a preliminary chapter in histories of mathematics or in books that offer an introduction to the spirit of the subject. Most of the presentations are theoretically and factually flawed. Here we show that it is time to revise these statements and urge toward an ethnomathematics appropriate to contemporary thought.

II

We choose the term "nonliterate" for people that are elsewhere called "primitive." In so doing, we raise an issue that is much more than quibbling about a word. "Primitive" as applied to people is a product of the theory of classical evolution.[1] Crucial in this largely nineteenth-century paradigm is the assumption that nonliterate peoples are the earliest living representatives along a straight evolutionary path that led from savagery to civilization in a series of predestined stages.[2] Nonliterate peoples were called "primitive" because they were thought to be original, early, ancient, primeval. It may be common knowledge that "primitive," when applied to the ethnographic present, is a theoretical anachronism but its use continues in mathematical texts. Thus, the ideas of living or recently extinguished nonliterate peoples are erroneously placed at the beginning of chronologically organized discussions of mathematics.

For the classical evolutionists, the mathematical thought of nonliterate peoples was confined to number. It could be nothing more if, as was the case, it was taken for granted that nonliterate peoples were living ancestors and that mathematics began with numeration. E. B. Tylor, the most highly regarded member of the school, devoted one chapter of his *Primitive Culture* to the "The art of counting."[3] Some twenty-two years later, the mathematician L. L. Conant wrote his influential book *The Number Concept: Its Origin and Development*.[4] The book, which is pervaded by the ideas of the classical evolutionists and in particular by the ideas of Tylor, continues to be followed today.[5] For current mathematicians too,[6] mathematical thought of nonliterate peoples centers on number using number words as data. Discussions emphasize body parts referred to in the number words, how words are formed, how high number words go, and whether different number words are used for different categories of objects. Records of counts, if they occur, are 1–1 assemblages of pebbles, notches on sticks, or knots on strings. Differences in ideas are associated with the needs of the different economies of hunting and gathering or herding or commerce. And sometimes nonliterate number comprehension is compared to a number sense found in other animals.[7] These notions require examination.

The fact that the same word is used for a number and a body part tells us nothing about number as number. In every language, words for numbers, just as words for anything else, are coined freely in several ways. If a word is adopted from an already existing word, it soon takes on a meaning appropriate to its new context. For example, when an English speaker says "a foot" in the context of measurement, no English hearer thinks he is thinking of a body part. If in a nonliterate culture, we hear a word for a body part in the context of number, there is no reason to presume otherwise. For speakers within both language communities, the circumstances of the coinage are irrelevant.

How high number words go reflects only how high people in a language community wish to count and is unrelated to intelligence or ability to formulate abstractions. The flexibility and expendability of all languages permit the addition of higher number words as and if they are needed.[8] The concept of counting is universal.[9] What is universally being recognized, and what the large variety of number words have in common, is that there is an equal amount from one to the next. In short, all number words are names given to 1, $1+1$, $1+1+1$, and so forth. New names are created when they are needed, but their presence or absence implies nothing new about the number concept. Kronecker's "God made the integers . . ." and the properties being captured by the Peano postulates are perhaps yet another way of saying the same thing.

The study of language is also mainly responsible for our insights into the diversity and richness of the ways in which the world is categorized in different cultures.[10] The use of different number words for different categories of objects is often erroneously cited in mathematical texts as evidence for the lack of an abstract concept of number. For example, Tsimshians are said to have seven distinct sets of number words with one set for flat objects and animals, one for round objects and divisions of time, one for men, one for long objects, one for canoes, and one for counting when no definite object is referred to.[11] Mathematicians should, but apparently do not, recognize that these categories are neither mutually exclusive nor exhaustive and wonder what word a Tsimshian uses for round flat objects or objects such as stars. The use of sets of number words or numerical classifiers is widespread and still eludes conclusive analysis.[12] Let us accept for the moment the simple statement that some groups have different number words for objects that are living and dead and diagram it as

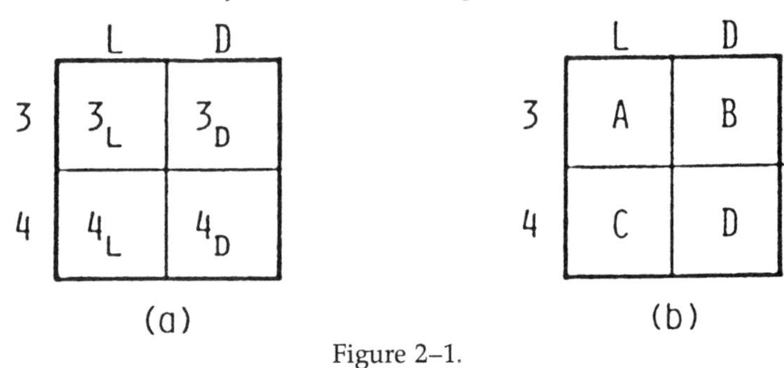

Figure 2–1.

in figure 2–1a. The concept of 3_L does not detract from either the concepts of 3 or the concept of living. It is their intersection. In figure 2–1b, only the labels are changed. Someone, not privy to either diagram, hearing only A, B, C, D might think them quite distinct. The hearer might brand them a product of concrete thinking, while in actuality, the ability to simultaneously carry both dimensions makes the idea rather subtle. In mathematical writing, Blackfeet are said to use different sets of numbers for the living and the dead.[13] All that can be concluded is that they insist on this fundamental opposition. We might speculate on the effect of such a distinction if introduced into applications of Western mathematics. Perhaps when a mathematical model is constructed from a realistic problem, it would be better to retain with the variables the knowledge that they refer to living or nonliving objects. They could be dealt with as variables with restricted domains as is the case with Diophantine equations.

Somewhere in the now conventional story of number, there are shifts from the area of language to the visual-tactile. Artifacts are introduced—pebbles, notched sticks, knotted cords. The often emphasized aspects of artifacts—the material from which they are made—is of no mathematical importance. Knotted cords, for example, can be 1–1 tallies, or they can be as sophisticated as the logical-numerical system of the Incas,[14] just as chalkboards can be records of 1—1 tallies or complex integrals. A 1—1 tally is a step-by-step matching process in which individual objects, be they pebbles, strokes, or whatever, are associated, one at a time, with other individual objects creating for some purpose a tractable set of signs. A number symbol, on the other hand, regardless of the medium, is a representation of whatever is subsumed by the meaning of the number. Having a set of number symbols implies having some organized concept of numbers and their relationships. Stressed as often as the material is the motivation for the creation and use of number symbols. It is depicted in terms of the property and business needs of the individual hunter or herder. Clearly, to be other than a private code, a symbolic system must be shared, spread, and reiterated by a group of users.

The conventional wisdom is well encapsulated in an anecdote that is repeated wholly or in part with such frequency that it must have special appeal. The anecdote tells of an exchange between a native African Demara sheepherder and someone else variously described as an explorer, trader, scientist, anthropometrist, or ethnologist. It is intended to show that the herder cannot comprehend the simple arithmetic fact that $2+2$(or 2×2)$=4$. It describes how the herder agrees to accept two sticks of tobacco for one sheep but becomes confused and upset when given four sticks of tobacco after a second sheep is selected.[15] Of course, the problem is not that the shepherd doesn't understand arithmetic, it is rather that the scientist/trader doesn't understand sheep. Sheep are not standardized units. Since the Demara herder finally agreed to the trade, his confusion could be attributed to the trader's willingness to pay an equal amount for the second, different animal.

Other than demonstrating two people talking past each other, the anecdote raises the issue of the difference between a mathematical concept and its application. When inferring mathematical ideas from concrete usage, one is always dealing with applications. It is recognized, for example, that $2+2=4$ does not apply when discussing combining volumes of gases or combining elements of overlapping sets.[16] Even more important, as was recognized as early as 1912 by Wertheimer,[17] the applicability of even the simplest of mathematical models becomes a question of cultural categorization. We say, for example, that 2 apples + 2 pears = 4 fruit, but 2 pants + 2 jackets = 2 suits. A while ago we might have said that 2 men + 2 women = 2 couples, but now we are careful to say that 2 men + 2 women = 4 people. Further, we believe there is meaning in 2 in. × 2 in. = 4 sq. in. but not in 2 apples × 2 apples = 4 sq. apples. A question that has been posed by Western scholars when trying to determine the universality of mathematics,[18] is whether or not 2 + 2 always equals 4. Once 4 is recognized as the name for 2 + 2, the question becomes, as it is here, whether or not the model applies. That question can have different answers in different cultures and even different answers within the same culture.[19]

While discussion of mathematical ideas of nonliterate peoples centers on number, the presumptions and conclusions extend to logic and thought. Nonliterate peoples are often explicitly characterized as simpleminded or childlike, as only capable of concrete thought and not of abstraction or generalization, as of lesser intelligence, as incapable of analytic thought, and as without formal reasoning or logic.[20] In any context these descriptions are heavily judgmental; in the context of mathematics, they are condemning.

To examine this characterization, we pick up the thread of intellectual history. For the classical evolutionists, the "psychic unity of mankind" was fundamental.[21] This tenet was challenged in the highly influential writings of Lévy-Brühl.[22] In place of "psychic unity," he erects a divided world of thought—nonliterate people are "prelogical"; we are logical, where logical is intended in the broad sense of all thought and thought processes. Once again going over the issue of number but now paying attention to thought, Lévy-Brühl reconsiders Conant's findings claiming that he can account for them for the first time in a satisfactory way. He says that Conant, believing 5 should be the most "natural" base, was puzzled by the diversity of bases he found in nonliterate cultures. But there is no puzzle, says Lévy-Brühl, if one understands that the minds of primitives do not function as ours do, that they are mystical rather than logical and that they know almost nothing of abstraction—how then, for them, can one base be more "natural" than another.[23] The line that Lévy-Brühl drew between himself and the classical evolutionist is not as hard and fast as he would have us believe. They needed the notion of psychic unity for their evolutionary hypothesis but contradicted it at almost every turn as the racial and class prejudices of their culture and times overshadowed their science.[24] For example, they characterized nonliterate peoples as childlike and meant two things by it. Used one way, childlike stood for an early stage in a human life, just as primitives were an early stage in the evolution of culture. Alternatively, childlike meant that primitives thought like children.[25] In effect, a Victorian prejudice of the evolutionists became philosophy in the works of Lévy-Brühl.[26]

Within a few years of the 1910 publications of *How Natives Think*, the ideas of Lévy-Brühl had entered the mathematical literature in a lengthy, uncritical manner,[27] they are found with force in Menninger, and continue a generation later in the works of others.[28] The myth of the childlike, prelogical primitive persists in spite of a host of anthropologists from Durkheim to Lévi-Strauss who have inveighed against it.[29] Why is this so? One answer is that there is political, social, economic, and ideational value in maintaining that most of the people in the world are our intellectual inferiors. Another answer is found in the belief that higher technology goes with higher intelligence. Its persistence within scholarly disciplines did not evaporate because it is difficult to deal with any statement so broad as to encompass all nonliterate peoples and so incisive as to pinpoint a mode of thought. Nevertheless, beginning about fifteen years ago, scholars with diverse perspectives have turned directly to the question.

One argument important for us addresses the specific ethnographic examples relating to number used by Lévy-Brühl. He noted, for example, that certain numbers and numbers that are multiples or divisors of them play an important role in Veddic religion, rituals, and legends. In this context, on occasion, 3, 7, or 9 are substituted for each other. From this, Lévy-Brühl concludes "this equivalence, an absurdity to logical thought, seems quite natural to prelogical mentality, for the latter, preoccupied with the mystic participation, does not regard these numbers in abstract relation to other numbers, or with respect to the arithmetical laws in which they originate."[30] But a recent field study of the numerical ideas and arithmetic abilities of the Kédang, who also have this type of substitution, enables a different conclusion.[31] Their use of numbers in practical arithmetic contexts and their use of numbers in non-arithmetic contexts do not contradict or detract from each other. When used in symbolic contexts, odd numbers are associated with life and even numbers with death. Substitutions within these classes are possible if circumstances require it. If, for example, a ceremonial period of four days is stipulated but cannot be met, two days will do but three would be a serious infringement. Four and two are members of the same class and so are equivalent in that sense in this context. The formation of these equivalence classes is, we think, instead an example of an abstract idea about number.

Other counters to Lévy-Brühl address logical processes specifically. None of the hundreds of languages studied so far lacks the ability to handle the logical connectors *and, not, or, if . . . then,* and *iff.*[32] Natural discourse, however, need not be explicitly structured as formal logic. Commonly held beliefs more often are presumed than stated as premises. Thus, some grasp of the beliefs of a given culture are needed so that the course of inference may be followed. An outsider, lacking such understanding, can too easily fall into the mistaken notion that what is heard is somehow irrational or illogical.[33] There is, as is posited by some philosophers, a universal "natural rationality" that enables learning through inferences made from experience with the diversity of conclusions stemming from different frameworks of concepts and beliefs in different cultures.[34] Further, the forms of human thought deemed logical are now viewed more broadly and include inferences about the plausibility of statements[35] and three-valued logic.[36] In a similar vein, such areas as heuristics, circumscription,[37] and fuzzy-logic[38] are being explored by computer scientists who are concerned with artificial intelligence and are trying to formalize human-like processes.

The use of syllogisms for investigating the reasoning ability of nonliterate people was recently reexamined by cognitive psychologists.[39] The study probed more deeply reasoning that respondents used to arrive at what were formerly viewed as unsatisfactory responses. Here is one example in which a Kpelle respondent would not reply to the question.

Question:	All Kpelle men are rice farmers. Mr. Smith is not a rice farmer. Is he a Kpelle man?
One part of the response:	If you know a person, if a question comes up about him you are able to answer. But if you do not know the person, if a question comes up about him, its hard for you to answer.[40]

Although the syllogism posed by the questioner has gone unanswered, the ability to reason and to think hypothetically has been demonstrated. What has also been demonstrated is that the Kpelle respondent and his Western questioner have different views on talking about people whom you do not know. Another question can be summarized as:

if A or B then C,
given not A and B,
is C true?

and the answer as:

> if A then Q,
> if B then not Q,
> if not Q then not C,
> given not A and B,
> the conclusion is not C.[41]

Here again the logic is fine but the respondent has explicitly substituted statements consistent with his world view. Comparing several similar studies with Kpelle, Vai, Yucatecans, and North Americans, it was concluded that the level of schooling seemed to be most significant in eliciting the responses expected by the Western questioner.[42] But this is as it should be since in formal Western-style education, one learns the school culture which emphasizes "playing along" with the questioner. Additionally, some psychologists are convinced that tests used on Western children and the developmental stages they define are invalid when used on adults and children in other cultures.[43] Piaget and others modified their positions to note that even within Western culture there are differences in results when questions more closely mirrored the life experiences of those tested.[44]

There is not one instance of a study or restudy that upon close examination supports the myth of the childlike primitive. What the studies do show might be summarized in these words: "Cultural differences in cognition reside more in the situations to which particular cognitive processes are applied than in the existence of a process in one cultural group and its absence in another."[45] No longer can nonliterate peoples be thought to be childlike. Indeed, the approach of many contemporary anthropologists is to think of themselves as children.[46] After all, an outsider, from the perspective of the people visited, is the one who is childlike and is often told so because he does not know, for example, how to address an elder, the proper way to hold a utensil or, perhaps, that different words *must* be used to count the living and the dead.

III

Mathematical ideas of nonliterate people must be drawn from ethnographic literature with the understanding that the task will not be easy. Most anthropologists were limited in their understanding of mathematics and have seldom asked relevant questions.[47] Ideas that might have been delved into more deeply or recorded more specifically by someone with an interest in mathematics, may well have not been seen for what they were. Fortunately, at least some anthropologists included information, sometimes even unbeknownst to them, that can be used, and some recent studies have been specifically concerned with related ideas. Also, in some cases, there are available ethnohistorical accounts as well as artifacts. The category mathematics is our own, and so we cannot expect to find anything so labelled by other peoples. Moreover, because they have no professional classes particularly devoted to the doing of mathematics, there is no explicit mathematics. Mathematical ideas will have to be found implicit in other areas and activities. Their context in a culture will depend on what the culture thinks about and on how it thinks about what it thinks about.

Specific studies within ethnomathematics require detailed investigation and analysis. As with any serious endeavor, the results will be more than anecdotal. Substantial studies are already available on, for example, the logical-numerical system of the Incas[48] and Maya calendrics[49] with their cultural ramifications. Here, however, we continue with instances isolated from their full cultural elaboration to indicate possibilities and problems. Expanding beyond number and numeracy, we contrast some impositions of order on space and cite some conceptual models for which the users are clearly aware of and concerned with logical structures and drawing inferences from them.

For a long time in Western culture, it was believed that our Euclidean geometry was describing truths about the physical world rather than being an elaboration of a mental construct. Our most fundamental concepts were then recognized as arbitrary, but some illustrations from other cultures will emphasize just how arbitrary they are. Points, lines, right angles, rectangles, and planar surfaces are essential to the world we have constructed around us. This, of course, influenced the course of Western mathematics, but mathematics, in turn, reinforced their importance through our art, architecture, measuring and mapping schemes, ways of seeing and describing, and even our aesthetic sense. What could be more natural than looking up at the sky, spotting particular stars, mentally connecting the star-points with straight line segments, and creating constellations that are seen by generation after generation? Apparently, in keeping with their spatial ideas, native Andean peoples see other constellations far more irregularly shaped made up of darker and lighter blotches (clouds of interstellar dust) in the sky.[50] Just before he died in the 1930s, Black Elk, an Oglala

Sioux, spoke about his life and thoughts. His statement about the circle (below, right) is presented in contrast to a statement about the line (below, left) which appeared in a recently highly lauded work by two American professors of mathematics. While they differ on the geometric form, the writers share their degree of conviction in the rightness of their ideas and support their view with nature, God, achievement of goals, and proper human development. Black Elk and the Sioux, however, were forcibly made to realize that their view was not shared by other cultures.

. . . in every human culture that we will ever discover, it is important to go from one place to another, to fetch water or dig roots. Thus, human beings were forced to discover—not once, but over and over again, in each new human life—the concept of the straight line, the shortest path from here to there, the activity of going directly towards something.

In raw nature, untouched by human activity, one sees straight lines in primitive form. The blades of grass or stalks of corn stand erect, the rock falls down straight, objects along a common line of sight are located rectilinearly. But nearly all the straight lines we see around us are human artifacts put there by human labor. The ceiling meets the wall in a straight line, the doors and windowpanes and tabletops are all bounded by straight lines. Out the window one sees rooftops whose gables and corners meet in straight lines, whose shingles are layered in rows and rows, all straight.

[continued on next page]

. . . I am now between Wounded Knee Creek and Grass Creek. Others came too, and we made these little gray houses of logs that you see, and they are square. It is a bad way to live, for there can be no power in a square.

You have noticed that everything an Indian does is in a circle, and that is because the Power of the World always works in circles, and everything tries to be round. In the old days when we were a strong and happy people, all our power came to us from the sacred hoop of the nation, and so long as the hoop was unbroken, the people flourished. The flowering tree was the living center of the hoop, and the circle of the four quarters nourished it. The east gave peace and light, the south gave warmth, the west gave rain, and the north with its cold and mighty wind gave strength and endurance. This knowledge came to us from the outer world with our religion. Everything the Power of the World does is done in a

[continued on next page]

The world, so it would seem has compelled us to create the straight line so as to optimize our activity, not only by the problem of getting from here to there as quickly and easily as possible but by other problems as well. For example, when one goes to build a house of adobe blocks, one finds quickly enough that if they are to fit together nicely, their sides must be straight. Thus, the idea of a straight line is intuitively rooted in the kinesthetic and the visual imaginations. We feel in our muscles what it is to go straight toward our goal, we can see with our eyes whether someone else is going straight. The interplay of these two sense intuitions gives the notion of straight line a solidity that enables us to handle it mentally as if it were a real physical object that we handle by hand.

By the time a child has grown up to become a philosopher, the concept of a straight line has become so intrinsic and fundamental a part of his thinking that he may imagine it as an Eternal Form, part of the Heavenly Host of Ideals which he recalls from before birth. Or, if his name be not Plato but Aristotle, he imagines that the straight line is an aspect of Nature, an abstraction of a common quality he has observed in the world of physical objects.[51]

circle. The sky is round, and I have heard that the earth is round like a ball, and so are all the stars. The wind, in its greatest power, whirls. Birds make their nests in circles, for theirs is the same religion as ours. The sun comes forth and goes down again in a circle. The moon does the same, and both are round. Even the seasons form a great circle in their changing, and always come back again to where they were. The life of a man is a circle from childhood to childhood, and so it is in everything where power moves. Our tepees were round like the nests of birds, and these were always set in a circle, the nation's hoop, a nest of many nests, where the Great Spirit meant for us to hatch our children.

But the Waischus (whitemen) have put us in these square boxes. Our power is gone and we are dying, for the power is not in us any more. You can look at our boys and see how it is with us. When we were living by the power of the circle in the way we should, boys were men at twelve or thirteen years of age. But now it takes them very much longer to mature.

Well, it is as it is. We are prisoners of war while we are waiting here. But there is another world.[52]

Sharply contrasting with both of these views are the Avilik (Inuit) spatial concepts studied by Edmund Carpenter, an anthropologist whose particular concern is the interconnectedness of media and thought.[53] He concludes that they do not separate space and time, but join them into a dynamic process in which space itself is never a static enclosure but "direction in operation." He relates this to their environment which he sees as vast with few permanent reference points. For orientation the Avilik rely on wind direction and changing snow and ice forms and relationships. Ramifications of this basic concept are seen in their art and ways of seeing and describing. In carving ivory, for example, a figure is sketched until the limit of the surface is reached and then it is turned over and the figure completed on the other side. Each independent figure has its own horizon and orientation. Sometimes groups of figures are all the same figure but shown from several different perspectives or they may depict an event that requires the passage of time. Also, if an Avilik wishes to show what is inside or behind something, he or she draws the objects as if they were transparent. Carpenter makes the analogy to "an engineer's sketch of the moving parts of an engine, all relevant elements are shown, in spite of the fact that they could never be observed from a single vantage point" or in a single moment.[54] The most important feature is that no single orientation seems to be assumed in drawing or viewing. Carpenter reports that the children poked fun at what they found to be his odd behavior of turning surfaces or his head in order to look at pictures. Some of the children, when taught our mode of perspective drawing in school, instead made the parallel lines converge as they approach the viewer, which served to open out instead of close the space, while others used the bottom of the paper as the closest visual points and higher positions for points more distant.

Continuing with the ordering and representation of space, we turn to the construction and use of a spatial model. It is an example of an application—the goal is to navigate from one place to another throughout the Caroline Islands, a chain that extends for about 1,500 miles. Its users do not believe it to be a statement of reality. As with most mathematical models, it is an abstraction from physical reality from which relations and logical implications can be derived in order to deal with all possible journeys rather than only journeys already taken. The model is used to organize fixed data, incorporate realistic cues, and make decisions accordingly.[55] Its use is standardized throughout the Caroline Islands and requires a long, intensive period of learning, much of which takes place in lectures on land. The lectures are augmented by diagrams sketched in the sand. The navigator's knowledge includes much more than this model, but the model is central to navigation. Leaving an island and getting started in the right direction, keeping a straight course by combining the angles of three sets of prevalent wave types, and homing in on the destination once within about twenty miles of it are also important, but the model is essential for knowing where you are when out of sight of any land.

The model consists, first of all, of a "compass" which is thirty-two points irregularly spaced around the horizon. The points are the rising and setting positions of different stars and, as part of a mental construct, do not depend on the visibility of the stars during a journey. The paths of these stars, other stars, and all known islands are positioned within this framework. The region is sufficiently latitudinally confined so that the star rising and setting points are essentially the same throughout. To each pair of origin and destination islands, there is associated a third island, well out of sight of the course, which is used as a reference point during a journey. The boat is conceived of as stationary and the island as moving. The direct course of the journey passes the sationary boat as the reference island passes consecutive star positions (figure 2–2).[56] Meanwhile, of course, tacking

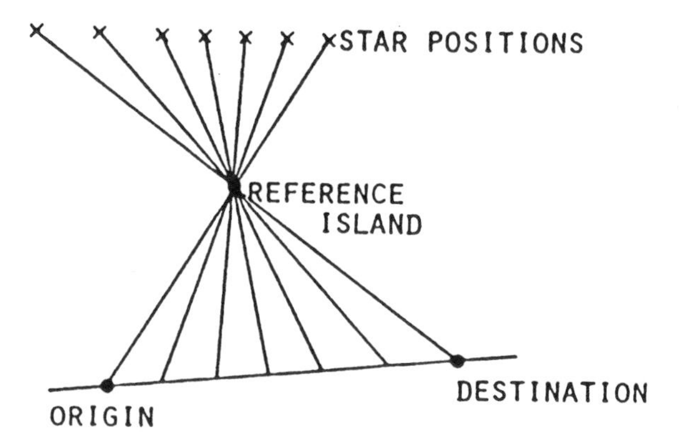

Figure 2–2. A journey.

and responding to particular circumstances are detours that need be correlated with the direct course.[57] With this model, the navigators know just where they are while sailing and are able to successfully reach landfall targets hundreds of miles from their starting point which subtend as little as 5°–7°.[58]

Ordering human relationships is certainly as fundamental as imposing order on space. In our culture, interest in the logical structuring of kinship is minimal; in others, it is a much more elaborated and dynamic element in daily life. There is, of course, no particular structure that is any more correct than any other. Each system contains an arbitrary set of rules which must be consistent and must lead to the desired outcomes which incorporate the values and world view of the culture. Most important is that the members of the culture share the model and constantly draw inferences from it about their relationships to whomever they encounter and the appropriate behavior towards them. The logic of relations, in mathematics, is more recent than the Aristotelean logic of classes and propositions. While it has no necessary connection to kin structuring, it is not uncommon to find Western-kin terms used as illustrative examples. For example, *spouse of* is used to illustrate a symmetric relationship, *mother of* to illustrate an asymmetric relationship, *father of* and *uncle of* are nonintersecting relations (with the exception of some few religious sects), and *grandfather of* is the relative product of *father of* and *mother of*.[59] The relations being stated explicitly are implicit in the definition of the kin relationships. More complex kin structures have more complex logical relations implicit in them.

Let us examine the kin relationships found among the Aranda of Australia. Each person is in one of eight marriage classes. Each must marry a person from a specified class and their children are in another class which depends on the class of the mother (figure 2–3). The

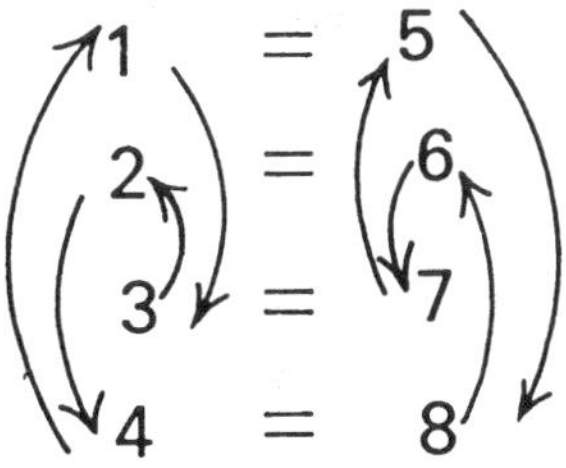

Figure 2–3. Aranda marriage classes: 1. An equal sign indicates marriage partners. The arrow points from the mother's class to the child's class.

representation in figure 2–3 raises an important problem. It conveys the rules but does not display the underlying logical structure. Each descriptive organization we create brings out those aspects of the structure we believe to be important. But it may be we are superimposing our own views while missing what is important to the Aranda. Looking at two additional representations, we will begin to see the logical richness of the system and some of the effects of using data collected, translated, and presented by others. Figure 2–4 dis-

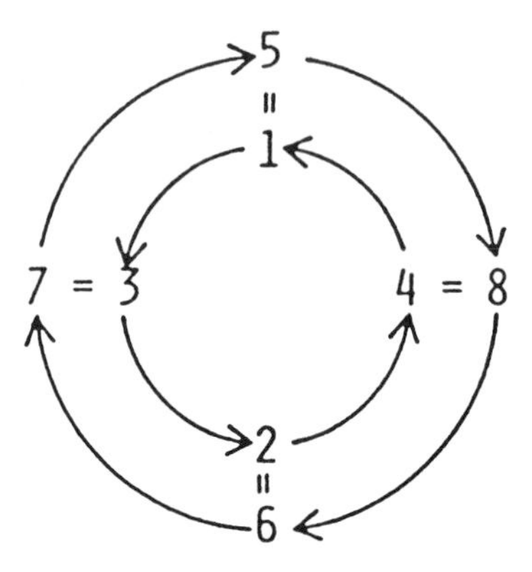

Figure 2–4. Aranda marriage classes: 2. An equal sign indicates marriage partners. The arrow points from the mother's class to the child's class.

plays the presence of two nonoverlapping matricycles, each of length four, and, although somewhat less blatently, four nonoverlapping particycles each of length two.[60] One of the primary features of the kin structure, which echoes an important feature of Aranda world view, and, in general, native Australian cosmology, is that in these cycles, past and future are drawn together into the present.[61] The author of the much more elaborate representation in figure 2–5 notes that the

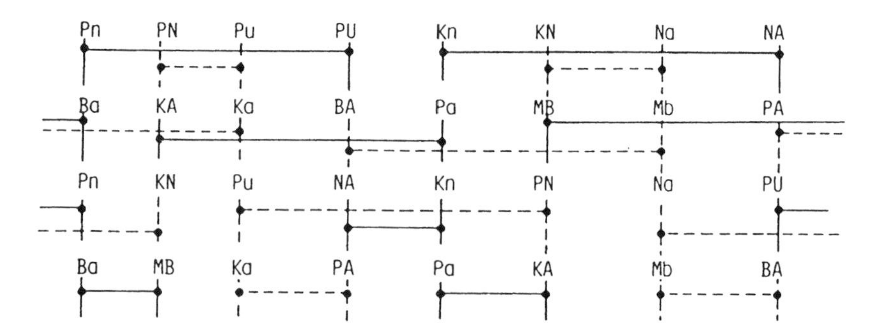

Figure 2–5. Aranda marriage classes: 3. Capitals indicate females, small letters indicate males. Terminating points of horizontal lines are marriage partners. Vertical lines connect parents to children.

right and left sides should be contiguous and the top and bottom should be contiguous. He further notes that the diagram should be on the surface of a torus.[62] Here the unified nature of time and both the matricycles are patricycles are made visual. The unified nature of blood ties is also made visible: the lesser blood ties on either side are drawn together into the closest blood ties. While the rules are stated in terms of the class of the person one marries, by implication, they also specify the individuals one cannot marry. A man, for example, cannot marry his mother, aunt, sister, first cousin, or daughter.

It would, of course, be possible for people to memorize rules and never conceive of the system abstractly or as a whole. There is, at least in one case, dramatic evidence that this is not so. In the 1920s, well before anthropologists had as sophisticated an understanding of kinship structures as they now do,[63] a young anthropologist, A. Bernard Deacon, was investigating the regulation of marriage in Ambrym, New Hebrides. Another who had studied it without conclusion noted later that "class systems of relations were but little known" and what there was in the scholarly literature contained too many details and too few diagrams obscuring the simple main principles.[64] The drama of Deacon's findings was in the way he learned of it: ". . . the older men explained the system to me perfectly lucidly, I could not explain it to anyone better myself. It is perfectly clear that the natives (the intelligent ones) *do* conceive of the system as a connected mechanism which they can represent by diagrams [which they drew for Deacon on the sand]. . . . The way they could reason about relationships from their diagrams was absolutely on a par with a good scientific exposition in a lecture room."[65] Deacon, as is clear from other materials he collected, had a broader view than many of his colleagues and an interest in information that others ignored. His letter to his mentor continues: "I have collected in Malekula, too, some cases of a remarkable mathematical ability. I hope, when I get my material together, to be able to prove that the native is capable of pretty advanced abstract thought."[66] Most unfortunately, less than a month later, Deacon was stricken with blackwater fever and died. His edited field notes, however, were published posthumously.

In the explanation given to Deacon,[67] the elder first drew three long lines arranged as equally spaced spokes of a wheel to represent men from each of three *bwelem*. The three men married, which was indicated by appending a short line to each of the long lines. Each couple had a boy and a girl who are of the same *bwelem* as the father, but the other "line," so longer lines, coming together at the center, were added and the children placed on the other side of them. The diagram, so far, is shown in figure 2–6.[68] The wives were from another *bwelem*, as indicated by arrows added to show where they came from.

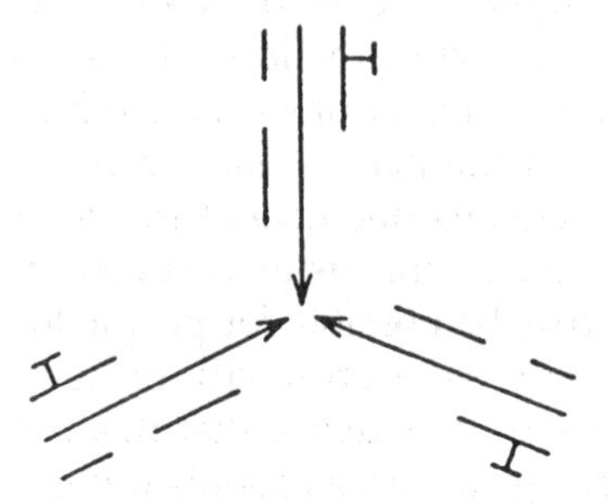

Figure 2–6. Marriage rules in Ambrym: 1. For explanation see text.

But, the elder emphasized, marriage moves in both directions. The males on the diagram yet without wives would, therefore, marry the sisters of the men who had married their sisters. This was shown by adding three short lines and arrows from them. The final diagram is figure 2–7.[69] Two aspects of the diagram that are particularly notewor-

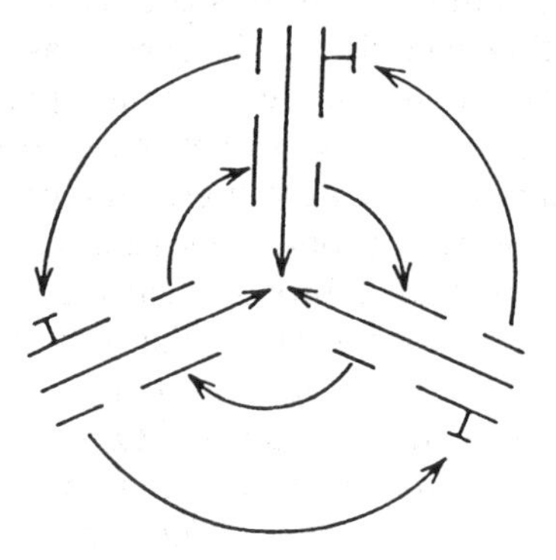

Figure 2–7. Marriage rules in Ambrym: 2. For explanation see text.

thy are its parsimony and the early interchanging of the positions of the male and female symbols in the first and second "lines" of each *bwelem*, which results in a final diagram without any crossing lines. But even more important mathematically is that a conceptual model and the relationships between its elements have been explicitly, consistently and effectively transformed into a planar model. The system itself has interesting features: the "line" of a woman and her female descendants remain the same, but their *bwelem* form a cycle of length three with the cycle for one "line" being a permutation of the cycle for the other; whereas the *bwelem* of a man and his male descendants remain the same, but their "lines" form a cycle of length two. Again, the system does not permit the marriage of a man and his mother, sister, aunt, or daughter. One of the original sources of confusion about the marriage rules was that the people of Ambrym were referring to the class of an individual, while the anthropologists were taking it concretely to mean the individuals themselves. In the description, for example, the word mistaken for blood sisters actually means classificatory sisters, which is made clear by the symbols and their placement in a diagram. The fact that logical inferences are drawn from the model was born out by the elder explaining, before being asked, that two or three kin words were misleading; they did not exactly reflect the model. One word, for example, was used by a man for his wife's mother and his father's sister, although they are not in the same class. Another elder explained the matricycles and patricycles explicitly with a different diagram on the sand. He placed three stones "to form the apices of an equilateral triangle"[70] saying each represented a *bwelem,* and that a woman of the first married a man of the second, her daughter in the second married a man of the third, and her daughter's daughter of the third married a man of the first again who would be in the line of the first woman's father. In yet another explicit statement, it was explained that the mother's mother's mother "came back" to a man's *bwelem* and "line."

IV

With this interplay of conceptual models and diagrams, cycles superimposed on other cycles, and rules and their logical implications, we come to the close of our discussion. Number systems, spatial ordering, and kinship structures are of particular interest because they are fundamental. Elsewhere we have briefly described interest of several cultures in tracing closed figures without lifting the finger or crossing lines.[71] Whether a children's puzzle, part of a myth about death or speculation on being able to take a Sunday walk on Königsberg's bridges, the fact that many share this concern can only make our Western solutions richer. Ethnomathematics is not a part of the history of Western mathematics although we will, of necessity, need to use Western terminology in discussing it. As Westerners, we are confined in what we can see and what we can express to ideas in some way analogous to our own. The ideas of non-Westerners belong, as do ours, in the global and ongoing history of mathematics always keeping in mind that there is no single linear ordering and no necessary route that all must follow. At the very least, ethnomathematics can lead to an appreciation of the intellectual endeavors of others.

The late Raymond L. Wilder, the primary mathematical spokesman for the importance of relating mathematics and culture, used his understanding to describe the processes of mathematical development in the West.[72] Claudia Zaslavsky increased our sensitivity to developments in Africa.[73] Some other mathematicians and philosophers, such as Keyser, Kline, Spengler, and Wittgenstein, also realized that mathematics has a cultural context but stopped short of probing other cultures.[74] As time passed, our culture and world view gave rise to different philosophies of mathematics, none of which seems quite satisfactory now.[75] An understanding of what is universal and what is not, a better understanding of the mathematical ideas of nonliterate peoples, and acceptance of the fact that they are not our early history are essential to the emergence of a philosophy of Western mathematics fitting our times and our culture.

Notes

1. R. L. Carneiro, "Classical evolution," in *Main currents in cultural anthropology*, ed. by R. Naroll and F. Naroll (Englewood Cliffs, NJ: 1973), 57–121, p. 73; M. J. Herskovits, "A geneology of ethnological theory," in *Context and meaning in cultural anthropology*, ed., M. E. Spiro (New York: 1965), 403–15, p. 408.

2. P. Bagby, *Culture and history* (Berkeley: 1963), 18; R. L. Beals and H. Hoijer, *An introduction to anthropology* (New York: 1965), 711.

3. E. B. Tylor, *Primitive culture* (Boston: 1874), i, 240–72. Tylor says that counting began with gestures using fingers and toes. At that stage, humans could count no further than four. Later, with the realization that words existed for hands and feet, words for numbers came into being. Still later, it was found that a system based on five (one hand) was scanty, one based on twenty (hands and toes) was cumbersome, and so the base ten was adopted, and so on. At every juncture in the story, Tylor points to a culture, from native Australians and native Americans through Polynesians and native Africans and, of course, up to and ending with us.

4. L. L. Conant, *The number concept: Its origin and development* (New York: 1896).

5. C. B. Boyer, *A history of mathematics* (New York: 1968); R. Dubisch, The nature of number (New York, 1952); H. W. Eves, *An introduction to the history of mathematics* (New York: 1st edn., 1953; 5th edn., 1982); G. Ifrah, *Histoire universelle des chiffres* (Paris: 1981).

6. An exception to the classical evolutionary paradigm is found in A. Seidenberg, "The diffusion of counting practices," *University of California publications in mathematics*, iii (Berkeley: 1960), 215–300; idem, "The ritual origin of counting," *Archive for history of exact sciences* ii (1962): 1–40. Seidenberg draws on diffusionism, the major contending nineteenth-and early twentieth-century scheme. According to him, nonliterate peoples invented nothing whatsoever in the way of number, let alone in other areas of mathematical thought.

7. Some or all of these ideas are in: A. D. Aleksandrov, "A general view of mathematics," in *Mathematics: Its content, methods, and meaning*, edited by A. D. Aleksandrov, A. N. Kolmogorov, M. A. Lavrent'ev (Cambridge, MA: 2nd edn., 1969; original Russian edn., 1956), i, 1–64, pp. 7–10; P. Beckmann, *A history of* π *(pi)*, 2nd edn. (New York: 1971), 10–12; C. B. Boyer, *op. cit.* (ref. 5), 3–5; W. F. Brett, E. B. Feldman and M. Sentlowitz, *An introduction to the history of mathematics, number theory and operations research* (New York: 1974), 11–13; L. N. H. Bunt, P. S. Jones, and J. D. Bedient, *The historical roots of elementary mathematics* (Englewood Cliffs, NJ: 1976), 2–3; N. A. Court, *Mathematics in fun and earnest* (New York: 1961; original edn., 1935), 64–66; T. Dantzig, *Number* (New York: 4th edn., 1967; original edn., 1930), 1–21; M. de Villiers, *The numeral-words: Their origin, meaning, history, and lesson* (Cape Town: 1923), 16, 46–49; R. Dubisch, *op. cit.* (ref. 5), 4–9; H. W. Eves, *op. cit.* (ref. 5, 1982), 2–5, 16; G. Flegg, *Numbers, their history and meaning* (New York: 1983), 8–46; G. Ifrah, *op. cit.* (ref. 5), 9–107; M. Kline, *Mathematics in Western culture* (New York: 1972; original edn., 1953), 13–14; E. E. Kramer, *The nature and growth of modern mathematics* (Princeton, NJ: 1982), 5; K. Menninger, *Number words and number symbols* (Cambridge, MA: 1969; original edn., 1957), 9–12, 33–40, 196–7, 223–56, 297; C. D. Miller and V. E. Heeren, *Mathematical ideas* (Glenview, IL: 2nd edn., 1973), 1–4; C. D. Miller and V. E. Heeren, *Mathematics an everyday experience* (Glenview, IL: 1976), 1; N. Myers, *The math book* (New York: 1975), 1–4; D. E. Smith, *History of mathematics* (New York: 1958; original edn., 1923), i, 6–14; D. J. Struik, *A concise history of mathematics* (New York: 2nd rev. edn., 1948), 2–5; B. L. van der Waerden and G. Flegg, *History of mathematics, counting* (Milton Keynes, England: Bucks, 1975), pt. 1, 12, 40; pt. 11, 42–45.

8. C. F. Hockett, *Man's place in nature* (New York: 1973), 138.

9. K. Hale, "Gaps in grammar and culture," in *Linguistics and anthropology: In honor of C. F. Voegelin*, edited by M. D. Kinkade, K. L. Hale, O. Werner

(Lisse, The Netherlands: 1975), 295–316; Z. Salzmann, "A method for analyzing numerical systems," *Word*, v (1950): 78–83.

10. M. R. Crick, "Anthropology of knowledge," *Annual review of anthropology*, xi (1982): 287–313, pp. 288–98.

11. R. Dubisch, *op. cit.* (ref. 5).

12. R. Burling, "How to choose a Burmese number classifier," in *Context and meaning in cultural anthropology*, edited by M. E. Spiro (New York: 1965), 243–64.

13. A. Seidenberg, *op. cit.* (ref. 6, 1960), p. 275.

14. M. Ascher and R. Ascher, *Code of the Quipu: A study in media, mathematics, and culture* (Ann Arbor, MI: The University of Michigan Press, 1981).

15. L. L. Conant, *op. cit.* (ref. 4), p. 4; N. A. Court, *op. cit.* ref. 7), 65–66; M. de Villiers, *op. cit.* (ref. 7), 48–49; H. W. Eves, *In mathematical circles*, reprinted as part of *The other side of the equation* (Boston: 1973; original, 1969); 5; G. Flegg, *op. cit.* (ref. 7), p. 19; M. Kline, *op. cit.* (ref. 7), p. 14; K. Menninger, *op. cit.* (ref. 7), p. 34.

16. M. Kline, *Mathematics, a cultural approach* (Reading, PA.: 1962), 581; L. Wittgenstein, *Remarks on the foundations of mathematics*, edited by G. H. von Wright, R. Rhees, and G. E. M. Anscombe, translated by G. E. M. Anscombe (Oxford, England: 1956), 14e.

17. M. Wertheimer, "Numbers and numerical concepts in primitive peoples," in *A source book of gestalt psychology*, edited by W. E. Ellis (London: 1955), 265–73.

18. E. g., S. Restivo, "Mathematics and world view: Otto Spengler's analysis of numbers and culture" (Paper presented at International Society for the Comparative Study of Civilizations Conference, Bloomington, Indiana, May 1981).

19. For a somewhat different insight into the persistence of this story of the Demara, we go to its source, F. Galton, *The narrative of an explorer in tropical South Africa* (London: 1853). A cousin of Darwin and a peer and friend of E. B. Tylor, Galton was considered a genius by many because of the breadth and originality of his thinking. He is included among the earliest of modern statisticians. His great interest in numbers and in measuring mental differences in people were related to an idea he advocated to which he gave the name "eugenics." Discussions of Galton can be found in: S. J. Gould, *The mismeasure of man* (New York: 1981), 75–77; and J. R. Newman, "Commentary on Sir Francis Galton," in *The world of mathematics*, edited by J. R. Newman (New York: 1956), 1167–72. Galton's 1853 African travelogue provides a fascinating glimpse into European reactions upon just meeting others so different from themselves. Now, 125 years later, those reactions can hardly be used as scientific evidence about nonliterate peoples.

20. A. D. Alexsandrov, *op. cit.*, (ref. 7), p. 7; L. Brunschvicg, *Les étapes de le philosophie mathématique* (Paris: 1972; original edn., 1912), 7; T. Dantzig, *op. cit.* (ref. 7), p. 4–6; M. de Villiers, *op. cit.* (ref. 7), 46–47; R. Dubish, *op. cit.* (ref. 7), 3–4, 7, 11; W. C. Eels, "Number systems of the North American Indians," *American mathematical monthly*, 20 (1913): 263–72, 293–9; H. W. Eves, *op. cit.* (ref. 15), 5–6; G. Ifrah, *op. cit.* (ref. 5), 11–12; C. J. Keyser, *Mathematics as a culture clue and other essays* (New York: 1947), 33–34; M. Kline, *op. cit.* (ref. 7), 13; R. W. Marks, "Introduction," in *The growth of mathematics*, edited by R. W. Marks (New York: 1964), 1–4, p. 1; K. Menninger, *op. cit.* (ref. 7), 9–10; D. E. Smith, *op. cit.* (ref. 7), 7; M. Wertheimer, *op. cit.* (ref. 17), p. 267; A. N. Whitehead, "Mathematics as an element in the history of thought," in *The growth of mathematics*, edited by R. W. Marks (New York: 1964), 7–24, pp. 7–8.

21. The exact formulation of psychic unity varies from text to text, but usually within the first ten pages there is a statement to the effect that the mental capacities of human beings throughout the world are the same. For example, contrast L. H. Morgan, *Ancient society* (New York: 1877), 8, 17–18, with E. B. Tylor, *op. cit.* (ref. 3), p. 6, 33.

22. For a discussion of the influence of Lévy-Brühl, see J. Cazeneuve, *Lucien Lévy-Brühl* (New York: 1972).

23. L. Lévy-Brühl, *How natives think* (London, England: 1926; original French edn., 1910).

24. R. L. Carneiro, *op. cit.* (ref. 1), 90–93.

25. E. B. Tylor, *op. cit.* (ref. 3), 31, 246, 272; J. Lubbock, *Prehistoric times* (London, England: 1865), 569; and idem, *The origin of civilization and the primitive condition of man: Mental and social condition of savages* (New York: 1870), 4–5.

26. Through the years Lévy-Brühl modified his opinions, although his posthumously published work shows that his position on number remained the same. See L. Lévy-Brühl, *The notebooks on primitive mentality* (New York: 1975), 143–4. Others followed Lévy-Brühl dressing his ideas in new clothes. See, for example, the substitution of "incomplete logic" for "prelogical" and "pre-operative" for "childlike" in C. R. Hallpike, *Foundations of primitive thought* (Oxford, England: 1979).

27. See L. Brunschvicg, *op. cit.* (ref. 20).

28. K. Menninger, *Zahlwort and Ziffer* (Göttingen, Germany: 1957); idem, *op. cit.* (ref. 7), and others, e.g., Ifrah, *op. cit.* (ref. 5).

29. E. Durkheim, *The elementary forms of religious life* (New York: 1915), 270–1; C. Lévi-Strauss, *The savage mind* (Chicago: 1966).

30. L. Lévy-Brühl, *op. cit.* (ref. 23), p. 221.

31. R. H. Barnes, "Number and number use in Kédang, Indonesia, *Man*, xvii (1982): 1–22.

32. J. F. Hamill, "Trans-cultural logic: Testing hypotheses in three languages," in *Discourse and inference in cognitive anthropology*, edited by M. D. Loflin and J. Silverberg (The Hague: 1978), 19–43, p. 31; M. D. Loflin, "Discourse and inference in cognitive anthropology," ibid., 3–16, p. 4.

33. J. Silverberg, "The scientific discovery of logic: The anthropological significance of empirical research on psychic unity (inference-making)", ibid., 281–95.

34. S. B. Barnes, "Natural rationality: A neglected concept in the social sciences," *Philosophy of the social sciences*, vi (1976): 115–26.

35. E. Hutchins, "Reasoning in Triobriand discourse," *Quarterly newsletter of the Laboratory of Comparative Human Cognition*, 1(1979): 13–17.

36. D. E. Cooper, "Alternative logic in 'primitive thought,'" *Man*, 10 (1975): 238–56.

37. J. McCarthy, "Circumscription—a form of non-monotonic reasoning," *Artificial intelligence*, xiii (1980): 27–39.

38. L. A. Zadeh, "Coping with the imprecision of the real world: An interview," *Communications of the Association for Computing Machinery*, 27 (1984): 304–11.

39. In the early 1930s, syllogisms were used to investigate the reasoning abilities of nonliterate people in Uzbekistan, Central Asia. For a discussion of this see A. R. Luria, *Language and cognition*, edited by J. V. Wertsch (Washington, DC: 1981). Luria believed syllogisms could "serve as a model for those techniques of language that make logical thinking possible" (ibid., 201). Frequently, responses were in terms of direct personal experience instead of the premises stated or there was a refusal to answer when the respondent had no knowledge of the subject. Comparing this with other responses, it was concluded that logical thinking is a concomitant of the more developed socio-economic systems.

40. S. Scribner, "Modes of thinking and ways of speaking: Culture and logic reconsidered," in *Thinking—readings in cognitive science*, edited by P. N. Johnson-Laird and P. C. Watson (New York: 1977), 483–500, p. 490.

41. Ibid., 487.

42. Ibid., 486.

43. M. Cole, "An ethnographic psychology of cognition," in P. N. Johnson-Laird and P. C. Watson, *op. cit.* (ref. 40), 468–82; M. Cole, J. Gay, J. A. Glick and D. W. Sharp, *Cultural context of learning and thinking* (New York: 1971); M. Cole and P. Griffin, "Cultural amplifiers reconsidered," in *The social foundations of language and thought*, edited by D. R. Olsen (New York: 1980), 343–64;

Laboratory of Comparative Human Cognition, "Cognition as a residual category in anthropology," *Annual review of anthropology*, vii (1978): 51–69.

44. P. N. Johnson-Laird and P. C. Watson, "A theoretical analysis of insight into a reasoning task", in P. N. Johnson-Laird and P. C. Watson, *op. cit.* (ref. 40), 143–58; J. Piaget, "Intellectual evolution from adolescence to adulthood", ibid., 158–65.

45. M. Cole, J. Gay, J. A. Glick, D. W. Sharp, *op. cit* (ref. 43), 233. For a similar view stated in a different way, see C. Lévi-Strauss, "The structural study of myth," *Journal of American folklore*, lxviii (1955): 428–44, p. 444.

46. G. E. Marcus and D. Cushman, "Ethnographies as texts," *Annual review of anthropology*, xi (1982): 25–69, p. 42.

47. Focus on number was difficult to broaden as the questions put to nonliterate peoples by ethnologists remained essentially the same. For example, the 1951 edition of *Notes and queries*, British Association for the Advancement of Science, London, contains essentially the same questions as the arithmetic section written for the 1874 edition of the same work. This book is a general guide to anthropological fieldwork.

48. M. Ascher and R. Ascher, *Code of the Quipu databook* (Ann Arbor, MI: 1978); idem, *op. cit.* (ref. 14).

49. D. H. Kelley, *Deciphering the Maya script* (Austin, TX: 1976).

50. G. Urton, *At the crossroads of the earth and sky: An Andean cosmology* (Austin, TX: 1981).

51. P. J. Davis and R. Hersh, *The mathematical experience* (Boston: 1981), 158–9.

52. J. G. Neihardt, *Black Elk speaks* (Lincoln, NB: 1961), 198–200.

53. E. Carpenter, "Eskimo space concepts," *Explorations*, 5 (1955): 131–45.

54. Ibid., 143.

55. K. G. Oatley, "Mental maps for navigation," *New scientist*, lxiv (1974): 863–6; K. G. Oatley, "Inference, navigation, and cognitive maps," in P. N. Johnson-Laird and P. C. Watson, *op. cit.* (ref. 40), 537–47.

56. W. H. Goodenough, "Native astronomy in Micronesia: A rudimentary science," *The scientific monthly*, lxxiii (1951): 105–10; W. H. Alkire, "Systems of measurement on Woleai Atoll, Caroline Islands," *Anthropos*, lxv (1970): 1–71. Figure 2-2 follows the diagrams in ibid., 53 and in K. G. Oatley, *op. cit.* (ref. 55, 1974), p. 866.

57. It is usually difficult to sail in the direction one wishes to go. For example, it is impossible to sail directly into a wind. Tacking is sailing back and forth at angles to the wind. Changing from one tack to another across the wind requires changing the setting of the sails from one side of the boat to the

other. Because of their design, this is harder to do on the boats of the Pacific navigators, as their entire mast with sails must be moved to the other end of the boat. The tacks by the Pacific navigators are sometimes, therefore, as long as 120 miles. For a fuller discussion see T. Gladwin, *East is a big bird—navigation and logic on Pulwat Atoll* (Cambridge, MA: 1970).

58. K. G. Oatley, *op. cit.* (ref. 55).

59. C. J. Keyser, *op. cit.* (ref. 20), p. 223.

60. Barry Alpher, personal communication 1979.

61. W. E. H. Stanner, "The dreaming", in *Australian signposts*, edited by T. A. G. Hungerford (Melbourne, Australia: 1956), 51–65.

62. Figure adapted from T. G. H. Strehlow, *Aranda traditions* (Melbourne, Australia: 1947), 174.

63. P. Courrège, "Un modèle mathématique des structures élémentaries de parenté," in *Anthropologie et calcul*, ed. by P. Richard and R. Jaulin (Paris: 1971), 126–81.

64. J. Layard, *Stone men of Malekula: Vao* (London: 1942), 98.

65. A. B. Deacon, *Malekula, a vanishing people in the New Herbrides*, edited by C. H. Wedgewood (London: 1934), pp. xxii–iii.

66. Ibid., p. xxiii.

67. A. B. Deacon, "The regulation of marriage in Ambrym," *Journal of the Royal Anthropological Institute*, lvii (1927): 325–42.

68. Ibid., 331.

69. Ibid., 332.

70. Ibid., 329.

71. M. Ascher and R. Ascher, *op. cit.* (ref. 14), 160–4. We plan to discuss these figures at greater length in another paper.

72. R. L. Wilder, *Mathematics as a cultural system* (New York: 1981).

73. C. Zaslavsky, *Africa counts* (Westport, CT: 1979; original edn., 1973).

74. C. J. Keyser, *op. cit.* (ref. 20); M. Kline, *op. cit.* (ref. 16); M. Kline, *op. cit.* (ref. 7); O. Spengler, *The decline of the West* (New York: 1926), i; L. Wittgenstein, *op. cit.* (ref. 16).

75. S. MacLane, "Mathematical models: A sketch for the philosophy of mathematics," *American mathematical monthly*, lxxxviii (1981): 462–72.

Section II

Uncovering Distorted and Hidden History of Mathematical Knowledge

Arthur B. Powell and Marilyn Frankenstein

Freire (1970, 1973) insists that in our struggle toward human liberation, the "culture of silence" represents a major obstacle. Through its mechanisms, the oppressed participate in their own domination by internalizing views of oppressors and by not speaking or otherwise acting against those oppressive views. In the United States, this culture of silence surrounding mathematical knowledge is fueled by the ideology of "aptitudes"—the deep-seated belief that "a difference in essence among human beings . . . predetermines the diversity of psychic and mental phenomena" (Bisseret, 1979, p. 2). Particular individuals, and various communities believe, speak and act as if they have nonmathematical minds. In the case of women, researchers have explored the structural, emotional, and culturally conditioned cognitive factors that lead women to believe men have more mathematical "aptitude." Tobias (1978) and Dowling (1990) discuss research into hidden messages about gender, race, and class in the content and images of mathematics textbooks; Beckwith (1983) summarizes studies of media influence on children's perception of alleged superior mathematics abilities of boys. Ernest (1976) concludes that these beliefs are "the result of many subtle (and not so subtle) forces, restrictions, stereotypes, sex roles, parental-teacher-peer group attitudes, and other cultural and psychological constraints" (p. 11). We argue that another significant reason for the silence around mathematics of so many women and people of color proceeds from the widespread myths presented in Western "his-stories" of mathematics.

The prevailing Eurocentric, and male-centric, myth, expressed and believed by many Western mathematicians, such as Kline (1953), is that:

> [mathematics] finally secured a firm grip on life in the highly congenial soil of Greece and waxed strongly for a short period. . . . With the decline of Greek civilisation, the plant remained dormant for a thousand years . . . when the plant was transported to Europe proper and once more imbedded in fertile soil (pp. 10–11).

This and other myths permeate so deeply the history of mathematics that even the images of mathematicians presented in textbooks, such as Euclid, who lived and studied in Alexandria, are "false portraits . . . which portray them as fair Greeks not even sunburned by the Egyptian sun." There are no actual pictures of Euclid and no evidence to suggest that he was not a black Egyptian (Lumpkin, 1983, pp. 104–105/reprinted here as chapter 5). Joseph (1987/reprinted here as chapter 3) discusses the cosmopolitan, racially diverse nature of Alexandrian society, "a meeting place for ideas and different traditions . . . [involving] continuing cross-fertilisation between different mathematical traditions, notably the algebraic and empirical traditions of Babylonia and Egypt interacting with the geometric and anti-empirical traditions of classical Greece" (p. 18). African, Egyptian, Alexandrian society created the environment in which some of its citizens (and probably their students)—for example, Euclid, Archimedes, Apollonius, Diophantus, Ptolemy, Heron, Theon, and his daughter Hypatia—contributed to the development of mathematics.

We gain further insight into why such myths were created and perpetuated, denying communities and cultures their history, when we examine how racism, sexism, and philosophical or ideological perspectives have impacted academic research, the historiography of mathematics, and foundational issues of mathematics itself. For example, European scholars arbitrarily, yet purposefully, changed the date of the origination of the Egyptian calendar from 4241 to 2773 B.C., claiming that, "such precise mathematical and astronomical work cannot be seriously ascribed to a people slowly emerging from neolithic conditions" (Struik, 1967, pp. 24–25, quoted in Lumpkin, 1983, p. 100).[1] For another example, the name of a key researcher in the theory of the elasticity of metals—the research which made possible the construction of such remarkable engineering feats as the Eiffel Tower—was not listed among the seventy-two scientists whose names are inscribed on that structure. They are all men, and the contribution of Sophie Germain remains unrecognized (Mozans, quoted in Osen, 1974, p. 42). This is just a small example of a much larger historical picture that obliterated knowledge that, in spite of sexism, women did contribute to the mathematical sciences.

Often the historiography of mathematics reveals philosophical and even ideological biases. Yet histories that almost exclusively privilege the mathematics traditionally taught in a "Western" conception neglect important variants. Within "Western" schools of thought on the foundations of mathematics and, in particular, controversies surrounding the calculus, historians and mathematicians "fail" to recognize that radically distinct philosophical and political perspectives generate decidedly different ideas about the nature of the calculus. Aside from professional mathematicians, others worried and wrote seriously about the calculus and its philosophical and theoretical foundations. One such thinker was Karl Marx, whose philosophical ideas on the calculus and alternative theoretical formulations were only first discussed in English by Struik (1948/reprinted here as chapter 8).[2] Still in the current North American effort to update and to create a lean structure and lively pedagogy for the calculus, we still find no account of Marx's ideas on the calculus.

We gain additional insight into the complexity of the eurocentric myth when we note that, although Euclid is adamantly described as "Greek," Ptolemy (circa A.D. 150) whose work dominated astronomy until replaced by Copernicus's theory around 1543, is often described as "Egyptian." Ptolemy's more "practical," applied work could be contrasted to Euclid's more "theoretical" contributions (Lumpkin, 1983, p. 105). Classically, dominant cultures employ a supposed intellectual hierarchy to differentiate their valuation of the products of the oppressed as practical in contrast to their own "theoretical" products. Harris (1987/reprinted here as chapter 10) shows that this distinction continues to denigrate women's knowledge. Diop (1991) discusses a number of cases in which European scholars used this practical-theoretical hierarchy to deny the sophisticated mathematical knowledge of the ancient Egyptians. For example, in discussing the Egyptian formula for the surface area of a sphere ($s = 4\pi r^2$ demonstrated in problem 10 of the Papyrus of Moscow), Diop shows how Peet (1931) "lets his imagination run its course" in a "particularly whimsical effort" to avoid attribution of this mathematical feat to the Egyptians. Instead, Peet tries to demonstrate that problem 10 represents the formula for the surface of a half-cylinder, knowledge which is consistent with the less sophisticated mathematics he believed the Egyptians understood:

> The conception of the area of a curved surface does not necessarily argue a very high level of mathematical thought so long as that area is one which, like that of the cylinder, can be directly translated into a plane by rolling the object along the ground. (quoted in Gillings, 1972, p. 198)

To transform this information in the Papyrus, Peet, "who does not recoil from this difficulty," explains that only one datum[3] is given in problem 10 because the diameter and height of the cylinder were equal, so the one datum represents both values! In addition, Peet supposes that the scribe in charge of recopying the Papyrus must have made a mistake and omitted a statement about the second missing datum (Diop, p. 253). Further, Diop points out that even Gillings, who argued forcefully for the sophisticated mathematical knowledge of the ancient Egyptians, gets caught up in the practical-theoretical dichotomy. After accepting the interpretation of problem 10 as the formula for the curved surface area of a hemisphere, 1500 years ahead of Archimedes, Gillings speculates that:

> Whether the scribe stumbled upon a lucky close approximation or whether their methods were the results of considered estimations over centuries of practical applications, we cannot of course tell. . . . [From murals and other art, one can conclude that] the art of the basket maker or weaver must have been one of some consequence in the Egyptian economic world. When one is weaving baskets which are roughly hemispherical one requires a quantity of material for the circular plane lid that is about half that required for the basket itself. Since the calculation of the area of a circle was a common place operation to the scribes (problem 50 of the Rhind Mathematical Papyrus), over a period of years it could have come to be equally commonplace that the curved area of the hemispherical basket was double that of the circular lid (pp. 200–201).

Diop comments on how absurd it is to think that solely empirical observation, without any theoretical reasoning, could lead to such complex mathematical knowledge. Finally, Diop (1991) remarks how curious it is that

> [i]f the ancient Egyptians were merely vulgar empiricists who were establishing the properties of figures only through measuring, if the Greeks were the founders of rigorous mathematical demonstration, from Thales onwards, by the systemization of "empirical formulas"

> from the Egyptians they would not have failed to boast about such an accomplishment (p. 255).

Bernal (1992/reprinted here as chapter 4) argues forcefully against the denial of the centrality of ancient Egyptians in the development of mathematical and scientific knowledge. For instance, he cites research that "the notion that the Egyptians were better geometers [than the Mesopotamians] fits both with their unparalleled architectural achievement and with their reputation among the Greeks as the founders of geometry and their teachers in it" (p. 605). In another instance, Bernal cites another scholar, de Santillana, who argues against the skepticism with which contemporary historians of science treat ancient Greek writers who claimed Egyptian mathematics and astronomy were superior to their own: "We are asked to admit, then, that the greatest mathematician of Greece [Eudoxos] learned Egyptian and tried to work on astronomy in Egypt without realizing that he was wasting his time" (p. 606).

In another example, Bernal (1987, pp. 272–280) details how this practical-theoretical split was used, in yet another feat of intellectual gymnastics, to resolve the "tensions" around the discoveries of the mathematical knowledge embedded in the structure of the pyramids. If the Greeks were the first "true" mathematicians, how could European scholars explain that such extraordinary mathematical precision, including measurements that lead to important relations such as π, ϕ, and Pythagoras's triangle, had been built into the pyramids by the ancient African Egyptians (described by classical Greek scholar Herodotus as having black skins and woolly hair)?[4] Using sarcasm, Bernal describes how this tension

> is made still more unbearable by the fact that the Greeks had been told about many of the Pyramids extraordinary features and that they believed the Egyptians to have been the first mathematicians and astronomers. Finally, there is the problem that so many of the Greek mathematicians and astronomers had studied in Egypt (p. 277).

Bernal further shows how the simplest resolution—"believe the Greeks and accept . . . that there was an . . . 'axial age' around 3000 B.C.," followed a few centuries later by a sophisticated knowledge of mathematics, built into the pyramids, retained by later Egyptians and passed on by them to visiting Greeks—was "not available to conventional scholars at the height of imperialism" (p. 278). The rejection of this simpler solution persisted in spite of the fact that there is nothing to back the alternative hypothesis—that the Greeks achieved a sudden, qualitative intellectual breakthrough in the fourth century B.C.—"approximating to the actual achievements of the Pyramids and the consistent ancient tradition of a superior Egyptian mathematics" (p. 278). The foundation supporting the alternative "Greek hypothesis" was the argument that the mathematical knowledge embedded in the Pyramids were "chance qualities that had remained totally unsuspected to the constructors . . . [purely the result of] intuitive and utilitarian empiricism" (Lauer, quoted in Bernal, 1987, p. 277–278)—practical, not theoretical.

The above discussion provides some detailed examples of the interaction and intersection of racism[5] and sexism as well as philosophy and ideology with intellectual elitism, which, in part is fueled by the different values attributed to practical and to theoretical work. For the dichotomy in work and value assigned to theory are what Anderson (1990/reprinted here as chapter 14) theorizes as key factors in the alienation that results from capitalist modes of production which "distances people from their creative source and their creativity . . . and allows capital to extract more surplus value from human labor and gain more control over our minds and socio-political activities" (p. 352). Instead, if we understand the creation and development of mathematics as inextricably linked to the material development of society, we can correct and uncover its hidden history.

In ancient agricultural societies, the needs for recording numerical information that demarcated the times to plant, gave rise to the development of calendars such as that on the Ishango bone, approximately 25,000 years old (Marshack, 1991), found at a fishing site of Lake Edwards in Zaire (Zaslavsky, 1983, pp. 111–112; De Heinzelin, 1962, June). And, as African women, for the most part, were the first farmers,[6] they were most probably the first people involved in the struggle to observe and understand nature, and therefore, to contribute to the development of mathematics (Anderson, 1990, p. 354). Then, as societies evolved, the more complex mathematical calculations that were needed to keep track of trade and commerce gave rise to the development of place-value notation by Babylonians (circa 2000 B.C.) (Joseph, 1987, p. 27/reprinted here as chapter 3). And this continues to the present day when for example, military needs and funding drive the development of artificial intelligence (Weizenbaum, 1985).

Notes

1. Lumpkin goes on to report that new discoveries caused Struik to reconsider. In a personal communication to her, he states that "[a]s to mathematics, the Stonehenge discussions have made it necessary to rethink our ideas of what Neolithic people knew. Gillings (1972) has shown the ancient Egyptians could work with their fractions in a most sophisticated way."

2. Translated and published in English in 1993, Marx's *Mathematical Manuscripts* are available from New Park of London, England. Years after Struik's article, Gerdes (1983/1985) wrote the first popular account of portions of Marx's manuscripts.

3. Only one datum, the diameter, is needed in the formula for the surface of a sphere; both the diameter and height are needed in the formula for the surface of a cylinder.

4. See Diop (1991, pp. 103–108) for archeological evidence and hieroglyphic analysis to support the thesis that the ancient Egyptians were black peoples descended from Southern Africans who migrated north.

5. The Eurocentric myth is tenacious, pernicious, and silencing, distorting perspective and inducing myopic vision. This ideological message of the dominant culture becomes internalized by the oppressed. In the United States, for example, a distinguished mathematician, one of the first African-Americans to earn a Ph.D. in mathematics, in a prestigious publication of the Mathematical Association of America, a vehicle for reform in collegiate mathematics education, stated that within the Black community "the tradition of mathematics and science study has *never* been very strong" (Newman, 1989, p. 4, emphasis added). This position disregards the historical contributions of Africans and African Americans to mathematics. His statement also ignores the structural role that the political economy of the United States plays in the mathematical underdevelopment of African Americans. Further, along with hundreds of others, such as Thomas Fuller, Benjamin Banneker, Elbert Frank Cox, Marjorie Lee Browne, and Evelyn Boyd Granville, this mathematician is a sterling example of how, in spite of social and institutional racism, the African-American community produces contributors to the mathematical sciences.

For social and biographical information on Fuller, see Fauvel and Gerdes (1990) and, on Banneker, see Bedini (1972). Concerning biographical information on Cox, Browne, Granville, and Newman, see Newell, Gibson, Rich, and Stubblefield (1980). Also see Giles-Giron (1991) for sketches of Browne, Granville, and Cox.

6. In summarizing recent paleontological and genetic evidence, Stringer (1990) contends that "modern demographic patterns most probably began

with the dispersal of early modern humans from Africa within the past 100,000 years" (p. 101).

References

Anderson, S. E. (1990). Worldmath curriculum: Fighting eurocentrism in mathematics. *Journal of Negro Education* 59(3): 348–359.

Beckwith, J. (1983). Gender and math performance: Does biology have implications for educational policy? *Journal of Education* 165: 159–173

Bedini, S. A. (1972). *The life of Benjamin Banneker*. New York: Scribner.

Bernal, M. (1987). *Black Athena: The afroasiatic roots of classical civilization*. Vol. 1. London: Free Association.

———. (1992). Animadversions on the Origins of Western Science. *ISIS*, 83: 596–607.

Bisseret, N. (1979). *Education, class language and ideology*. London: Routledge and Kegan Paul.

De Heinzelin, J. (1962). Ishango. *Scientific American* 206(6): 105–116.

Diop, C. A. (1991). *Civilization or Barbarism: An authentic anthropology*. New York: Lawrence Hill Books.

Dowling, P. (1990). Some notes towards a theoretical model for reproduction, action, and critique. In R. Noss, A. Brown, P. Drake, P. Dowling, M. Harris, C. Hoyles, and S. Mellin-Olsen. (Ed.), *Proceedings of the First International Conference: Political Dimensions of Mathematics Education: Action & Critique*. London: Institute of Education: University of London.

Ernest, J. (1976). Mathematics and sex. *American Mathematical Monthly* 83: 595–614.

Fauvel, J., and Gerdes, P. (1990). "African slave and calculating prodigy: Bicentenary of the death of Thomas Fuller." *Historia Mathematica* 17: 141–151.

Freire, P. (1970). *Pedagogy of the oppressed*. New York: Seabury.

———. (1973). *Education for critical consciousness*. New York: Seabury.

Gerdes, P. (1985). *Marx demystifies calculus*. Trans. B. Lumpkin. Vol. 16. Minneapolis: Marxist Educational Press. (Original published in 1983 as *Karl Marx: Arrancar o véu misterioso à matemática*. Maputo: Universidade Eduardo Mondlane.)

Giles-Giron, J. Black pioneers in mathematics: Brown, Granville, Cox, Clayton, and Blackwell. *Focus: The newsletter of the Mathematical Association of America*, (Jan–Feb, 1991): p. 18, 21.

Gillings, R. J. (1972). *Mathematics in the time of the pharaohs*. Cambridge, MA: MIT Press.

Harris, M. (1987). "An example of traditional women's work as a mathematics resource." *For the Learning of Mathematics* 7(3): 26–28.

Joseph, G. G. (1987). Foundations of eurocentrism in mathematics. *Race & Class* 28(3): 13–28.

Kline, M. (1953). *Mathematics in western culture*. New York: Oxford.

Lumpkin, B. (1983). Africa in the mainstream of mathematics history. In I. V. Sertima (Ed.), *Blacks in science: Ancient and modern* (pp. 100–109). New Brunswick, NJ: Transaction.

Marshack, A. (1991). *The roots of civilization*. Revised Edition. Mount Kisco, NY: Moyer Bell.

Marx, K. (1983). *Mathematical manuscripts*. London: New Park.

Newell, V. K., Gipson, J. H., Rich, L. W., and B. Stubblefield. (Ed.). (1980). *Black mathematicians and their works*. Ardmore, PA: Dorrance.

Newman, R. (1989). Minorities and mathematics: A case for algebra. *UME Trends* (August): p. 4.

Osen, L. M. (1974). *Women in mathematics*. Cambridge, MA: MIT.

Stringer, C. B. (1990). The emergence of modern humans. *Scientific American* (December): 98–104.

Struik, D. J. (1948). Marx and mathematics. *Science and Society* 12(1): 181–196.

Weizenbaum, J. (1985). Computers in uniform: A good fit? *Science for People* 17(1 and 2), 26–29.

Zaslavsky, C. (1983). The Yoruba number system. In I. V. Sertima (Ed.), *Blacks in science: Ancient and modern* (pp. 110–126). New Brunswick, NJ: Transaction.

Chapter 3

Foundations of Eurocentrism in Mathematics

George Gheverghese Joseph

Editors's comment: George Gheverghese Joseph, a mathematical statistician, provides an important challenge to Eurocentricism through this overview and critical analysis of the dominant historiography of mathematics. Joseph has heavily revised his original article of the same name that appeared in *Race and Class*, 28 (3): 13–28, in 1987, published in England by the Institute of Race Relations. He further elaborates issues raised in this chapter, including specific examples of the mathematics of different cultures, in *The Crest of the Peacock: Non-European Roots of Mathematics* (London: Penguin Books, 1992) and in *Multicultural Mathematics* (London: Oxford University Press, 1993).

Introduction

There exists a widespread Eurocentric bias in the production, dissemination, and evaluation of scientific knowledge. And this is in part a result of the way people perceive the development of science over the ages. For many Third World societies, still in the grip of an intellectual dependence promoted by European dominance during the past two or three centuries, the indigenous scientific base that may have been innovative and self-sufficient during precolonial times is neglected or often treated with a contempt that it does not deserve. An understanding of the dynamics of precolonial science and technology of these societies is essential in formulating a strategy of meaningful adaptation of the indigenous forms that remain to present-day scientific and technological requirements.

Now an important area of concern for anti-racists is the manner in which European scholarship has represented the past and potentialities of nonwhite societies with respect to their achievement and capabilities in promoting science and technology. The progress of Europe and its cultural dependencies[1] during the last four hundred years is perceived by many as inextricably—or even causally—linked with the rapid growth of science and technology during that period. In the minds of some, scientific progress becomes a uniquely European phenomenon that can be emulated by other nations only if they follow a specifically European path of social and scientific development.

Such a representation of societies outside the European cultural milieu raises a number of issues that are worth exploring, however briefly. First, recent studies of India, China, and parts of Africa, contained, for example, in the work of Dharampal (1971), Needham (1954), and Van Sertima (1983), would suggest the existence of scientific creativity and technological achievements long before the incursions of Europe into these areas. If this is so, we need to understand the dynamics of precolonial science and technology in these and other societies and to identify the material conditions that produced these developments.

Second, there is the issue of who "makes" science and technology. In a material and non-elitist sense, each society, impelled by the pressures and demands of its environment, has found it necessary to create a scientific base to cater for its material requirements. The perceptions of what are the particular requirements of a society would vary according to time and place, but it would be wrong to argue that the capacity to "make" science and technology is a prerogative of one culture alone.

Third, if one attributes all significant historical developments in science and technology to Europe, then the rest of the world can impinge only marginally either as an unchanging residual experience to be contrasted with the dynamism and creativity of Europe or as a rationale for the creation of academic disciplines congealed in subjects such as development studies, anthropology, orientalism, sinology, and indology. These subjects then serve as the basis from which more elaborate Eurocentric theories of social development and history are developed and tested.

One of the more heartening aspects of academic research in recent years is that the shaky foundations of these "adjunct" disciplines are being increasingly exposed by scholars, a number of whom originate from countries that provide the "raw materials" of these disciplines. In a recent contribution, Edward Said (1985) points to a number of examples of "subversive" analyses, inspired by similar impulses as his seminal anti-orientalism critique (1978), which are aimed at nothing less than the destruction of the existing Eurocentric paradigmatic norms. For example the growing movement towards promoting a form of indigenous anthropology that sees its primary task as questioning, redefining and, if necessary rejecting particular ideas that grew out of colonial experience in Western anthropology, is well examined in Fahim (1982). In a similar vein, I propose to show that the standard treatment of the history of non-European mathematics exhibit a deep-rooted historiographic bias in the selection and interpretation of facts, and that mathematical activity outside Europe had as a consequence been ignored, devalued, or distorted.[2]

The Historical Development of Mathematical Knowledge

The "Classical" Eurocentric Trajectory

Most histories of mathematics that were to have a great influence on later work were written in the late nineteenth or early twentieth century. During that period, two contrasting developments were taking place that had an impact both on the content and the balance of the books produced on both sides of the Atlantic. Exciting discoveries of ancient mathematics on papyri in Egypt and clay tablets pushed back the origins of written mathematical records by at least 1,500 years. But a far stronger adverse influence was the culmination of European domination in the shape of political control of vast tracts of Africa and Asia. Out of this domination arose the ideology of European superiority which permeated a wide range of social and economic activities, with traces to be found in the histories of science that emphasized the unique role of Europe in providing the soil and spirit of scientific discovery. The contributions of the colonized were ignored or devalued as part of the rationale for subjugation and dominance. And the developments in mathematics before the Greeks—notable in Egypt and Mesopotamia—suffered a similar fate, dismissed as unimportant to the later history of the subject. In his book, *Black Athena* (1987), Martin Bernal has shown how the respect for ancient Egyptian science and civilization, shared by ancient Greece and pre-nineteenth-century Europe alike, was gradually eroded, leading eventually to a Eurocentric model with Greece as the source and Europe as the inheritor and guardian of the Greek heritage.

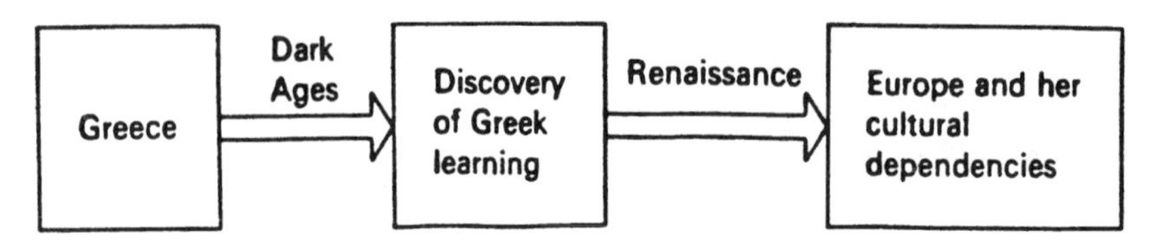

Figure 3–1.

Figure 3–1 presents the "classical" Eurocentric view of how mathematics developed over the ages. This development is seen as taking place in two sections, separated by a period of stagnation lasting over a thousand years; Greece (from about 600 B.C. to A.D. 300) and post-Renaissance Europe from the sixteenth century to the present day. The intervening period of inactivity was the "Dark Ages"—a convenient label that expressed post-Renaissance prejudices about its immediate past and the intellectual self-confidence of those who saw themselves as the true inheritors of the "Greek miracle" of two thousand years earlier.

Two passages, one by a well-known historian of mathematics writing at the turn of the century and the second by a contemporary writer whose books are still widely referred to on both sides of the Atlantic, show the durability of this Eurocentric view and its imperviousness to new evidence and sources:

> The history of mathematics cannot with certainty be traced back to any school or period before that of the Ionian Greeks (Rouse Ball, 1908, p. 1). . . .
>
> [Mathematics] finally secured a firm grip on life in the highly congenial soil of Greece and waxed strongly for a short period. . . . With the decline of Greek civilization the plant remained dormant for a thousand years . . . when the plant was transported to Europe proper and once more imbedded in fertile soil (Kline, 1953, pp. 9–10).

The first statement is a fair summary of what was popularly known and accepted as the origins of mathematics then, except for the neglect of the early Indian mathematics contained in the *Sul-*

basutras (*The Rules of the Cord*), belonging to the period between 800 and 500 B.C., which would make them at least as old as the earliest known Greek mathematics. Thibaut's translation of these works, published around 1875, were known to historians of mathematics at the turn of the century.

The second statement, however, ignores a substantial body of research evidence pointing to the development of mathematics in Mesopotamia, Egypt, China, pre-Columbian America, India, and the Arab world. Mathematics is perceived as an exclusive product of white men and European civilizations. And that is the central message of the Eurocentric trajectory described in figure 3–1.

But this comforting rationale for European dominance became increasingly untenable for a number of reasons. First, there is the fulsome acknowledgment given by ancient Greeks themselves of the intellectual debt they owed to the Egyptians and Babylonians. There are scattered references from Herodotus (fl. 450 B.C.) to Proclus (fl. A.D. 400) of the knowledge acquired from the Egyptians in fields, such as astronomy, mathematics, and surveying, while some commentators even considered the priests of Memphis to be true founders of science.

To Aristotle (fl. 350 B.C.), Egypt was the cradle of mathematics. His teacher, Eudoxus, one of the notable mathematicians of the time, had studied in Egypt before teaching in Greece. Even earlier, Thales (d. 546 B.C.), one of the earliest and greatest of Greek mathematicians, were reported to have travelled widely in Egypt and Mesopotamia and learnt much of their mathematics from these areas. Some sources even credit Pythagoras (fl. 500 B.C.) with having travelled as far as India in search of knowledge, which may explain some of the close parallels between Indian and Pythagorean philosophy and religion.[3]

A second reason why the trajectory described in figure 3–1 was found to be untenable arose from the combined efforts of archaeologists, translators, and interpreters who unearthed evidence of a high level of mathematics practised in Mesopotamia and in Egypt at the beginning of the second millennium B.C., providing further confirmation of Greek reports. In particular, the Babylonians (a generic term that is often used to describe all inhabitants of ancient Mesopotamia) had invented a place value number system, knew different methods of solving quadratic equations (which would not be improved upon until the sixteenth century A.D.) and knew the relationship between the sides of a right-angled triangle which came to be known as the "Pythagorean theorem."[4]

The neglect of the Arab contribution to the development of European intellectual life in general and mathematics in particular is another serious drawback of the "classical" view. The course of European cultural history and the history of European thought are inseparably tied up with the achievement of Arab scholars during the Middle Ages and their seminal contributions to mathematics, natural sciences, medicine, and philosophy. In particular, we owe to the Arabs in the field of mathematics the bringing together of the technique of measurement, evolved from its Egyptian and Babylonian roots to its final form in the hands of Greeks and Alexandrians, with the remarkable instrument of computation (our number system), which originated in India, and the supplementing of these strands with a systematic and consistent language of calculation which came to be known by its Arabic name, "algebra." An acknowledgment of this debt by certain books contrast sharply with a failure to recognize other Arab contributions to science.[5]

Finally, in discussing the Greek contribution, there is a need to recognize the difference between the classical period of Greek civilization (i.e. from 600 to 300 B.C.) and the post-Alexandrian dynasties (i.e. from the third century B.C. to the third century A.D.). In early Eurocentric scholarship, the Greeks of the ancient world were perceived as ethnically homogeneous and originating from areas which were mainly within the geographical boundaries of present-day Greece. It was part of the Eurocentric mythology that from the mainland of Europe had emerged a group of people who had created, virtually out of nothing, the most impressive of all civilization of ancient times. And from that civilization had sprung not only the cherished institutions of the present-day Western culture, but also the mainspring of modern science. The reality, however, is more complex.

The term "Greek," when applied to times before the appearance of Alexander (356–323 B.C.), really refers to a number of independent city-states, often at war with one another, but exhibiting close ethnic or cultural affinities and, above all, sharing a common language. The conquests of Alexander changed the situation dramatically, for at his death his empire was divided among his generals, who established separate dynasties. The two notable dynasties from the point of view of mathematics were the Ptolemaic dynasty of Egypt and the Seleucid dynasty which ruled over territories that included the earlier sites of the Mesopotamian civilization. The most famous center of learning and trade became Alexandria in Egypt, established in 332 B.C. and named after the conqueror. From its foundation, one of its most striking features was its cosmopolitanism—part Egyptian, part Greek, a liberal sprinkling of Jews, Persians, Phoenicians, and Babylonians, and even attracting scholars and traders from as far away as India. A lively contact was maintained with the Seleucid dynasty. Alexandria thus became the meeting-place for ideas and different traditions. The character of Greek mathematics began to change slowly, mainly as a result of the continuing cross-fertilization between different mathematical traditions, notably the algebraic and empirical basis of Babylonian and Egyptian mathematics interacting with the geometric and anti-empirical traditions of the early Greek mathematics. And from this mixture came some of the greatest mathematicians of antiquity—notably Archimedes and Diophantus. It is, therefore, misleading to speak of Alexandrian mathematics as Greek, except in so far as the term shows that Greek intellectual and cultural traditions served as the main inspiration and the Greek language as the medium of instruction and writing in Alexandria. In that sense, our use of the term "Greek" is closely analogous to the use of the term "Arab" to describe a civilization which contained a number of ethnic and religious groups, but all of whom were imbued with the Arabic culture and language.

A Modified Eurocentric Trajectory

Figure 3–2 below takes on board some of the objections raised about the "classical" Eurocentric trajectory. The figure acknowledges that there is some awareness of the existence of mathematics before the Greeks, and of their debt to earlier mathematical traditions, notably those of Babylonia and Egypt. But this awareness is all too likely to be tempered with dismissive rejections of their importance compared to Greek mathematics: "the scrawling of children just learning to write as opposed to great literature" (Kline, 1962, p. 14).

The differences in character of the Greek contribution before and after Alexander are recognized to a limited extent in figure 3–2 by the separation of Greece from the Hellenistic world (in which the Ptolemaic and Seleucid dynasties became the crucial instruments of mathematical creation). There is also some acknowledgement of the Arabs, but mainly as custodians of Greek learning during the Dark Ages in Europe.[6] Their role as transmitters and creators of knowledge is ignored. So are the contributions of other civilizations—notably those of China and India. They are perceived either as borrowers from Greek sources, or having made only minor contributions to mainstream mathematical development (i.e. developments eventually culminating in modern mathematics).[7] More recently, histories of mathe-

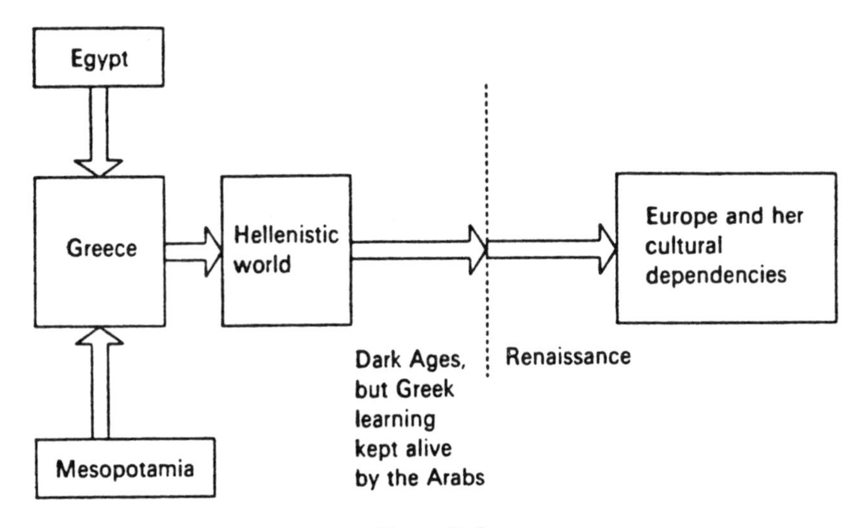

Figure 3–2.

matics carry separate chapters, serving as "residual" dumps, entitled "Oriental" mathematics or "Indian/Chinese" mathematics, which are of marginal relevance to the mainstream themes pursued in these books. This marginalization of non-European mathematics is reflected in the nature of the scholarships that characterizes the treatment of these subjects in successive textbooks. An openness to more recent research findings, especially in the case of Indian and Chinese mathematics, is sadly missing. As a consequence, paraphrases of the contents of earlier texts or quotes from individuals whose scholarship or impartiality have been seriously questioned are reproduced in each succeeding generation of textbooks.[8]

Figure 3–2 therefore remains a flawed representation of how mathematics developed: it contains a series of biases and remains quite impervious to new evidence and arguments. With minor modifications, it remains the model to which many recent books on the history of mathematics conform. It is interesting that a similar Eurocentric bias exists in other disciplines as well: for example, diffusion theories in anthropology and social geography indicate that "civilization" spreads from the Center (i.e., "Greater" Europe) to the Periphery (i.e., to the rest of the world). And the theories of modernization or evolution developed within some Marxist frameworks are characterized by a similar type of Eurocentrism. In all such conceptual schemes, the development of Europe is seen as a precedent for the way in which the rest of the world will follow—a trajectory whose spirits is not dissimilar to the one suggested in figures 3–1 and 3–2.

An Alternate Trajectory for the Dark Ages

If we are to construct an unbiased alternative to figures 3–1 and 3–2, the guiding principle should be to recognize that different cultures in different periods of history have contributed to the world's stock of mathematical knowledge. Figure 3–3 presents such a trajectory of mathematical development, but confines itself to the period between the fifth and fifteenth centuries A.D.—the period represented by the arrow labelled in figures 3–1 and 3–2 as the "Dark Ages" in Europe. The choice of this trajectory as an illustration is deliberate: it serves to highlight the variety of mathematical activity and exchange between a number of cultural areas that went on while Europe was in deep slumber.

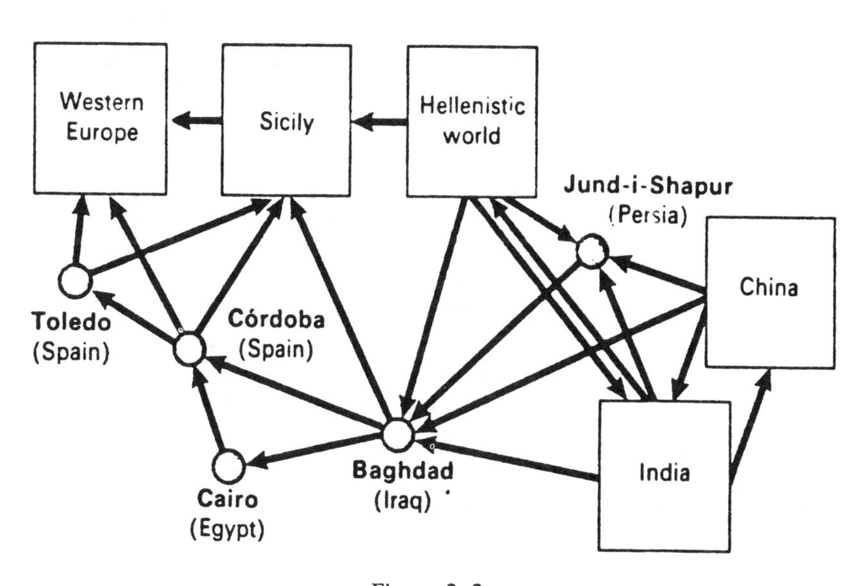

Figure 3–3.

The role of the Arabs is brought out in figure 3–3. Scientific knowledge which originated in India, China, and the Hellenistic world was sought out by Arab scholars and then translated, refined, synthesized, and augmented at different centers of learning, starting with Jund-i-shapur in Persia around the sixth century (even before the coming of Islam), and then moving to Baghdad, Cairo and finally to Toledo and Cordoba in Spain, from where this knowledge spread into Western Europe. Considerable resources were made available to the scholars through the benevolent patronage of the caliphs—the Abbasids (the rulers of the Eastern Arab Empire with its capital at Baghdad) and the Ummayads (the rulers of the Western Arab Empire with its capital first at Damascus and later at Cordoba).

The role of the Abbasid caliphate was particularly important for the future development of mathematics. The caliphs, notably al-Mansur (754–775), Harun al-Rashid (786–809) and al-Mamun (A.D. 809–833), were in the forefront of promoting the study of astronomy and mathematics in Baghdad. Indian scientists were invited to Baghdad. When Plato's Academy was closed in 529, some of its scholars found refuge in Jund-i-shapur, which a century later became part of the Arab world. Greek manuscripts from the Byzantine Empire, the translations of the Syriac schools of Antioch and Damascus, the remains of the Alexandrian library in the hands of the Nestorian Christians at Edessa were all eagerly sought out by Arab scholars, aided and abetted by the rulers who had control over or access to men and materials from the Byzantine Empire, Persia, Egypt, Syria and places as far east as India and China.

Caliph al-Mansur built at Baghdad a *Bait al-Hikma* (House of Wisdom) which contained a large library for manuscripts collected from various sources; an observatory which became a meeting place of Indian, Babylonian, Hellenistic, and probably Chinese astronomical traditions; and a university where scientific research continued apace. A notable member of this institution Mohammed ibn-Musa al-Khwarizmi (fl. A.D. 825) wrote two books which were of crucial importance to the future development of mathematics. One of the books, the Arabic text of which is extant, is entitled *Hisab al-djabr wa-al muqabala* (which may be loosely translated as the *Calculation by Restoration and Reduction*). The title refers to the two main operations in solving equations: "restoration," the transfer of negative terms from one side of the equation to the other, and "reduction," the merging of like terms on the same side into a single term. In the twelfth century the book was translated into Latin under the title *Liber Algebrae Et Almucabola*, thus giving a name to a central area of mathematics.

Al-Khwarizmi wrote a second book, of which only a Latin translation is extant: *Algorithmi De Numero Indorum*, which explained the Indian system of numeration. While al-Khwarizmi was at pains to point out the Indian origin of this number system, subsequent translations of the book attributed not only the book but also the numerals to the author. Therefore in Europe any scheme using these numerals came to be known as an algorism or later algorithm (a corruption of the name of al-Khwarizmi) and the numerals themselves as Arabic numerals.

Figure 3–3 shows the importance of two areas of southern Europe in the transmission of mathematical knowledge to Western Europe. Spain and Sicily were the nearest points of contact with Arab science and had been under Arab hegemony, Cordoba succeeding Cairo as the center of learning during the ninth and tenth centuries. Scholars from different parts of Western Europe congregated in Cordoba and Toledo in search of both ancient and contemporary knowledge. It is reported that Gherardo of Cremona (c. 1114–1187) went to Toledo after its recapture by the Christians, in search of Ptolemy's *Almagest*, an astronomical work of great importance produced in Alexandria during the second century A.D. He was so taken by the intellectual activity there that he stayed over twenty years, during which he is reported to have copied or translated eighty manuscripts of Arab science or Greek classics, which were later disseminated across Western Europe. Gherado was just one of a number of European scholars, including Plato of Tivoli, Adelard of Bath, and Robert of Chester, who flocked to Spain in search of knowledge.

The main message of figure 3–3 is that it is dangerous to characterize mathematical development solely in terms of European developments. The darkness that was supposed to have descended over Europe for a thousand years before the illumination that came with the Renaissance did not interrupt mathematical activity elsewhere. Indeed, the period saw great activity in other parts of the world, a discussion of which will be found in Joseph (1992).

There are two additional features of mathematical knowledge that figure 3–3 could serve to highlight. First, it is not generally recognized that practically all topics taught in school mathematics today are directly derived from the mathematics originating outside Western Europe before the fifteenth century A.D. The failure to recognize this fact is partly a function of the heavily Eurocentered nature of school curricula and partly due to the unwarranted neglect of history (and particularly non-Eurocentric history) of mathematics in a typical mathematics classroom. Second, figure 3–3 shows the one-way traffic of mathematical knowledge into Western Europe up to the fifteenth century. Thus, the Arab mathematical renaissance between the eighth and twelfth centuries shaped and determined the pace of developments in the subject for the next five hundred years.

The Anatomy of Eurocentric Bias

The Eurocentric historiography of mathematics exhibits certain features which may explain the biases that result. First, there is a general disinclination to locate mathematics in a materialist base and thus link its development with economic, political, and cultural changes. Second, there is a tendency to perceive mathematical pursuits as confined to an elite, a few who possess the requisite qualities or gifts denied to the vast majority of humanity. This is a view prevalent even today in the classroom and thus determines what is taught and who benefits from learning mathematics. Third, there is a widespread acceptance of the view that mathematical discovery can only follow from a rigorous application of a form of deductive axiomatic logic, which is perceived as a unique product of Greek mathematics. As a consequence "intuitive" or empirical methods are dismissed as of little relevance in mathematics. Finally, the presentation of mathematical results must conform to the formal and didactic style following the pattern set by the Greeks over 2,000 years ago. And, as a corollary, the validation of new additions to mathematical knowledge can only be undertaken by a small, self-selecting côterie whose control over the acquisition and dissemination of such knowledge through journals has a highly Eurocentric character today.

As an illustration of how the features listed above can create Eurocentric bias, let us examine the status ascribed to mathematical pursuits which do not conform to the criteria mentioned in the last paragraph, notably in Egypt and Mesopotamia before the emergence of Greek mathematics.

A commonly expressed view is that, before the Greeks, there was no mathematics in the sense of the characteristic intellectual activity which goes under that name today. The argument goes: pre-Greek mathematics had neither a well-defined idea of "proof" nor any perception of the need for proof. Where the Egyptians or Mesopotamians were involved in activities which could be described as "mathematics," these activities were purely utilitarian, such as the construction of calendars, parcelling out land, administration of harvests, organization of public works (e.g., irrigation or flood control), or collection of taxes. Empirical rules were devised to help undertake these activities, but there is no evidence of any overt concern with abstractions and proofs which form the core of mathematics. In any case, the argument continues, the only evidence that we have to assess the mathematics of these two civilizations amounts to little more that the exercises that school children of today are expected to work out, which merely involve the application of certain rules or procedures; they are hardly "proofs" or results which have universal application.

The word "proof" has different meanings, depending on its context and the state of development of the subject. To suggest that because existing documentary evidence does not exhibit the deductive, axiomatic, logical inference characteristic of much modern mathematics, these cultures did not have an idea of proof, would be misleading. Generalizations about the area of a circle and the volume of a truncated pyramid are found in Egyptian mathematics. Checking the correctness of a division by a subsequent multiplication or verifying the solutions of different types of equation by the method of substitution are found in Babylonian mathematics. A method in common use in Europe until about a hundred years ago for solving linear equations is generally known as the method of "false position."[9] This method was in common use to solve practical problems such as finding the potency of beer or obtaining optimal feed mixtures for cattle and poultry in Egyptian and Babylonian mathematics. As Gillings (1972) has argued, Egyptian "proofs" are rigorous without being symbolic, so that typical values of a variable are used and generalization to any other value is immediate. Or again, generalizations of the methods used in solving problems contained in the Ahmes papyrus (c. 1650) and the Moscow papyrus (c. 1850 B.C.) two of the most important mathematical documents from Egypt, involve applications of the same procedure to one example after another. To illustrate, consider one of the "lesson texts" dating back to the time of the first Babylonian Dynasty of Hammurabi (c. 1700 B.C.), translated and interpreted by Neugebauer (1935). For the sake of simplicity, I have converted the quantities expressed in base 60 (i.e., sexagesimal) system to our base 10 (i.e., decimal system).

Problem
Length (*ush*), breadth (*sag*). I have multiplied length and breadth, thus obtaining the area (*asha*). Then I added to the area the excess of length over breadth: 183 (was the result). Then I added length and breadth: 27. Required (to obtain) length, breadth and area.

Solution
Given: 27 and 183, the Sums.
Result: 15 length, 12 breadth, 180 area.

Method
One follows this method: [step 1] $27 + 183 = 210$; $2 + 27 = 29$
Take one half of 29 and square it: [step 2] $(14.5)^2 = 210.25$

Subtract 210 from the result: [step 3] $210.25 - 210 = 0.25$
Take the square root of 0.25 [step 4] Square root of $0.25 = 0.5$
Then, length $= 14.5 + 0.5 = 15$
breadth $= (14.5 - 0.5) - 2 = 12$
area $= 15 \times 12 = 180$

Solution using present-day notation
Let length $= x$ and breadth $= y$. Then the problem is solved by evaluating the two equations:

$$xy + x - y = 183 \quad (1a)$$

$$x + y = 27 \quad (1b)$$

Now define a new variable y^* such that $y^* = y + 2$
Then (1a) and (1b) can be re-written as:

$$xy^* = 27 + 183 = 210 \quad (2a)$$

$$x + y^* = 2 + 27 = 29 \quad (2b)$$

[Note: The transformations of (1a) and (1b) to (2a) and (2b) are shown by step1]
The general system of equations of which (2a)–(2b) is a particular case is:

$$xy^* = p$$
$$x + y^* = s$$

And the solution is:

$$x = 1/2s + w \quad (3a)$$

$$y^* = 1/2s - w \quad (3b)$$

where $w =$ square root of $[(1/2s)^2 - p$
Substituting $p = 210$, $s = 29$ gives $w = 0.5$, which can then be used to evaluate $x = 15$, $y = y^* - 2 = 12$ and area $= 180$.

What the Babylonian method involved was the step-by-step application of the general formula, expressed in modern algebraic symbolism, given in (3a) and (3b) to numbers. The Sumerian symbols *ush* and *sag*, for length and width respectively, serve the same purpose as our algebraic symbols x and y. And instead of providing a formula for the solutions of this type of problem, the Babylonians gave one example after another, just as an elementary school textbook may do today to ensure that the method is correctly applied. Such a demonstration may be as effective as formal "proofs" in problems of this nature.

This problem is also indicative of the level of sophistication reached by Babylonian mathematics. To dismiss such a work as "scrawlings of children just learning to write" (Kline, 1962) is more a reflection of the author's prejudices than an objective assessment of the real quality of such mathematics.

A further criticism levelled against Egyptian and Babylonian mathematics is that their mathematics was more a practical tool than an intellectual pursuit. This criticism is symptomatic of a widespread attitude, again originating with the Greeks,[10] that mathematics lacking a utilitarian bent is in some sense a finer or better mathematics. This attitude has even percolated right across the mathematics curriculum in schools and colleges.[11] As a consequence, there is both a sense of remoteness and irrelevance associated with the subject among many who study it, and an ingrained elitism among those who teach it. This elitism is translated at a classroom level into a view, often implicit and not spoken, that real mathematics as opposed to "doing sums" is an activity suited for a select few—which when extended provides the broader argument that mathematics is a unique product of European culture. Thus, elitism in the classroom is ultimately linked to the form of intellectual racism that I have described as Eurocentrism.

Countering Eurocentrism in the Classroom

The foregoing analysis illustrates the need to confront and then counter Eurocentrism in mathematics. A commonly expressed view of the educational establishment in this country [England, editors's note] is that while a correction of the Eurocentric bias in history may be a worthwhile exercise, it has little relevance to mathematical activities within the classroom. I have stated elsewhere why I think this is a misconceived view and how an unbiased historical perspective can enrich the quality of mathematical activity in the classroom as well as provide a valuable input into anti-racist education generally. (Joseph, 1984; 1985; 1986; 1994, and Nelson et al., 1993). It would be useful to restate these arguments in the context of the themes explored here.

First, mathematics is shown to have flourished all over the world, with its internal logic providing a point of convergence for different mathematical traditions, without being constrained by geography, gender[12] (see Osen, 1974) or race. Yet within this unity there is an interesting diversity which could serve to entertain and educate at the same time. By bringing to the attention of the students differences in the language and structure of counting systems found across the world, by showing how different calendars and eras operate, or by examining different spatial relations contained in, say, traditional African designs, Indian *rangoli* patterns and Islamic art, they could serve both as useful examples of applied mathematics as well as increase their awareness of cultural diversity.[13]

Second, a historical approach may, if handled carefully, provide a useful materialistic perspective in evaluating contributions made by different societies. The implied myth of the "Greek" miracle in explaining the origins of mathematics will give way to a more balanced assessment of the nature of early mathematical accomplishments. Thus, the Ishango bone, found on a fishing site by the banks of Lake Edward in Zaire dating back about twenty thousand years, was first thought of as a permanent numerical record of unknown objects. A closer study of the notches on it revealed that it may have been a six-month calendar of the phases of the moon.[14] Similarly, an American *quipu* found in Peru was first thought of as an art object consisting of an intricate pattern of woven knots. But it was later recognized that the artefact contained the record of a whole population census, where the knots of varying sizes stood for different numerical magnitudes and different color coding used to show characteristics such as sex and age. In a predominantly pastoral or simple agricultural economy, such ingenious devices were invented to satisfy the main mathematical requirements—the recording and preservation of such information as was required to keep track of the passage of time or predict seasons for planting seed or the coming of rains. But as societies evolved, mathematical demands became more varied and sophisticated, leading, for example, to the discovery of the place value notation by Babylonians (c. 2000 B.C.) for more complex computations, and the eventual adoption 3,000 years later, where mechanical contrivances such as the abacus or rod numerals were no longer sufficient, of our number system (developed by the Indians about 2,000 years ago), when written calculations became essential for trade and commerce. Both the Babylonian invention and the Indian numerals were momentous discoveries at the time, but are taken for granted today.

Finally, if we accept the principle that teaching should be tailored to children's experience of the social and physical environment in which they live, mathematics should also draw on these experiences, which would include in contemporary Britain the presence of different ethnic minorities with their own mathematical heritage. Drawing on the mathematical traditions of these groups, showing that these cultures are recognized and valued, would also help to counter the entrenched historical devaluation of them. Again, by promoting such an approach, mathematics is brought into contact with a wide range of disciplines, including art and design, history and social studies, which it conventionally ignores. Such a holistic approach would serve to augment, rather than fragment, a child's understanding and imagination.

Notes

1. The term "cultural dependencies" is used here to describe those countries—notably the United States, Canada, Australia, and New Zealand—which are mainly inhabited by populations of European origin or with similar historical and cultural roots. For the sake of brevity, the term "Europe" is used from now on to include these areas as well.

2. A concise and meaningful definition of mathematics is virtually impossible. In the context of this article, the following aspects of the subject are particularly highlighted. Mathematics is a global activity which has developed into a worldwide language, with a particular kind of logical structure. It contains a body of knowledge relating to number and space, and prescribes a set of methods for reaching conclusions about the physical world. And it is an intellectual activity which calls for both intuition and imagination in reaching conclusions. Often it rewards the creator with a strong sense of aesthetic satisfaction.

3. These parallels include: (*a*) a belief in transmigration of souls; (*b*) the theory of five elements making up matter; (*c*) the reasons for not eating beans; (*d*) the structure of the religio-philosophical character of the Pythagorean fraternity which resembled Buddhist monastic orders; and (*e*) the contents of the mystical speculations of the Pythagorean school which bear a remarkable resemblance to *Upanishads*. According to Greek tradition, Pythagoras, Thales, Empedocles, Anaxagoras, Democritus, and others undertook journeys to the East to study philosophy and science. While it is farfetched to assume that all these individuals reached India, there is a strong historical possibility that some of them became aware of Indian thought and science through Persia.

4. The statement and demonstration of the so-called Pythagorean theorem is found in varying degrees of detail all over the world. A variety of evidence is now available on the widespread practical use of the theorem among the Babylonians (c1800–1600 B.C.). The Chinese provided a proof of the theorem in their oldest extant mathematical text entitled *Chou Pei* (c. 1000–800 B.C.). The earliest known Indian work on geometry, *Sulbasutra* (c. 800–600 B.C.), contains a discussion of the theorem and its application to construction of altars. It is also worth noting that even though the theorem is universally associated with the name of Pythagoras, there is no evidence that Pythagoras

had either stated or proved the theorem. The earliest Greek proof, which is still to be found in school geometry texts was given by Euclid (fl. 300 B.C.).

5. They include: (*a*) an earlier description of pulmonary circulation of the blood by ibn al-Nafis, usually attributed to Harvey, though there are records of an even earlier explanation in China; (*b*) the first known statement of the refraction of light by ibn al-Haytham, usually attributed to Newton; (*c*) the first known scientific discussion of gravity by al-Khazin, again attributed to Newton; (*d*) the first clear statement of the idea of evolution by ibn Miskawayh, usually attributed to Darwin; and (*e*) the first exposition of the rationale underlying the "scientific method" found in the works of ibn Sina, ibn al-Haytham and al-Biruni but usually credited to Francis Bacon.

6. In a review article, Nisbet (1973) has pointed out how much the myth of a Renaissance occurring in Europe between the fifteenth and sixteenth centuries has persisted, in spite of overwhelming evidence to indicate that there was continuous intellectual development taking place in Europe from the twelfth century.

7. Chinese, Japanese, or Mayan mathematics are often ignored on the grounds that they fall outside the main line of mathematical development that culminated in the European advance of the subject. In the case of other traditions, Eurocentric histories remain largely silent. A notable exception is a recent book by Katz (1993). Its coverage is global and its examination of non-Western mathematical traditions is clear and thorough.

8. One individual who is frequently quoted by historians as an authority on Indian mathematics is G. R. Kaye who was in the service of the Raj at the turn of the century. His interpretations both with regard to dating certain mathematical documents (notably *Sulbasutra* and the *Bakhshali Manuscript*), which he generally tended to put much later than other scholars, usually on fairly flimsy grounds, as well as his tendency to attribute anything significant in Indian mathematics to a Greek origin, have been criticized by notable scholars of ancient Indian mathematics (see for example Datta and Singh, 1935 and 1938, Sarasvati Amma 1979, Srinivasiengar 1967) without apparently making much impression on those who continue to write histories of mathematics in Europe and her cultural dependencies. Hellonocentrism still prevails. For example, Pingree (1981) has prepared a chronology of Indian astronomy notable for the absence of any Indian presence and the ever presence of Greek influences!

9. To solve for x in equation $x + \frac{x}{5} = 24$, the method of false position involves arguing that if $x = 5$, then $x + \frac{x}{5} = 24$ will equal 6. So to obtain the required 24, we need to multiply 6 by 4. Or the correct x value is 20.

10. An important distinction running right across Greek thought has been *arithmetica,* the study of the properties of pure numbers, and *logistica* the use of numbers in practical applications. The cultivation of the latter discipline was to be left to the slaves. A legend has it that when Euclid (fl. 300 B.C.) was asked what was to be gained from studying geometry, he told his slave to toss a coin at the inquirer.

11. There is, however, a discernible movement towards "utilitarian" mathematics in the modern classroom. So the tide may be turning.

12. The contribution of women mathematicians has also been neglected in standard histories of early mathematics, except for the occasional mention of Hypatia (d. A.D. 415) whose cruel death at the hands of a Christian mob is taken by some to represent the end of Alexandrian mathematics.

13. It is not my intention here to enter the controversy regarding the precise meaning of the culture. The relationship between a people who possess a culture and the culture itself is highly complex and very germane to the point under discussion. The term "culture" is used here in an anthropological sense to describe a collection of customs, rituals, beliefs, tools, mores, and so forth, possessed by a group of people who may be related to one another by factors such as a common language, geographical contiguity, or class.

14. Marshack (1972) has argued, on the basis of a close fit observed between the numbers in each group of notches and the astronomical lunar periods, that the Ishango bone offers possible evidence of one of man's earliest intellectual activities, devising sequential notation based on a six-month lunar calendar for activities such as tattooing, decorating, gaming, or ceremonial festivals.

References

Ball, W. W. R. (1908). *A short account of the history of mathematics.* Reprint. New York: Dover, 1960.

Boyer, C. R. (1978). *A history of mathematics.* Princeton, NJ: Princeton University.

Datta, B. B. and Singh, A. N. (1962). *History of hindu mathematics.* 2 Vols. Bombay: Asia Publishing House.

Eves, H. (1983). *An introduction to the history of mathematics.* 5th ed. Philadelphia: Holt, Rinehart and Winston.

Fahim, H. (1982). *Indigenous anthropology in non-western countries.* Chapel Hill, NC: University of North Carolina Press.

Gillings, R. J. (1972). *Mathematics in the time of the pharaohs*. Cambridge, MA: MIT.

Joseph, G. G. (1984). "The multicultural dimension." *Times Educational Supplement, Mathematics Extra* (5 October).

———. (1985). A historical perspective. *Times Educational Supplement, Mathematics Extra* (11 October).

———. (1986). A non-Eurocentric approach to school mathematics. *Multicultural Teaching* 4 (2): 13–14.

———. (1987). Foundations of Eurocentrism in mathematics. *Race and Class* 28, (3): 13–28.

———. (1992) *The crest of the peacock: Non-European roots of mathematics*. London: Penguin.

———. (1994). The politics of anti-racist mathematics. *European Education Journal*. Vol. 26: 67–74.

———. (1994). Different ways of knowing: Contrasting styles of argument in Indian and Greek mathematical traditions. In *Mathematics, Education and Philosophy: An International Perspective*. Edited by P. Ernest. London: Falmer, pp. 194–204.

———. (1995). 'Cognitive encounters in India during the age of imperialism. *Race and Class* 36 (3): pp. 39–56.

Katz, V. J. (1993). *A history of mathematics: An introduction*. New York: Harper Collins.

Kline, M. (1962). *Mathematics: A cultural approach*. Reading, MA: Addison Wesley.

———. (1953). *Mathematics in western culture*. New York: Oxford University.

Marshack, A. (1972). *The Roots of Civilisation*. London: Weidenfeld and Nicolson.

Needham, J. (1954). *Science and civilisation in China*. Vol. 1. Cambridge, England: Cambridge University.

Nelson, D. Joseph, G. G., Williams, J. (1993). *Multicultural mathematics*. Oxford, England: Oxford University.

Neugebauer, O. (1935). *Mathematics Keilschrift-text*. Vol. 2. Berlin: Springer Verlag.

Nisbet, R. (1973). 'The myth of the renaissance'. *Comparative Studies in Society and History* 15 (4 October).

Osen, L. M. (1974). *Women in mathematics*, Cambridge MA: MIT.

Pingree, D. (1981). History of mathematical astronomy in India. In *Dictionary of Scientific Biography.* Vol. 15 (Suppl.); C. C. Gillespie, ed. New York: Charles Scribner's Sons.

Said, E. (1985). Orientalism reconsidered. *Race and Class*. Vol. 26, No. 2.

———. (1978). *Orientalism*. New York and London: Vintage Books.

Sarasvati Amma, T. A. (1979). *Geometry in ancient and medieval india*. Delhi: Motilal Banarsidass.

Scott, J. F. (1958). *A history of mathematics from antiquity to the beginning of the nineteenth century*. London: Taylor and Francis.

Srinivasiengar, C. N. (1967). *The history of ancient indian mathematics*. Calcutta: World Press.

Van Sertima, I. (1983). *Blacks in science.* New Brunswick, NJ: Transaction Books.

Chapter 4

Animadversions on the Origins of Western Science

---

*Martin Bernal**

Editors's comment: Martin Bernal, a professor of government and Near Eastern studies, reviews and then challenges specific claims made for the originality of Greek science, such as the alleged lack of original or abstract ideas among the Egyptians. As in *Black Athena: The Afroasiatic Roots of Classical Civilization*, New Brunswick, NJ: Rutgers University (1987), he provides an invaluable account of how Eurocentric ideology has distorted the history of the development of scientific and mathematical knowledge. This chapter first appeared in *ISIS*, 83: 596–607, in 1992.

I spent the first fifty years of my life trying to escape from the shadow of my father John Desmond Bernal, and hence, among other things, from science and the history of science. Therefore, the trepidation that is proper for anyone who is neither a scientist nor a historian of science writing for *Isis* is multiplied manyfold in my case. Nevertheless, I am grateful for the invitation to put forward my views on the origins of Western science.

Any approach to this question immediately stumbles over the definition of "science." As no ancient society possessed the modern concept of "science" or a word for it, its application to Mesopotamia, Egypt, China, India, or Greece is bound to be an arbitrary imposition. This lack of clarity is exacerbated by the clash between historians, like David Pingree, who are concerned with "sciences" as "functioning systems of thought" within a particular society and those who apply

---

* I could not have begun, let alone completed, this paper without many years of patient help and encouragement from Jamil Ragep, who, it should be pointed out, is far from accepting all my conclusions.

transhistorical standards and see "science" as "the orderly and systematic comprehension, description and/or explanation of natural phenomena . . . [and] the tools necessary for the undertaking including, especially, mathematics and logic."[1] I should add the words "real or imagined" after "natural phenomena."

Pingree denounces the claims of what he calls "Hellenophilia" that "science" is an exclusively Greek invention owing little or nothing to earlier civilizations and that it was passed on without interference to the Western European makers of the "scientific revolution." Puzzlingly, the work of Otto Neugebauer—and his school, including Pingree himself—on the extent and sophistication of Mesopotamian astronomy and mathematics and Greek indebtedness to it, as well as M. L. West's demonstration of the Near Eastern influences on the pre-Socratic cosmologies, appears to have left this kind of thinking unscathed.[2]

There are still defenders of the claim that "Thales [seen as a Greek] was the first philosopher scientist" the word "scientist" being used here in the positivist sense. According to G. E. R. Lloyd, the Greeks were the first to "discover nature," "practice debate," and introduce such specifics as the study of irrational numbers (notably $\sqrt{2}$ and geometrical modeling for astronomy. Lloyd sees the discovery of nature as "the appreciation of the distinction between 'natural' and 'supernatural,' that is the recognition that natural phenomena are not the products of random or arbitrary influences but regular and governed by determinable, sequences of cause and effect."[3] However, it is clear that at least by the second millennium B.C. Mesopotamian astronomy and Egyptian medicine, to take two examples, were concerned with regular and, if possible, predictable phenomena with relatively little supernatural involvement.[4]

It is true that Egyptian medicine contained some religion and magic. At one point even the "scientific" Edwin Smith Papyrus on surgery turns to magical charms. However, E. R. Dodds and others have shown how isolated the natural philosophers' criticism was against the widespread Greek belief in the efficacy of magic.[5] Even Hippocratic medicine, which is generally regarded as highly rational, was institutionally centered on the religious cult of Asclepius and his serpents, which laid great emphasis on the religious practice of incubation. Both the cult and the practice, incidentally, had clear Egyptian roots.[6]

On the question of the alleged uniqueness of Greek "scientific" debate, as we can see from those in Gilgamesh, "debates" are at least as old as literature. Some, such as the "Dispute between a man and his *Ba,"* which dates back to Middle Kingdom Egypt, contain quite profound philosophy. It is also clear that different Mesopotamian, Syrian, and Egyptian cities had not merely different gods but distinct cosmogonies, most of which involved abstract elements or forces without cults, of which the priesthoods of the others were aware. There were also attempted and actual syncretizations, suggesting that there had been debates.[7] This situation resembles that plausibly reconstructed for the cosmological disputes of the pre-Socratics.

Later Greek philosophical and scientific debates clearly owed a great deal to the Sophists, who came from the Greek tradition of "persuasion," with its close association with legal disputes. Oratory, persuasion, and justice are highly valued in nearly all cultures, but, interestingly, they received particular emphasis in Egypt. The central scene in Egyptian iconography is the judicial weighing of the soul of the dead person, and the legal battle between Horus and Seth is a central episode in its mythology. One of the most popular Egyptian texts was that of *The Eloquent Peasant,* which its most recent translator into English, Miriam Lichtheim, describes as "both a serious disquisition on the need for justice and a parable on the utility of fine speech."[8]

I have written elsewhere on the centrality of the image of Egyptian justice to both Mycenaean and Iron Age Greece, and there is no doubt that Greeks of the Classical and Hellenistic periods saw Egyptian law as the ultimate basis of their own. As Aristotle wrote at the end of the *Politics*: "The history of Egypt attests the antiquity of all political institutions. The Egyptians are generally accounted the oldest people on earth; and they have always had a body of law and a system of politics. We ought to take over and use what has already been adequately expressed before us and confine ourselves to attempting to discover what has already been omitted."[9]

While the first attestation of written law in Egypt comes from the tomb of Rekhmire in the fifteenth century B.C., there is no reason to doubt that it existed much earlier.[10] In any event, the Egyptian New Kingdom is sufficiently old by Greek standards. It is clear that what Aristotle was recommending had not hitherto been carried out. Nevertheless, it would seem likely that Aristotle was conventional in his belief that, even though Egyptian and Greek laws were very different in his own day, the true foundation of Greek law and justice lay in Egypt.

The emphasis on law is important both because of its promotion of argument and dialectic and because of the projections of social law into nature and the establishment of regularities.[11] There is no doubt that the Egyptian M3't (*Maat*: "truth," "accuracy," "justice") was central to both social and natural spheres in the same way as the Greek *Moira*, which derived from it. Similarly, it is clear that the Egyptians applied the "justice" of scales to social and legal life at least as early as the Middle Kingdom.[12]

To return to some of the specific claims made for the originality of Greek science, there is now no doubt that Babylonian scholars were concerned with $\sqrt{2}$ Pythagorean triples as well as having a good approximation of π. The Egyptian estimate for π was even more accurate. The standard use in land measurement of the diagonal of a square of one cubit was the so-called double *remen*, that is to say $\sqrt{2}$ the cubit.[13] Thus, the irrational number par excellence was employed in Egypt from the beginning of the second millennium B.C. at the latest; whether or not its irrationality was proved in Euclidean fashion, its use provides circumstantial evidence that Egyptian scribes were aware of the incommensurability of the side and diagonal.

Modern scholars have poured scorn on the widespread ancient tradition holding that Egyptians had known of the "Pythagorean" triangle. However, the very cautious Gay Robins and Charles Shute maintain that knowledge of it is shown by the use in Late Old Kingdom pyramids of a *seked* of $5\frac{1}{4}$ palms, which imposed "a half-base width to height of 3:4 and so could have been modeled on a 3:4:5 right angled triangle."[14]

I shall discuss the strong possibility that geometry, thought to be typically Greek, came from Egypt. However, at this point it would seem difficult to argue that before the second half of the fourth century B.C. any aspect of Greek "science"—with the possible exception of axiomatic mathematics—was more advanced than that of Mesopotamia or Egypt.

Was Neugebauer Right to Dismiss Ancient Traditions of Egyptian Science?

In this section, I should like to take it as given that R. O. Steuer, J. B. de C. M. Saunders, and Paul Ghalioungui have established not merely that Egyptian medicine contained considerable "scientific" elements long before the emergence of Greek medicine, but that Egyptian medicine played a central role in the development of Greek medicine.[15] Similarly, the work of Neugebauer and his school has made it impossible to deny that some Mesopotamian mathematicians and astronomers were "scientific" in the positivist sense and that Mesopotamian "science" in these areas was crucial to the creation of Greek mathematics and astronomy. However, I should like to challenge these scholars' dismissal of claims that there was an Egyptian mathematics that could have had a significant influence on Greek thinkers.

Despite his early passion for ancient Egypt and his considerable work on Egyptian astronomy, throughout his long life Neugebauer insisted that the Egyptians had no original or abstract ideas and that mathematically and scientifically they were not on the same level as the Mesopotamians. He claimed that the accurate astronomical alignments of the pyramids and temples in Egypt and the use of π and could all be explained as the results of practical knacks rather than of profound thought. An example of this approach is the following: "It has even been claimed that the area of a hemisphere was correctly found in an example of the Moscow papyrus, but the text admits also of a much more primitive interpretation *which is preferable.*"[16]

In his *Exact Sciences in Antiquity*, Neugebauer did not argue with the pyramidological school; he simply denounced it, recommending that those interested in what he admitted to be "the very complex historical and archaeological problems connected with the pyramids" read the books by I. E. S. Edwards and J. F. Lauer on the subject.[17]

While Edwards does not involve himself with the pyramidologists and their calculations, the surveyor and archaeologist Lauer did, in the face of opposition from Egyptologists, who were "astonished that we should give so much importance to the discussion of theories which have never had any credit in the Egyptological world." Lauer's work had a certain contradictory quality. He admitted that the measurements exhibited by the pyramids do have some remarkable properties; that one can find such relations as π , ϕ, and Pythagoras's triangle from them; and that these facts generally bear out the claims Herodotus and other ancient writers made for them. On the other hand, he denounced the "fantasies" of pyramidologists and claimed that the formulas according to which the pyramids were aligned and the extraordinary degree of sidereal accuracy they exhibited were purely the result of "intuitive and utilitarian empiricism."[18]

A conflict between the acceptance of the extraordinary mathematical precision of the Great Pyramid and a "certainty" that the Greeks were the first "true" mathematicians runs throughout Lauer's many writings on the subject. The strain is made still harder to bear by Lauer's awareness that some Greeks had been told about many of this pyramid's extraordinary features and that they believed the Egyptians to have been the first mathematicians and astronomers.

Moreover, there was the problem that so many of the Greek mathematicians and astronomers had studied in Egypt. Lauer's honest attempt to deal with these difficulties was the following:

> Even though up to now, no esoteric Egyptian mathematical document has been discovered, we know, if we can believe the Greeks, that the Egyptian priests were very jealous of the secrets of their science and that they occupied themselves, Aristotle tells us, in mathematics. It seems then reasonably probable that they had been in possession of an esoteric science erected, little by little, in the secrecy of the temples during the long centuries that separate the construction of the pyramids, towards the year 2800 [I should put it two hundred years earlier] to the eve of Greek mathematical thought in the sixth century B.C. As far as geometry is concerned, the analysis of buildings as famous as the Great Pyramid would take a notable place in the researches of these priests; and it is perfectly conceivable that they could have succeeded in discovering in it, perhaps long after their erection, chance qualities that had remained totally unsuspected to the constructors.[19]

The question of when Egyptians developed this sophisticated mathematical knowledge is not directly relevant to the topic of this article. However, apart from the precision and intricacy of many of the architectural constructions of the Old Kingdom, there is another argument for the existence of relatively "advanced" mathematics in the first half of the third millennium B.C. This is that although the two great mathematical texts that have survived, the Moscow and the Rhind papyri, come from the Middle Kingdom in the twentieth and nineteenth centuries B.C., some of the problems set in them use measures that belong to the Old Kingdom, which had been discarded by the later period.[20]

Lauer's solution still allowed some later Egyptians to have been capable of relatively advanced thought. He continued:

> For the whole length of the three thousand years of her history, Egypt thus, little by little, prepared the way for the Greek scholars who like Thales, Pythagoras, and Plato came to study, then even to teach, like Euclid at the school in Alexandria. But it was in their philosophic spirit, which knew how to draw from the treasure amassed by the technical positivism of the Egyptians, that geometry came to the stage of a genuine science.[21]

Even this degree of recognition was too much for Neugebauer. As he put it at one point: "Ancient science was the product of a very few men and these few happened not to be Egyptians." In 1981, he published his note "On the Orientation of Pyramids," in which he showed how accurate alignments could be made without sophisticated astronomy, simply by measuring and turning the shadow of a model pyramid or the capstone over a period of some weeks. There is no evidence, one way or the other, whether this was the method used, but it would seem plausible, if only because pyramids appear to have had solar rather than stellar cultic associations. Nevertheless, the requirement of what Neugebauer concedes to be "remarkable accuracy of . . . orientation of the Great Pyramid," a structure of extraordinary sophistication, indicates very serious religious and theoretical concerns.[22] Thus, Neugebauer's choice of the word "primitive" to describe the alignment seems inappropriate, the word is—as we shall see—indicative of his general opinion of the ancient Egyptians.

There is little doubt that this modern view of the Egyptians' lack of mathematics and science has been influenced by a distaste for the theology and metaphysics in which much of Egyptian—and Platonic—knowledge was embedded and by progressivist views that no one who lived so early could have been so sophisticated. It may also have been reinforced by assumptions, almost universal in the nineteenth and early twentieth centuries, that no Africans of any sort could have been capable of such great intellectual achievements.

An indication that such attitudes may have had an impact even on such a magnificent champion of liberalism and foe to racism as Neugebauer comes in one of his bibliographical notes, where the first book he recommended "for a deeper understanding of the background that determined the character of Egyptian arithmetic" was Lucien Lévy-Brühl's *Fonctions Mentales Dans Les Sociétés Inferieures.* Lévy-Brühl was far from the worst of his generation. Nevertheless he belonged to it, and it was appropriate that his work was translated into English as *How Natives Think.*[23]

Having said this, there is no doubt that Neugebauer had some substantial arguments to back his case. The strongest of these were his claims that none of the surviving mathematical papyri from pharaonic Egypt contained what he believed to be sophisticated calculations and that the Egyptians' systems of numbers and fractions were too crude for profound mathematical and astronomical thought of the kind that had been attributed to them. There are seven major arguments against this position.

1. The strong possibility that—pace Neugebauer—the surviving Egyptian papyri do contain "advanced" mathematics.

2. Parallels from Mesopotamia and Ptolemaic Egypt showing that one can not rely on the papyrological record to gauge the full range of pharaonic Egyptian "science."
3. The general agreement that Egyptian geometry was equal to or better than that of Mesopotamia, in conjunction with the conventional wisdom that one of the chief contributions of the Greeks to Mesopotamian "arithmetic" was geometric modeling, which suggests that the geometrical input may well have come from Egypt.
4. The coordination of sophisticated geometry and computation in Egypt with extraordinary practical achievements.
5. The Greek insistence that they learned mathematics—and medicine—not from Mesopotamia but from Egypt.
6. The Greek adoption of an Egyptian rather than a Mesopotamian calendar.
7. The facts that much of Hellenistic and Roman science took place in Egypt, not Greece, and that although they wrote in Greek some of its practitioners, including the astronomer Ptolemy, were Egyptian.

The first argument is buttressed by the fact that, as we have seen, Neugebauer preferred "more primitive interpretations" and therefore could have overlooked evidence of more sophisticated work. Thus, we must allow for the possibility that the surviving texts contain or refer to elements that are more sophisticated than he and some other twentieth-century historians of science have supposed. There is little doubt about the employment of irrational numbers, mentioned above, and the use of arithmetical and geometrical progressions in the Rhind Papyrus problems 40 and 79.[24]

The Soviet scholar V. V. Struve, who was the first to study the Moscow Mathematical Papyrus, was much more respectful than Neugebauer. He wrote, for instance, that "we must admit that in mechanics the Egyptians had more knowledge than we wanted to believe." He was convinced that this papyrus and the Rhind Mathematical Papyrus demonstrated a theoretical knowledge of the volume of a truncated pyramid, and he has been followed in this interpretation by later scholars. Given the many pyramids successfully constructed during the Old and Middle Kingdoms, this would not in itself seem unlikely. Archimedes, however, maintained in the third century B.C. that the volumes of pyramids were first measured by Eudoxos of Knidos a hundred years earlier.[25]

Here, as in some other instances, Archimedes was knowingly or unknowingly mistaken. Even so, it is possible that Eudoxos was the first to transmit the formulas to Greece. Eudoxos spent many years in Egypt and was reported to have learned Egyptian and to have made translations, some of which may well have come from the *Book of the Dead,* into Greek. As Giorgio de Santillana pointed out, it is unlikely that Eudoxos translated these texts merely for their entertainment value; it is much more probable that he believed that they contained esoteric astronomical information.[26] This raises the important suggestion that Egyptian religious and mystical writings and drawings may well contain esoteric mathematical and astronomical wisdom.

To return to earth with the particular case of the measurement of the surface area of either a semicylinder or a hemisphere in the Moscow Papyrus: Richard Gillings, who believes the measurement refers to the latter, describes the Egyptian operations and writes:

> If this interpretation . . . is the correct one then the scribe who derived the formula anticipated Archimedes by 1,500 years! Let us, however, be perfectly clear [that] in neither case has any proof that either $A_{\text{cylinder}} = \frac{1}{2}\pi dh$ or $A_{\text{hemisphere}} = 2\pi r^2$ been established by the Egyptian scribe that is at all comparable with the clarity of the demonstrations of the Greeks Dinostratos and Archimedes. All we can say is that, in the specific case in hand, the mechanical operations performed are consistent with these operations which would be made by someone applying these formulas even though the order and notation might be different.[27]

In general, it is clear that the specifically mathematical papyri give considerable indications of sophisticated operations. As Struve put it in the conclusion of his study of the Moscow Papyrus:

> These new facts through which the Edwin Smith and Moscow papyri have enriched our knowledge, oblige us to make a radical revision of the evaluation made up to now of Egyptian "science" [Wissenschaft]. Problems such as the research into the functions of the brain or the surface area of a sphere do not belong to the range of practical "scientific" questions of a primitive culture. They are purely theoretical problems.

Or earlier:

> The Moscow Papyrus . . . confirms in a striking way the mathematical knowledge of the Egyptian scholars and we no longer have any reason to reject the claims of the Greek writers that the Egyptians were the teachers of the Greeks in geometry.[28]

Objections by Neugebauer and others to Struve's specific interpretation of the surface area of a hemisphere have now been answered.[29] Similarly, as mentioned above, claims for the use of "Pythagorean" triangles and the sophistication of the measurement of the volume of the truncated pyramid have both survived earlier skepticism. If these bases of Struve's general case still stand, should one accept Neugebauer's dismissal of it?

Even if one were to concede Neugebauer's argument that the mathematics contained in these papyri is merely practical and primitive, there is the second argument: the strong likelihood that more sophisticated work was recorded on others that have not been preserved. Lauer raised the point that all reports indicate that the Egyptian priests were secretive about their writings; therefore there would have been few copies and the chances of their survival would have been slim. It should be emphasized that relatively few papyri of any kind have survived. This is very different from Mesopotamia, where the baked clay tablets are remarkably durable and hundreds of thousands of them have been discovered. The problem with Mesopotamian texts is not a lack of them but the difficulty of finding enough Assyriologists to read and publish them. Even here, however, there are gaps in what exists. Neugebauer points out that the "great majority" of mathematical tablets come from one of two periods, the Old Babylonian period—of two hundred years—in the first half of the second millennium B.C. and the Seleucid period. Continuities between the two sets of texts make it clear that sophisticated mathematics was carried out in the twelve or more centuries that intervened. However, there is no record of this.[30]

The situation is far worse in Egypt, and there is no doubt that most of the papyri written and all of those that have survived were texts used for teaching scribes techniques that were useful for practical accounting rather than "state of the art" advanced mathematics.[31] An instructive parallel can be seen in the Ptolemaic period. Many more mathematical papyri have been found from these few centuries than from the whole pharaonic period, yet none of these go beyond Book 1 of Euclid or give any indications of the extraordinary sophistication of the work we know from textual transmission to have been taking place in Hellenistic Egypt. Thus, the argument from silence, which should always be applied sparingly, should be used with particular caution in evaluating the absence of textual proof of advanced Egyptian mathematics.

Against this, it is argued that the few texts that do exist show a consistency of techniques and notation that makes it impossible for the Egyptians to have produced sophisticated mathematics. This brings us to the third argument against skepticism: Egyptian numerical notation may not have been as flexible and helpful as that of the Mesopotamians, but it was, if anything, better than that in the Greeks wrote their sophisticated formulas. There is no doubt that Egyptian mathematics was based on very simple principles; on the other hand, the existing papyri show that extraordinarily elaborate mathematical structures were erected upon them.

Neugebauer admits that while the Egyptians were not as good in their arithmetic as the Babylonians, their geometry was equally good; and if we are to believe other scholars' interpretations of the Moscow Papyrus, Egyptians were able to carry out geometrical operations that were beyond those of the Mesopotamians. The notion that the Egyptians were the better geometers fits both with unparalleled architectural achievements and with their reputation among Greeks as the founders of geometry and their teachers in it.[32]

Given this concern with geometry, it is not surprising that there are direct and indirect proofs that Egyptians relied on plans for their architectural constructions. Struve may have been exaggerating when he wrote, "The Egyptian plans are as correct as those of modern engineers."[33] Nevertheless, there is no reason to suppose that they were inferior to those of the Greeks and Romans.

According to the Egyptians, the tradition of making plans went back to Imhotep, at the beginning of the third dynasty, circa 3000 B.C., but most modern scholars have understood this claim merely as a mythical projection on to deified prototype of all architects. However, it is now proven that architectural plans were used during the Old Kingdom and that Imhotep did design the Step Pyramid and the elaborate complex of buildings around it. Furthermore, a nostracon found at the Step Pyramid does contain measurements for a vault.[34]

This coordination of geometry and computation with architecture constitutes the fourth argument against modern denials that the Egyptians possessed a superior mathematics. While the textual evidence for such knowledge can be construed as ambiguous, the case for it is greatly strengthened by the architectural evidence. In addition to the pyramids there were temples, granaries, and irrigation networks on huge scales that required extraordinary planning and the ability to visualize these structures in advance on writing or drawing surfaces.

The fifth reason for supposing that the Egyptians had sophisticated mathematics is that the Greeks said so. Writers on the subject were unanimous that Egyptian mathematics and astronomy were superior to their own and that while only two Greek mathematicians were supposed to have studied in Mesopotamia, the majority of Greek scientists, astronomers, and mathematicians had studied or spent time in Egypt.

These reports are treated with skepticism by modern historians of science, who *know* that there was no Egyptian science or mathematics worth studying. However, as de Santillana wrote about Eudoxos, who undoubtedly studied in Egypt: "We are asked to admit, then, that the greatest mathematician of Greece learned Egyptian and tried to work on astronomy in Egypt without realizing that he was wasting his time."[35]

There is little doubt that after the Assyrian and Persian conquests the mathematics and astronomy of Egypt drew from both Egyptian and Mesopotamian sources. However, the Greek belief that it was all Egyptian tradition strengthens the case that the native component was significant.

The sixth argument against the skeptics is the fact that the Greeks adopted an Egyptian rather than a Mesopotamian calendar. Apart from the greater convenience of the Egyptian calendar, this adoption is indicative of what seems to have been a wider Greek tendency to draw from nearby Egypt rather than more distant Mesopotamia.

The final argument is that in Hellenistic times, while Athens remained the center of Greek philosophical studies, nearly all "Greek" science took place in Egypt. This was partly the result of Ptolemaic patronage, but if we are to believe Greek and Roman sources, the "scientists" also drew and built on Egyptian wisdom. It is striking that Euclid worked in Egypt at the very *beginning* of the Ptolemaic period, that is to say a mere fifty years earlier Eudoxos had felt the need to learn Egyptian in order to study mathematics and astronomy. Thus, it would seem more accurate to view Euclid's work as a synthesis of Greek and Egyptian geometry than as an imposition of the Greek rational mind on muddled oriental thinking.

While it is true that Babylonian mathematics and astronomy flourished under the Seleucids, as already noted, most of the great "Greek" scientists wrote in Greek but lived in Egypt, and some indeed may have been Egyptian. For possible examples of this, there are the inventor Heron, the algebraist Diophantos and the astronomer Ptolemy, who was known in early Arabic writings as an Upper Egyptian.[36]

It seems to be generally accepted that the great Greek contribution to mathematics and astronomy was the introduction of geometric modeling, in particular the transposition of Mesopotamian arithmetical astronomical cycles into rotating spheres.[37] However, the Greeks themselves believed that geometry developed in Egypt, a view supported by Egyptian architectural sophistication and the mathematical papyri. Furthermore, those most responsible for the Greek view of the heavens as spinning spheres, Plato and Eudoxos, were reported to have spent time in Egypt and were known for their deep admiration of Egyptian wisdom.[38]

We have seen how particularly close Eudoxos's association was with Egyptian priests, and it was precisely Eudoxos who established the new astronomy of complex concentric spheres.

I believe that these seven arguments present a very strong case indeed that there were rich mathematical—particularly geometrical—and astronomical traditions in Egypt by the time Greek scholars came in contact with Egyptian learned priests. After the Assyrian conquest of Egypt, in the seventh century B.C., Egyptian mathematics and astronomy were substantially influenced by Mesopotamian "scientific" thought, a process which continued in Ptolemaic and Roman Egypt.[39] The Egyptian medical tradition appears to have been less affected by Mesopotamia. In general, the "scientific" triumphs of Hellenistic Egypt would seem to the result of propitious social, economic, and political conditions and the meeting of three "scientific" traditions, those of Egypt, Mesopotamia, and Greece. However, the two former were much older than the third, reaching back to the third millennium or beyond, and more substantial. It should also be noted that the point at which the Greeks "plugged into" Near Eastern "science" was Egypt; this was the reason that the Greeks always emphasized the depth and extent of Egyptian wisdom.

The arbitrariness of the application of the word "science" to ancient civilizations was noted at the beginning of this essay. I suppose, like Humpty-Dumpty we can use words more or less as we please. However, the only way to claim that the Greeks were the first Western scientists is to define "science" as "Greek science." If less circular definitions are used, it is impossible to exclude the practice and theory of some much earlier Mesopotamians and Egyptians.

Notes

1. For Pingree see "Hellenophilia versus the History of Science," a lecture originally presented at the department of history of science, Harvard University, 14 November 1990, and now published in this special section. For

the passage cited see G. E. R. Lloyd, *Early Greek Science: Thales to Aristotle* (New York/London: Norton, 1970), p. 1, quoting Marshall Clagett, *Greek Science in Antiquity: How Human Reason and Ingenuity First Ordered and Mastered the Experience of Natural Phenomena*, new corrected ed. (New York: Collier, Macmillan, 1962), p. 15.

2. See Otto Neugebauer, *The Exact Sciences in Antiquity* (New York: Dover, 1969); and M. L. West, *Early Greek Philosophy and the Orient* (Oxford: Clarendon, 1971)—pace John Vallance, "On Marshall Clagett's *Greek Science in Antiquity*," *Isis* (1990) 81:713–721, on p. 715. See also G. S. Kirk, "Popper on Science and the Presocratics," *Mind* (1960) 69:318–339, esp. pp. 327–328; and Kirk, "Common-Sense in the Development of Greek Philosophy," *Journal of Hellenic Studies* (1961) 81 105–117, pp. 105–106.

3. Lloyd, *Early Greek Science*, p. 8.

4. Neugebauer, *Exact Sciences in Antiquity*, (cit. n. 2), pp. 29–52; Paul Ghalioungui, *The House on Life: Per Ankh Magic and Medical Science In Ancient Egypt*, 2nd ed. (Amsterdam: H. M. Israel, 1973); and Ghalioungui, *The Physicians of Ancient Egypt* (Cairo Al-Ahram Center for Scientific translation, 1983).

5. E. R. Dodds, *The Greeks and the Irrational* (Berkeley/Los Angeles: Univ. of California Press, 1951); C. E. R. Lloyd, *Magic, Reason, and Experience* (Cambridge, England: Cambridge Univ. Press, 1979) pp. 10–58, 263–264; and Garth Fowden, *The Egyptian Hermes: A Historical Approach to the Later Pagan Mind* (Cambridge, England: Cambridge Univ. Press, 1986), pp. 81–82.

6. J. B. de C. M. Saunders, *The Transition From Ancient Egyptian to Greek Medicine* (Kansas: Lawrence Univ. Kansas Press, 1963), p. 12.

7. See E. A. Wallis Budge, *The Gods of the Egyptians: Studies in Egyptian Mythology*, vol. 1 (London: Methuen, 1904), pp. 282–281; and Marshall Clagett, *Ancient Egyptian Science*, vol. 1: *Knowledge and Order*, 2 pts. (Philadelphia: American Philosophical l Society, 1989), pp. 263–372. For an annotated translation of the "Dispute between a man and his *Ba*" see Miriam Lichtheim *Ancient Egyptian Literature*, 3, vol. 1: *The Old and Middle Kingdoms* (Berkeley/Los Angeles: Univ. of California Press, 1975), pp. 163–169.

8. Lichtheim, *Ancient Egyptian Literature*, vol. 1, p. 169.

9. Aristotle, *Politics*, 7.10, translated by Ernest Barker in *The Politics of Aristotle* (Oxford: Oxford Univ. Press, 1958), p. 304. See Martin Bernal, "Phoenician Politics and Egyptian Justice in Ancient Greece," in *Anfänge politischen Denkens in den Antire*, edited by Kurt Raaflaub (Munich: Historisches Kollegs, Kolloquien 24, 1993), pp. 241–261.

10. A. Theodorides, "The Concept of Law in Ancient Egypt," in *The Legacy of Egypt*, edited by J. R. Harris (Oxford: Clarendon Press, 1971), pp. 291–313; and Anne Burton *Diodorus Siculus Book 1: A Commentary* (Leiden: Brill 1972), pp. 219–225.

11. See Joseph Needham, *Science and Civilization in China*, vol. 2 (Cambridge: Cambridge Univ. Press, 1956), pp. 518–583.

12. Bernal "Phoenician Politics and Egyptian Justice," Clagett subtitles *Ancient Egyptian Science* "Knowledge and Order," (cit. n. 7), pp. xi–xii. These words are translations of the Egyptian rḫt and M3't, which he sees as The Egyptian "rudimentary" science. See The constant references to balances and plumblines as symbols of justice in *The Eloquent Peasant*, translated by Lichtheim, Ancient Egyptian Literature (cit. n. 7), vol. 1, pp. 170–182.

13. For the triples see the bibliography at the end of Olaf Schmidt, "On Plimton 322: Pythagorean Numbers in Babylonian Mathematics." *Centaurus* (1980) 4:4–13, on p. 13. On the Egyptian estimate for π see Richard Gillings, *Mathematics in the Time of the Pharaohs* (New York: Dover, 1972), pp. 142–143; and Gay Robins and Charles Shute, *The Rhind Mathematical Papyrus: An Ancient Egyptian Text* (London: British Museum Publications, 1987), pp. 44–46. On the double *remen* see Gillings, *Mathematics in The Time of the Pharaohs*, p. 208.

14. Gay Robins and Charles Shute, "Mathematical Bases of Ancient Egyptian Architecture and Graphic Art," *Historia Mathematica* (1985) 12:107–122, on p. 112. See also Beatrice Lumpkin, "The Egyptian and Pythagorean Triples." Ibid., (1980) 7:186–187, and Gillings, *Mathematics in the Time of the Pharaohs*, app. 5.

15. See Ghalioungui, *House of Life* Ghalioungui, *Physicians of Ancient Egypt* (cit. n. 4) R. O. Steuer, and J. D. de C. M. Saunders, *Ancient Egyptian and Cnidian Medicine: The Relationship of Their Aetiological Concepts of Disease* (Berkeley/Los Angeles: Univ. of California Press, 1959); and Saunders, *Transition from Ancient Egyptian to Greek Medicine* (cit. n. 6). Pace J. A. Wilson, "Medicine in Ancient Egypt, *Bulletin of Historical Medicine* (1962) 36(2):114–123; Wilson, "Ancient Egyptian Medicine," editorial in *Journal of the International College of Physicians*, sect. I, (June 1964) 41(6):665–673; and G. E. R. Lloyd, introduction to *The Hippocratic Writings* ed. G. E. R. Lloyd (London: Penguin, 1983), p. 13n. Even the skeptical Heinrich von Staden, who is very reluctant to concede Egyptian influences on Hellenistic medicine, admits that the study of pulses and their timing by water clocks, for which his subject Herophilus of Alexandria was famous probably came from the Egyptian tradition: von Staden, *Herophilus: The Art of Medicine in Early Alexandria* (Cambridge, England: Cambridge Univ. Press, 1989), p. 10.

16. Neugebauer, Exact Sciences in Antiquity (cit. n. 2), p. 78 (italics added). I shall challenge this interpretation below.

17. Ibid., p. 96.

18. J. F. Lauer. *Observations sur les pyramides* (Cairo: Institut Francais d'Archeologie Orientale, 1960), pp. 11, 10, 4–24 (here and elsewhere, translations are my own unless otherwise indicated).

19. Ibid, pp. 1–3. For a more skeptical view of this see Robins and Shute, "Mathematical Bases," (cit. n. 14), p. 109.

20. Robins and Shute, *Rhind Mathematical Papyrus,* (cit. n. 13), p. 58.

21. Lauer, *Observations sur les pyramides,* (cit. n. 18), p. 10.

22. Neugebauer, *Exact Sciences in Antiquity,* (cit. n. 2j. p. 91; and Neugebauer, "On the Orientation of Pyramids," *Centaurus* (1980) 24:1–3. Professor James Williams challenges Neugebauer on this in his *Fundamentals of Applied Dynamics* (New York: Wiley, 1996, pp. 43–44).

23. Neugebauer, *Exact Sciences in Antiquity,* p. 92; and Lucien Levy-Bruhl, *How Natives Think* (New York: Knopf, 1926).

24. Robins and Shute, *Rhind Mathematical Papyrus* (cit. n. 13), pp. 42–43, 56.

25. V. V. Struve, "Mathematischer Papyrus des staatlischen Museums der schönen Kunste in Moskau," *Quellen und Studien zur Geschichte der Mathematik* (Pt. A) (1930) I:184; and Paul Ver Eecke, *Les oeuvres completes d'Archimede* (Paris: Blanchard, 1960), p. xxxi. Among later scholar who have supported Struve's view (on pp. 174–176) see Gillings, *Mathematics in the Time of the Pharaohs* (cit. n. 13), pp. 187–194; and Robins and Shute, *Rhind Mathematical Papyrus,* p. 48.

26. Giorgio de Santillana, "On Forgotten Sources in the History of Science," in *Scientific Change: Historical Studies in the Intellectual, Social, and Technical Conditions for Scientific Discovery an Technical Invention, from Antiquity, to the Present,* edited by A. C. Crombie (London: Heinemann, 1962), pp. 813–828, on p. 814.

27. Gillings, *Mathematics in the Time of the Pharaohs* (cit. n. 13), p. 200.

28. Struve, Mathematischer Papyrus," (cit. n. 25), pp. 183, 185.

29. Gillings, *Mathematics in the Time of the Pharaohs* (cit. n, 13), pp. 194–201.

30. Neugebauer, *Exact Sciences In Antiquity* (cit. n. 2), p. 29.

31. Robins and Shute, *Rhind Mathematical Papyrus* (cit. n. 13), p. 58.

32. See Herodotos, 2:109; Diodoros Sikeliotes, 1:69.5, 81:3, 94:3; Aristotle, *Metaphysics* 1:1. (981b); Hero, *Geometria* 2; Strabo, 16:2, 24, and 17:1, 3; and Clement of Alexandria, *Stromateis* 1:74.2. See also Cheikh Anta Diop, *Civilization or Barbarism: An Authentic Anthropology,* translated by Yaa-Lengi Meema Ngemi (New York: Lawrence Hill, 1991), pp. 257–258.

33. Struve, "Mathematischer Papyrus" (cit. n. 25), pp. 16}165, on p. 165.

34. Sergio Donadoni, "Plan," in *Lexikon der Agyptologie,* edited by Wolfgang Helck and Eberhard vols. Vol. 4 (Wiesbaden: Harrassowitz, 1977–19U),

cols. 1058–1060. For my dating see Martin Bernal, *Black Athena: The Afroasiatic Roots of Classical Civilization*, Vol. 2: *The Archeological and Documentary Evidence* (London: Free Association Books; New Brunswick, NJ: Rutgers University, 1991), pp. 206–216.

35. De Santillana, "Forgotten Sources" (cit. n. 26), p. 814.

36. J. F. Weidler, *Historia astronomiae* (Wittenberg: Gottlieb, 1741), p. 177.

37. Pingree, "Hellenophilia versus the History of Science" (cit. n. 1).

38. See the bibliography in Whitney Davis, "Plato on Egyptian Art," *Journal of Egyptian Archaeology* (1979) 66:121–127, on p. 122, n. 3.

39. A clear example of this can be seen in the fragment discussed by Neugebauer in his exquisite swan song: Neugebauer. "A Babylonian Lunar Ephemeris from Roman Egypt," in *A Scientific Humanist: Studies in Memory of Abraham Sachs*, edited by E. Leichty, et al. (Occasional Publications of the Samuel Noah Kramer Fund, 9) (Philadelphia: University Museum, Univ. Pennsylvania, 1988), 301–304.

Chapter 5

Africa in the Mainstream of Mathematics History

Beatrice Lumpkin

Editors's comment: Beatrice Lumpkin, a mathematics educator, presents specific examples to demonstrate the centrality of African contributions to the development of mathematics knowledge. This chapter was first published in I. Van Sertima (Ed.), *Blacks in science: Ancient and modern*, New Brunswick, NJ: Transaction, pp. 100–109, in 1983. In a postscript, written for this volume, she includes a brief update, discussing recent scholarship and addressing criticisms made since her chapter was first published. Elsewhere, along with Dorothy Strong, she has applied these and other examples to the teaching of mathematics in *Multicultural Science and Math Connections: Middle School Projects and Activities* (Portland, ME: Walch, 1992).

Summary: For thousands of years, Africa was in the mainstream of mathematics history. This history began with the first written numerals of ancient Egypt, a culture whose African origin has been reaffirmed by the most recent discoveries of archaeology. With a longer period of scientific work than any other area of the world, progress in mathematics continued on the African continent through three great periods, ancient Egyptian, Hellenistic, and Islamic. The language changed from Egyptian to Greek to Arabic. But the tradition of African science continued, despite a change of language. The Renaissance in Europe was triggered by the science and mathematics brought to Spain and Italy by the Moors of North Africa. Although all peoples and continents have played a role in the history of mathematics, the contributions of Africa are still unacknowledged by European and North American historians.

One of the earliest examples of writing were the hieroglyphs on Narmer's palette, named for the first king of upper and lower Egypt, who was also known as Menes. The numerals used cited thousands of heads of cattle and thousands of prisoners, indicating that numerals and hieroglyphs already had a long history in Egypt.[1] It has recently been learned from the findings of the International Nubian Rescue Mission, which salvaged ancient artifacts and monuments before the Aswan Dam flooded the Nubian area, that pharaonic kings and hieroglyphic writing were known south of the first cataract generations before Menes.

These findings have reaffirmed the African origin of the great ancient civilization. From this region, in the interior of Africa, has come evidence of the earliest known cultivation of grain.[2] With this new evidence, the date of the Egyptian calendar must also be reconsidered. Originally thought to date back to 4241 B.C. when first analyzed by European scholars, its apparent date was arbitrarily changed to 2773 B.C. It was claimed that "Such precise mathematical and astronomical work cannot be seriously ascribed to a people slowly emerging from Neolithic conditions."[3] The internal evidence is consistent with either date, based on the Sothic cycle of about 1,468 years.[4] Struik, himself, has kept an open mind on light that could be shed by new discoveries. In correspondence with this author, he wrote: "As to mathematics, the Stonehenge discussions have made it necessary to rethink our ideas of what Neolithic people knew. Gillings has shown that the ancient Egyptians could work with their fractions in a most sophisticated way."[5]

The early beginnings of algebra and geometry in ancient Egypt are briefly covered in many history books. But the full scope and depth of ancient Egyptian mathematics have been largely overlooked because the first judgement of the European translators of the papyri dismissed this mathematics as "primitive."[6] It is only in the last ten years that a full-length study of this mathematics was published. It is often not realized that African contributions did not end with the ancient Egyptians but continued through the Hellenistic and Islamic Empires. Indeed, Africa continued in the mainstream of mathematics for thousands of years, right up to the European Renaissance.

Any unprejudiced view of world history must acknowledge that many different peoples and races on every continent have made great mathematical discoveries. The Maya[7] of Central America used a zero hundreds of years before A.D. 876,[8] its earliest known use in India. And the ancient Chinese, almost 2,000 years ago, solved systems of equations with a method similar to the modern elementary transformations of matrices.[9] But it was through Africa that the science, mathematics, and knowledge of the entire Eastern world reached Europe. This was true in the time of the classical Greeks and continued through the Middle Ages when Islamic scholars dominated the intellectual life of Europe, Africa, and the western and central parts of Asia.

Very different from the above is the theory of history now taught at most North American universities and accepted as "fact" by thousands of practicing mathematicians and teachers. They teach the purely European origin of mathematics. According to this version of history, mathematics began in Greece in the fifth century B.C. With the decline of the Greek Empire, no further progress was made until Europe, the true home of mathematics, was ready to advance again during the Renaissance.[10] To those, such as Kline, who dismiss all mathematics before the Greeks as less than "true" mathematics, George Sarton, the encyclopedist of science, replies: "It is childish to assume that science began in Greece. The 'Greek miracle' was prepared by millennia of work in Egypt, Mesopotamia and possibly other regions. Greek science was less an invention than a revival."[11]

Three great periods of African mathematics will be briefly considered in this article. They are: The ancient Egyptian mathematics of the pyramids, obelisks, and great temples,[12] the African participation in classical mathematics of the Hellenistic period, and the African participation in Islamic mathematics. Other periods and locales of mathematics on the African continent are not described although future research may well show that these played an important role in the history of mathematics. This vast subject, not covered here, includes the mathematical games so widespread in Africa,[13] the systems of measurement used in the African forest kingdoms, and the mathematics used in building the great stone complexes of Zimbabwe. Perhaps now that Zimbabwe has its own government, more information will become available and new chapters in the history of mathematics will be written.

The great accuracy of the dimensions of the pyramids[14] still gives rise to wonder. Geometry, literally the measurement of the land, required a high technology in addition to theoretical mathematics. The famous "rope stretchers" to whom Democritus compared himself, used special ropes, twisted of many fine strands to assure high stability and constant length. The accuracy of the Egyptian value for π (the constant ratio of circumference to diameter of any circle) was probably a result of theoretical analysis of "squaring a circle"[15] and confirmation of experiment and accurate measurement. The Egyptian value for π was 3.16, much closer to the modern 3.14 than the biblical value, 3.0.

Contrary to the reports that the ancient Egyptians did not derive any general principles and limited themselves to specific examples, many of the problems in the mathematical papyri ended with general statements. For example, in the papyrus written by the scribe Ahmose (Rhind Mathematical Papyrus 61B) the solution was followed by: Behold! Does one according to the like for every uneven fraction which may occur. Gillings lists over ten such statements in this papyrus.[16]

In his book, *Mathematics in the Time of the Pharaohs,* Gillings tries to discover methods that may have been used by the ancient Egyptian scribes to derive their often amazing results. By approaching the subject without prejudice, with the keenness of a mathematical detective, Gillings investigated the terse clues left by the scribes and has revealed some unsuspected achievements. These include formulas for the summation of arithmetic and geometric series and the measurement of the area of a curved surface. An efficient irrigation technology, efficient central administration, and the skill of the farmers of ancient Egypt made possible a large food surplus, enough to support the mathematicians, teachers, and other intellectuals. In turn, these ancient African mathematicians contributed to production by developing methods of measuring the land through formulas for the areas of rectangles, triangles, circles, and even the area of a curved dome. Properties of similar triangles were known and some trigonometry, the equivalent of our co-tangent, helped assure a constant slope for the faces of pyramids.

Their very system of measurement shows knowledge of some "Pythagorean" triads. An area measured in cubits was doubled if cubits were replaced by double remens while the shape remained similar.[17] Alone of the ancients, the Egyptians knew the correct formula for the volume of a truncated pyramid, thus stumping the modern experts who wonder "How did they do it?" First and second degree equations were solved by the method of false position, a method that continued in use up to this century. But perhaps it was in their use of arithmetic and geometric series that the Egyptians' work has almost a modern ring. Indeed the formula used for the sum of n terms of an arithmetic series is the equivalent of one we use today,

$$s = \frac{n}{2}(2a + (n - 1)d).$$

Sometimes these ancient problems seemed pure mathematical fun. For what practical significance could there have been in problem

79 of the Ahmose papyrus which seems to anticipate this Mother Goose rhyme by 3,500 years,

> "As I was going to St. Ives,
> I met a man with 7 wives.
> Each wife had 7 sacks,
> Each sack had 7 cats . . . "

Compare with:

> Houses 7
> Cats 49
> Mice 343
> Spelt 2,401;
> (ears of grain)
> Hekats 16,807
> (measures of grain) [18]

Egyptian fractions were, perhaps, the most important application of mathematics in those times because they were used for the extensive bookkeeping needed for large public works such as pyramid construction. These were unitary fractions using 1 as the numerator. For example, instead of 2/5 they wrote the equivalent 1/3 + 1/15. These fractions remained in use in Europe until fairly recent times when they were replaced by the decimal fractions which the Moors had introduced. With their fractions, the Egyptians could add, multiply, divide, and take square roots.

Alexandrian Mathematics

Egyptian contributions to science and mathematics did not end with the conquest by the Macedonian, Alexander the Great. Attracted by the great wealth and learning of Egypt, Alexander, in 332 B.C., ordered the construction of Alexandria, a city which became the intellectual center of the Greek speaking world. In Alexandria, the products and ideas of the city-states of North Africa, Asia Minor, Greece, India, and China mingled and took firm root on African soil. A great museum and library attracted the best scholars and educated many generations of Egyptian students.

It was in the fourth century before our era[19] that Greek mathematicians developed the deductive, axiomatic method, establishing the logical foundation on which mathematics so proudly rests today. As Struik wrote, "This again may be connected with the fact that mathematics had become a hobby of a leisure class which was based on slavery, indifferent to invention, and interested in contemplation."[20]

Of course, no modern scholar has tried to belittle this great Greek accomplishment because it rested on an economic base of slavery. Contrast this with the case of Egypt, where slavery played a much lesser role. Yet Hollywood movies and popular texts claim that the greatness of the pyramid period is lessened because slave labor was used.

Up to the fourth century B.C., according to Neugebauer, Greek mathematics was similar to, and no doubt an outgrowth of Egyptian and Babylonian. He cautions that "if modern scholars had devoted as much attention to Galen or Ptolemy as they did to Plato and his followers, they would have come to quite different results and they would not have invented the myth about the remarkable quality of the so-called Greek mind to develop scientific theories without resorting to experiments or empirical tests."[21]

It was in Egypt that Hellenistic mathematics reached its peak. Struik attributes this flowering of mathematics to the central position that Egypt occupied during the Ptolemaic period as the intellectual and economic center of the Mediterranean world.[22] Who were the people of Alexandria? They were the African people of Egypt with a few immigrants from Greece, western Asia and neighboring African countries. Sarton reminds us that "Greek emigrants were too few in pre-Christian times and too little interested in science and scholarship to affect and change Eastern minds."[23] The ruling class, itself, was mixed from the first days of Alexandria because Alexander, the Macedonian, ordered his officers to marry and mix with the local population.

Nonetheless, although no pictures have come down to us of any of the great men and women of Alexandria, false portraits have been published which portray them as fair Greeks, not even sunburned by the Egyptian sun. This misleading practice is decried by George Sarton, in an article on "Iconographic Honesty" in which this dean of science history declares "I do not believe there is a single ancient scientist of whose lineaments we have any definite knowledge; thus to publish "portraits' of Hippocrates, Aristotle or Euclid is, until further notice, stupid and wicked."[24]

In the case of Euclid, best known of the Alexandrian mathematicians, there is not a shred of evidence to suggest that he was anything other than Egyptian. Euclid's fame is based on his thirteen major texts, *The Elements,* a strictly logical deduction of theorems from accepted definitions and axioms. For over 2,000 years these books dominated the teaching of mathematics to the delight of mathematicians and the discomfiture of students. In a similar manner, *The Almagest,* written by another Egyptian, Claudius Ptolemy, c. A.D. 150, dominated astronomy until finally replaced by Copernicus' theory of a sun-centered planetary system, c. 1543.

The Almagest (the greatest in Arabic) contains in its thirteen books the foundations of spherical trigonometry, a catalogue of 1,028 stars and the epicycle system of an earth-centered astronomy. By some peculiar racial reasoning, Ptolemy is often described as Egyptian only because his work was of a practical, applied nature, differing in this respect from the strictly theoretical work of Euclid. The fact is that both were Alexandrians and therefore it is highly probable that they were Africans. In Ptolemy's time, Alexandria was already 400 years old and very much a part of Egypt.

Of Heron, another Alexandrian who wrote *Metrica* on geometric measurement, and *Pneumatica,* a book about machines, Howard Eves says "There are reasons to suppose he was an Egyptian with Greek training."[25] Another great mathematician of that time, Diophantus, of the Alexandria of the third century, continued the tradition of Egyptian algebra. His *Arithmetica* on number theory marks the author as a genius in his field[26] and introduced brief symbols to simplify algebraic expressions (syncopation) in place of the long, wordy formulations then in use (rhetorical algebra.)

All of these Alexandrian mathematicians wrote their books in Greek. Their use of Greek makes them no more European than the use of English by Nigerians today changes that nationality. To this very incomplete list of Egyptian mathematicians who worked in Alexandria must be added Theon and his daughter Hypatia, whose memory still inspires women to become mathematicians.

African Mathematics During the Islamic Empire

In the lengthy period between the decline of the Greek and Roman Empires to the eve of the Renaissance, a period of almost 1,000 years, Europe disappears from the mainstream of the history of mathematics. That is, with the exception of Moorish mathematicians in

Spain and Italy who came from North Africa and brought with them 4,000 years of African-Asian mathematics. In Moorish Spain, "Cordoba, in the tenth century, was a great centre of learning, where one could walk for several miles in a straight line by the light of the public lamps."[27] A whole series of new inventions became available during this period: steel, silk, porcelain, and paper. African papyrus paper was still used and appreciated for its fine qualities, but pulp paper from China was coming into wider use.[28]

Struik, who has read the Islamic mathematicians in the Russian translations (unfortunately much of this work is still not available in English) stresses their continuity of culture under Arabic rule. "The ancient native civilizations had even a better chance to survive under this rule than under the alien rule of the Greeks."[29]

In North Africa, the Arabs, as the Greeks before them, intermarried with the African people of these countries and quickly absorbed the culture and learning of Egypt. The rapid physical expansion of the Islamic Empire had its intellectual parallel in the exchange of knowledge among Egypt, Persia, India, and China.

From the eighth century until the fifteenth, Arabic was the language of mathematics and science. About 773 Al-Fazari translated the Indian Siddhanta to Arabic, popularizing the Hindu decimal system; the zero is believed to have come into use later. Thabit ibn Qurra and his school produced excellent Arabic translations of Euclid, Appolonius, Archimedes, Ptolemy, and Theodosius and made important additions of their own (826–901). To this day, most of these classics are known to us only through the Arabic translations, the original Greek versions having been lost. But one book, more than any other, was the vehicle for introducing Europe to Islamic algebra and the Hindu-Arabic numerals and arithmetic, al-Khwarizmi's *Al-jabr wa'l muqabalah.* From the author's name we get the common mathematical term, "algorithm." From the title, al-jabr, we get the modern term, "algebra."

Some of the greatest scholars of this time came to Egypt to work, where they could enjoy support for the full scope of their research. Among them was the outstanding mathematician-physicist Ibn al-Haytham (Alhazen). Although born in Basra, his productive life was spent in Egypt, pioneering in optics and geometry.[30] Ibn al-Haytham is the first of three Islamic mathematicians who opened the door to non-Euclidean geometry. Through his work and that of his successors, Umar al-Khayyami and Nasir al-Din al-Tusi, European mathematicians hundreds of years later were inspired to create new geometries.

Islamic Mathematics Reaches Europe

The main routes of transmission of Islamic learning were from North Africa to Spain, also to Sicily and southern Italy, where Moorish rule lasted for many generations. The Europeans who appear in the mathematical history of the time had studied with Islamic scholars. Constantine the African (d. 1087), a merchant from Carthage, brought a precious cargo of manuscripts to Salerno where a school was founded to translate and study the Arabic works. Adelard of Bath (1116–1142) made a long voyage to Arab countries and translated Arab classics into Latin. Fibonacci (Leonardo Pisano) (1170–1240) got his start in mathematics during his long residence in the North African coastal city where his father was a merchant.[31]

In short, as summarized by Haskins, "The full recovery of this ancient learning, supplemented by what the Arabs had gained from the Orient and from their own observation, constitutes the scientific renaissance of the Middle Ages."[32]

Interrupted Progress

The period which follows the European Renaissance and brings us down to the modern era saw the pillaging of Africa, Asia, and the Americas by European colonialism. Slavery depopulated Africa and drastically interrupted African progress. But it is no mere academic exercise to reconstruct without prejudice the thousands of years of history during which Africa contributed to the mainstream of mathematics. World science will become much richer when the former colonial peoples take their place, once more, in the mainstream of mathematics and science.

Notes

1. Sir Alan Gardiner, *Egyptian grammar* (Griffith Institute: Oxford, 1927, 1978), 5.

2. Wendorf, Schild, El Hadidi, Close, Kobusiewicz, Wieckowska, Issawi, Haas, "Use of barley in the Egyptian late paleolithic," *Science* 28, no. 4413

(Sept. 79), V. 205: 1341–47. (Paper has been withdrawn by the authors. See update below.)

3. Dirk J. Struik, *A concise history of mathematics* (New York: Dover, 1967), 24–5.

4. Carl B. Boyer, *A history of mathematics* (New York: Wiley), 1968, 12. Also see Richard A. Parker, "Egyptian astronomy, astrology, and calendrical reckoning," in *Dictionary of Scientific Biography*. New York: Scribners, vol. XV), 708.

5. Letter dated April 4, 1978. Gillings is the author of *Mathematics in the time of the pharaohs*, cited in ref. 15 below.

6. Morris Kline, *Mathematics, a cultural approach* (Reading, MA: Addison Wesley, 1962), 14.

7. J. Eric, S. Thompson, *The rise and fall of Maya civilization* (Norman, OK: Univ. of Oklahoma Press, 1954), p. 158. For possible African influence in Olmec and Maya civilizations, see lvan VanSertima, *They came before Columbus* (New York: Random House, 1976).

8. Boyer, *A history of mathematics*, p. 235.

9. Ibid., p. 219.

10. Morris Kline, *Mathematics in western culture* (New York: Oxford Press, 1953), 23.

11. George Sarton, *A history of science* (Cambridge, MA: Harvard Univ. Press, 1959), ix.

12. Beatrice Lumpkin, "The pyramids, ancient showcase of African science and technology," in *Journal of African Civilizations*, vol. 2, nos. 1 and 2: 10–26.

13. Claudia Zaslavsky, *Africa counts: number and pattern in African culture* (New York: Lawrence Hill Books, 1979), section 4.

14. I. E. S. Edwards, *The pyramids of Egypt* (Middlesex, Eng.: Penguin, 1961), 118.

15. Richard J. Gillings, *Mathematics in the time of the pharaohs* (Cambridge, MA: MIT Press, 1975), 141–5.

16. Ibid., p. 233.

17. Ibid., p. 208.

18. Chace, Arnold B. *The Rhind mathematical papyrus*, reprinted by the National Council of Teachers of Mathematics, Reston, VA: 1979, and again in 1986.

19. Otto Neugebauer, *The exact sciences in antiquity* (N.Y.: Dover, 1957, 1969), 148.

20. Dirk J. Struik, *A concise history of mathematics*, pp. 48, 49.

21. Otto Neugebauer, *Exact sciences*, p. 152.

22. Dirk J. Struik, *A concise history of science*, pp. 49, 50.

23. George Sarton, *A history of science*, p. 4.

24. George Sarton, "Iconographic honesty," *Isis* 30 (1939): 226.

25. Howard Eves, *An introduction to the history of mathematics* (New York: Holt, Rinehart and Winston, 1964, 1969), 159.

26. Ibid., p. 159.

27. Henry James J. Winter, *Eastern science*, (London: John Murray, 1952), 62.

28. John D. Bernal, *Science in history*, (New York: Cameron, 1954), 195.

29. Dirk J. Struik, *A concise history of mathematics*, p. 69.

30. Aldo Mieli, *La science arabe*, (Leiden: E.J. Brill, 1938), 105.

31. Aldo Mieli, *La science arabe*, p. 243.

32. Charles Homer Haskins, *Studies in the history of medieval science* (Cambridge, MA: Harvard U., 1927), 3.

Postscript

When this article first appeared in September 1980, it received a favorable response from many who wanted to restore Africa to its rightful place in history. In the years since 1980, there has been a growing awareness of African contributions to mathematics and science.

Several articles in this volume have contributed to this increased awareness. However, there has also been much adverse criticism of the article's main theme, that Africa was in the *mainstream* of mathematics history, and that school mathematics owes a lot to ancient African mathematicians. This concept seems to have touched a nerve among those who insist on staying with the Eurocentric approach that "mathematics began in Greece."

In view of recent scholarship, some additions and corrections to this article are appropriate at this time. Although the summary of the article states that the history of mathematics in Africa began "with the first written numerals of ancient Egypt," mathematics in Africa started much earlier. Even if history is taken in the narrow sense of *written* history, numbers were recorded in Africa long before the development of Egyptian numerals. The discovery of tally marks on a fossilized baboon bone, found in Border Cave between South Africa and Namibia, pushes the date of number records back to at least 35,000 B.C.E.[1] Long before 3100 B.C.E., the approximate date of the earliest known Egyptian numerals,[2] the Ishango bone was carved in the Lakes region of Central Africa. Recent work dates the Ishango fossil to between 23,000—18,000 B.C.E. The bone is inscribed with tallies that show a complex array of values.[3]

Wendorf and Schild et al., who published a paper on a South Egyptian location for the "earliest known cultivation of grain," have withdrawn their paper. They now believe that the domesticated barley they found was a later intrusion into the 14,000-year old site. However, the age of the harvesting tools reported in their paper remains unchanged, even if the grain that was harvested was probably wild. Moreover, other investigators at the same site have found evidence that grain was made into a mash for infants 19,000 years ago, possibly for weaning.[4]

Since the article was published, there have been many attempts to refute Sarton's position for a mostly Egyptian origin of the population of Alexandria. It has been claimed that early Alexandria was strictly a Greek enclave. But the question has not been addressed, "Why was the scientific center of the Mediterranean world established in Egypt instead of Greece?" It is this author's opinion that the synthesis of the new Alexandrian science was a synthesis of cultures, and that the genetic origin of the people of Alexandria is irrelevant to this issue. The magnetic pull of ancient Egyptian culture on Greek scholars has been well established.

This update will cite some sophisticated mathematical achievements that were not included in the original article. These concepts occur in the context of construction plans and bookkeeping applications, not in the "mathematical" papyri. In the sense that this mathematics was practiced outside the formal school setting, it could be called "Ethnomathematics." Perhaps that explains why the achievements described below do not appear in the "History of Mathematics" textbooks. However, one cannot help but wonder what part racial prejudice played in their omission from textbooks. The following examples were well known to founders of Egyptology such as Petrie, Gardiner, Reisner, Borchardt, and Scharff.

1. Rectangular Coordinates, c. 2700 B.C.E.

An architect's plan on a limestone ostracon, found at the Saqqara pyramid complex, is dated c. 2700 B.C.E. This artifact shows an architect's drawing for a curved section of a temple roof. The drawing shows horizontal coordinates spaced 1 cubit apart. For these horizontal coordinates, heights are given for points which define a curve. The curve in the sketch exactly matches the curve of the temple roof near the spot where the ostracon was found. The vertical coordinates are given in Egyptian numerals that are easy to read. Start at the upper left. Under the outstretched arm, the hieroglyph for cubits, the number three is shown as **I I I**. Under the 3 cubits, 3 palms and 2 fingers are shown. Since there are 7 palms/cubit, and 4 fingers/palm, this coordinate is equivalent to 98 fingers.

This architect's plan shows the earliest known use of rectangular coordinates. It is possible that the concept of coordinates grew out of the Egyptian use of square grids to copy or enlarge artwork, square by square. It needs just one short, important step from the use of square grids to the location of points by coordinates. Clarke and Engelbach recognized the architect's plan as "of great importance."[5] But no history textbook has picked up this "important" example.

Figure 5–1. Architect's diagram giving coordinates for a curve, c. 2700 B.C.E., Saqqara, du Service XXV. Clarke and Engelbach, 1930, from Annales du Service XXV.

2. The Egyptian Zero Symbol, Two Applications

It is true that a zero placeholder was not used (or needed) in the Egyptian hieroglyphic or hieratic numerals because these numerals did not have positional value. But the zero concept has many other applications. Only a few historians such as Boyer[6] and Gillings have reported any use of the zero concept in ancient Egypt. But Gillings added, "Of course zero, which had not yet been invented, was not written down by the scribe or clerk; in the papyri, a blank space indicates zero."[7] In this statement, Gillings was repeating the conventional wisdom. He was not aware that Egyptians had, in fact, invented a zero symbol. The ancient Egyptian zero symbol was the same as the hieroglyph for beauty, an abstraction of a human windpipe, heart, and lungs. The consonantal values were nfr; the vowel sounds are unknown.

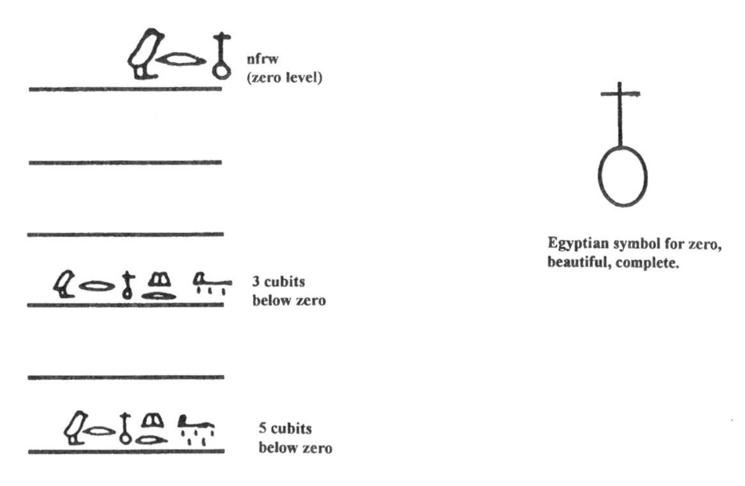

Figure 5–2. Diagram of leveling lines from a tomb at Meidum.

Horizontal leveling lines were used to guide the construction of pyramids and other large structures. These massive stone structures required deep foundations and careful leveling of the courses of stone. One of these leveling lines, sometimes at pavement level, was used as a reference for the other levels, and was labeled "nfr," or "zero." Other horizontal leveling lines were spaced 1 cubit apart and labeled as "1 cubit above nfr," "2 cubits above nfr," or "1 cubit, 2 cubits, 3 cubits," and so forth, below nfr. Here zero was used as a reference for a system of directed or signed numbers.[8]

A number of examples of these leveling lines are still visible at pyramid and tomb sites. In 1931, George Reisner described the zero reference for leveling lines at the Mycerinus (Menkure) pyramid at Giza built c. 2600 B.C.E. He gave the following list collected earlier by Borchardt and Petrie from their study of Old Kingdom pyramids.[9]

nfrw	zero (Note the w suffix added to nfr for grammatical reasons.)
m tp n nfrw	zero line
hr nfrw	above zero
md hr n nfrw	below zero

Zero Balance in Bookkeeping

The same nfr symbol was also used to express zero remainders in a monthly account sheet from the Middle Kingdom dynasty 13, c. 1770 B.C.E. The bookkeeping record looks like a double entry account sheet with separate columns for each type of goods. At the end of the month, the account was balanced. For each item, income was added, then disbursements were totaled. Finally, the disbursement total for each column was subtracted from total income for the column. Several columns had zero remainders, shown by the nfr symbol.[10] The Egyptian use of the same symbol for two different applications of the zero concept is of more than passing interest. Also, the ancient Egyptian penchant for making tables and organizing data is worthy of note.

3. Egyptian Cipherization of Numerals

Carl Boyer credited the Egyptians with introducing the idea of cipherization when they invented hieratic numerals. Hieratic script, generally speaking, was a cursive form of the hieroglyphs. Hieratic numerals, however, were different from the hieroglyphic numerals. The hieratic numerals used ciphers, as Boyer explained, "for each of the first nine integral multiples of integral powers of ten." He called the hieratic numeral system, "decimal cash-register cipherization," referring to old-style cash registers, which sent up a flag for each decimal place. Boyer added that, "The introduction by the Egyptians of the idea of cipherization constitutes a decisive step in the development of numeration."[11]

For example, a number such as 19,607 written in hieroglyphs would require twenty-three symbols, but in hieratic would need only four symbols. The four hieratic ciphers would be the equivalent of 10,000; 9,000; 600; and 7. (See the last line in Figure 5-3.) Both the hieroglyphic and the hieratic numeral systems did not use positional value and did not need a zero "placeholder." Ionic numerals, as Boyer showed, also used "decimal cash-register cipherization," yet another example of Egyptian influence on Greek mathematics.

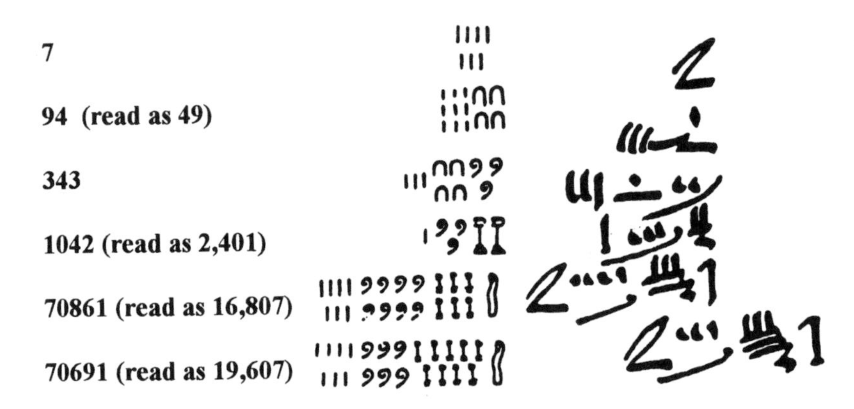

Figure 5–3. *Center*: Hieroglyphic numerals. *Right*: Hieratic Numerals. Read Egyptian numerals from right to left.

Adapted from Arnold Buffum Chace, *The Rhind mathematical papyrus* (Reston, VA: NCTM, 1979), 137.

Notes

1. J. Bogashi, K. Naidoo, and J. Webb. "The oldest mathematical artefact," in *Mathematical Gazette* 71: 294.

2. Bruce Williams and Thomas J. Logan. "The Metropolitan Museum Knife Handle and aspects of pharaonic imagery before Narmer," *Journal of Near Eastern Studies*, vol. 46, no. 4 (1987): 245–48.

3. Claudia, Zaslavsky, *Africa Counts* (Westport, CT: Lawrence Hill, 1979), 19. Her discussion of the Ishango bone is based on: De Heinzelin, Jean. "Ishango," in *Scientific American*, vol. 206 (June 1962): 105–114; and, Alexander Marshack, *The Roots of Civilization*, (Mt. Kisco, NY: Mayer, Bell, 1991).

4. Gordon C. Hillman, "Late Palaeolithic plant foods from Wadi Kubbaniya in Upper Egypt: dietary diversity, infant weaning and seasonality in a riverine environment," in *Foraging and Farming, the Evolution of Plant Exploitation,* edited by David C. Harris and Gordon C. Hillman (London: Unwin Hyman, 1989), 230.

5. Somers Clarke, and R. Engelbach, *Ancient Egyptian Construction and Architecture* (New York: Publisher, 1990, orig. 1930), 52–3.

6. Carl B. Boyer, *A history of mathematics* (New York: Wiley, 1968), 18.

7. Gilllings, *Mathematics in the time of the pharaohs,* p. 228.

8. Dieter Arnold, *Building in Egypt* (New York: Oxford University, 1991), 17.

9. George A. Reisner, *Mycerinus, the temples of the third pyramid at Giza,* (Cambridge, MA: Harvard University, 1931), 76–77.

10. Alexander Scharff, "Ein Rechnungsbuch des Königlichen Hofes aus der 13. Dynastie (Papyrus Boulaq Nr. 18)," *Zeitschrift für Ägyptische Sprache und Altertumskunde* 57 (1922): 58–9, 5**, 8**, 12**.

11. Carl B. Boyer, "Fundamental Steps in the Development of Numeration," in *Isis* 35 (1944): 153–68. All quotations in this paragraph are from pages 157–58.

Section III

Considering Interactions Between Culture and Mathematical Knowledge

Arthur B. Powell and Marilyn Frankenstein

In his educational practice, Freire initiates the process by first considering who creates culture. This is done to clarify that all people, whether literate or illiterate, are cultural actors. Toward this end, he emphasizes an anthropological concept of culture. In the following quote he indicates signposts of a definition of culture:

> the distinction between the world of nature and the world of culture; the active role of men [and women] *in* and *with* their reality; the role of mediation which nature plays in relationships and communications among men [and women]; culture as the addition made by men [and women] to a world they did not make; culture as the result of men's [and women's] labor, of their efforts to create and re-create; the transcendental meaning of human relationships; the humanist dimension of culture; culture as systematic acquisition of human experience (but as creative assimilation, not as information-storing); the democratization of culture; the learning of reading and writing as the key to the world of written communication. In short, the role of man [and woman] as Subject[s] in the world and with the world (1973, p. 46).

The salient points for our discussion are that cultural products are the creation of people and that transformations of nature are made by all people. Mathematics is a cultural product and, therefore, is created by humans in the interconnected midst of culture.[1] The interactions are dialectical: people's daily practice, language, and ideology effect and are effected by their mathematical knowledge. Bishop (1990), and other mathematics educators, reviewing anthropological studies and investigating mathematical activities in different cultures, view mathematics as a "pan-cultural phenomenon . . . a symbolic technology, developed through engaging in various [integrated] environmental activities" which can be classified as

> counting: the use of a systematic way to compare and order discrete objects . . . locating: exploring one's spatial environment, and conceptualising and symbolising that environment, with models, maps, drawings, and other devices . . . measuring: quantifying qualities like length and weight, for the purposes of comparing and ordering objects . . . designing: creating a shape or a design for an object or for any part of one's spatial environment . . . playing: devising, and engaging in, games and pastimes with more or less formalised rules that all players must abide by . . . explaining: finding ways to represent the relationships between phenomena (pp. 59–60).

Further Gattegno (1970, 1988) argues that mental functionings, or structures, needed to learn to speak a language are akin to those used in doing mathematics. The implication that he draws is that anyone who succeeds in learning a language has already mathematized his or her linguistic domain and, therefore, capable of mathematizing other domains. Not only does he posit that mathematics is a birthright, but also that mathematical structures are developed through a specific cultural activity: learning to speak a language.

Observers of less universal cultural contexts also narrate how people acquire "unschooled" knowledge of mathematics. Considerable research documents that unschooled individuals, in their daily practice, develop accurate strategies for performing mental arithmetic. For example, the Dioula, an Islamic people of Côte D'Ivore, have traditionally engaged in mercantile activities. Ginsburg, Posner, and Russell (1981) discovered that unschooled Dioula children develop similar competence in mental addition as those who attended school. These researchers hypothesize that this is a result, at least in part, of the daily experiences of children working in marketplaces. Further, studying Brazilian children who worked in their parents' markets, Carraher, Carraher and Schliemann (1985) conclude that "performance on mathematical problems embedded in real-life contexts was superior to that on school-type word problems and context-free computational problems involving the same numbers and operations" (p. 21). Through interviews with the youngsters, these investigators learned that in the marketplace the children reasoned by mental calculations, whereas in the formal test they usually relied on paper and pencil, school-taught algorithms.

> Mistakes often occur as a result of confusing the algorithms. Moreover, there is no evidence, once the numbers are written down, that the children try to relate the obtained results to the problem at hand in order to assess the adequacy of their answers. . . . The results [of this study] support the thesis . . . that thinking sustained by daily human sense can be—in the same subject—at a higher level than thinking out of context (p. 27).

We do not interpret this work as suggesting that these youngsters *cannot* do school mathematics. Carraher, Carraher and Schliemann (1985) conclude just that the school mathematics curriculum should *start* from the mathematical knowledge that the children already have (p. 28). Further, Ginsburg (1982) reflects on this issue cross-culturally:

> although culture clearly influences certain aspects of cognitive style (i.e., linguistic style), other cognitive systems seem to develop in a uniform and robust fashion, despite variation in environment or culture. Children in different social classes, both black and white, develop similar cognitive abilities, including basic aspects of mathematical thought (pp. 207–208).

In the same study, he also concludes that "upon entrance to school virtually all children possess many intellectual strengths on which education can build. . . . Elementary education should therefore be organized in such a way as to build upon children's already existing cognitive strengths." He further argues that the reason why poor children do not do "well" in school

> may include motivational factors linked to expectations of limited economic opportunities, inadequate educational practices, and bias on the part of teachers . . . [therefore] reform efforts must not be limited to the psychological remediation of the poor child. They must also focus on teaching practices, teachers, and the economic system (pp. 208–209).

Analyzing studies on interactions between culture and cognition as well as ethnographic data of adults in the United States engaged in supermarket and weight-watching activities, Lave (1988) argues against considering mathematical knowledge and context separately. Rather, she theorizes that "activity-in-setting [is] seamlessly stretched across persons-acting" and that the context often shapes the mathematical activity, becoming the calculating device, rather than merely the place in which the mathematical calculations are applied (p. 154). Scribner (1984) found that this occurs when dairy workers invent their own units (full and partial cases) to solve, on-the-job, problems of product assembly. In another example, Lave (1988) describes how a shopper who found, in a bin, a surprisingly high-priced package of cheese investigated for error by searching through the bin for a similar-sized package and checking to see whether there was a price discrepancy (p. 154). If instead, the problem were solved as a textbook problem rather than as a calculation shaped by the setting, the shopper would have divided weight into price and compared that quotient with the price per pound printed on the label. Lave uses, in both senses of the term, the phrase "dissolving problems" for discussing what happens in practice. Mathematics problems "disappear into solution with ongoing activity rather than "being solved." Such transformations pose a challenge to scholastic assumptions concerning the bounded character of math problem solving as an end in itself" (p. 120).

Lave then theorizes about the societal reasons why so many shoppers attend to arithmetic. School mathematics, she contends, is filled with shopping applications, so that money becomes a value-free, "natural" term, just a form of neutral school arithmetic. When adults go shopping their choices are first made qualitatively. That is, an item may be the best buy mathematically but is rejected because the package is too big to fit on their pantry shelf. However, they fall back on arithmetic calculations when there is no other criteria for choice. This provides a basis for believing that their decision is rational and objective. Thus, as Lave argues, "price arithmetic contributes more to constructing the incorrigibility of 'rationality' than to the instrumental elaboration of preference structures" (p. 158).[2]

Focusing on the linguistic construction of this kind of mathematical rationality, Walkerdine (1988) argues that a key way in which this "mathematically precise," positivist rationality gets constructed is through the suppression of the multiple meanings of lexical terms in different practices. One set of meanings, one path to cognitive development is chosen as true, as "normal." By analyzing transcripts of children using and learning basic mathematical concepts such as size relations, she indicates how the meaning of these mathematical terms is shaped by social relations constituting the practice in which those words are used. Schools, she contends, ignore these multiple significations and, therefore, make judgments about the conceptual development of children which produce a particular set of behaviors and which then are considered *the* rational path to intellectual development. For instance, in asking children questions that involve comparing the sizes of "daddy, mommy and baby bear," teachers tend to ignore how the power relations in their families add another signification to the terms "big, bigger, biggest." For instance, in the life of a child, daddy may be the biggest physically but mommy may be the biggest power figure. Children may answer the school mathematical question incorrectly because the "bear story" context is confused with the meanings of size in their family practices, not because they are "unready" to learn the size concepts of the curriculum. Formal academic mathematics is "built precisely on a bounded discourse in which the practice operates by means of suppression of all aspects of multiple signification. The forms are stripped of meaning, and the mathematical signifiers become empty" (p. 97). Walkerdine goes on to suggest that to learn school mathematics, children must learn to treat all applications, all practices as undifferentiated aspects of a value-free, neutral, and rational experience.

Even other supposedly more value-free mathematical concepts are shaped by specific philosophical and ideological orientations. For example, Martin (1988/reprinted here as chapter 7) cites Forman who analyzed how the intense antagonism to "rationality" which existed in the German Weimar Republic after World War I resulted in a particular interpretation of a mathematical construction.

> Forman suggests that this pressure led the quantum physicists to search for . . . a mathematical formalism which could be interpreted as non-causal. In crude terms, the acausal Copenhagen interpretation and its associated mathematical framework was adopted because they looked good publicly. . . . In the decades since the establishment of the orthodox or Copenhagen interpretation, a number of alternative interpretations have been put forth. Some of these use the same mathematical formulations, but interpret their physical significance differently, while others use different mathematical formulations to achieve the same or different results. . . . [So] the interpretation of the equations of quantum theory as supporting indeterminism was not *required* by the equations themselves. Furthermore, it seems possible that many of the achievements of the theory might have been accomplished using a somewhat different mathematical formulation which could well have been *difficult* to interpret inderterministically (pp. 210–211).

On other occasions, philosophy and ideology have prompted variant interpretations of fundamental mathematical concepts and techniques. For instance, the dialectics and historical materialism of Karl Marx, along with his project to elaborate the principles of political economy, between 1873 and 1881, led him to study, criticize, and develop an alternative theoretical foundation for the differential calculus (Marx, 1983). Struik (1948/reprinted here as chapter 8) was the first to reveal to the English speaking world that Marx engaged in this mathematical struggle. Marx's critique of prevailing methods for deriving the derivative of a function was twofold: (1) the derivative of a function was always present before the actual differentiation occurred, and (2) none of the methods accounted for the dialectical nature of motion and change to which a function is subjected in the process of differentiation (Powell, 1986, p. 120). Out of touch with professional mathematicians and unaware of Cauchy's work on the calculus and limits, Marx overcame his critique of the theoretical foundations of the calculus by developing both a conceptual formulation and a technique for differentiation that captured symbolically the vexing problematic that was the impulse behind the method of Newton and Leibniz: motion and change. Indeed, his discoveries, stimulated and informed by his philosophical and ideological framework, represented rediscoveries and, in some instances, anticipated future conceptual and philosophical developments (Gerdes, 1985; Powell, 1986).

Grounded in a cultural praxis—the conceptual, mathematical description of dynamics—Marx attempted to undergird the calculus with a cultural construct—dialectics—which was part of the philosophical and ideological perspective of an identifiable cultural group. Mathematical knowledge seems unconnected to cultural context since, in isolation and at historical moments distant from their genesis, particular mathematical ideas, such as the derivative, may appear detached from a specific cultural interpretation or application. Ideas, however, do not exist independent of social context. Moreover, as some critical theorists and realist philosophers remind us, our categories, concepts, and other ideas are essentially dependent on objective reality or nature. In a critique of anti-realist epistemology, Johnson (1991) rightly argues that

> . . . all these (social) things are materialized in, and dependent on, that which is essentially mind-independent, namely: the natural world. The very human activity of "cutting up" the world into [for example] hammers and chairs presupposes a world of naturally existing things (like trees and iron ore) capable of being fashioned into tools. In other words, the essential independence goes one way: *nature is essentially independent of mind, but mind (and all its products) is not essentially independent of nature* (1991, p. 25).

The social and intellectual relations of individuals to nature or the world and to such mind-dependent, cultural objects as productive forces influence products of the mind that are labeled mathematical ideas. Further, though there are recognized philosophical variants to the foundations of mathematics, the seemingly non-ideological character of mathematics is reinforced by a history which has labeled alternative conceptions as "non-mathematical" (Bloor, 1976, as cited in Martin, 1988, p. 210).

Notes

1. Rotman (1988) refers to Hegel and Marx to explain why both mathematicians and popular culture thinks of mathematical objects and "the semiotic basis of mathematical persuasion" (p. 29) as "'out there,' waiting independently of mathematicians, to be neither invented nor constructed nor somehow brought into being by human cognition, but *discovered* as planets and their orbits are discovered" (p. 29). He argues that Hegel elaborates how

> human products frequently appear to their producers as strange, unfamiliar, and surprising; that what is created need bear no obvious or transparent markers of its human (social, cultural, historical, psychological) agency, but on the contrary can, and for the most part does, present itself as alien and prior to its creator.
>
> Marx, who was interested in the case where the creative activity was economic and the product was a commodity, saw in this masking of agency a fundamental source of social alienation, whereby the commodity appeared as a magical object, a fetish, separated from and mysterious to its creator; and he understood that in order to be bought and sold commodities had to be fetishized, that it was a condition of their existence and exchangeability within capitalism. Capitalism and mathematics are intimately related: mathematics functions as the grammar of techno-scientific discourse which every form of capitalism has relied upon and initiated. So it would be feasible to read the widespread acceptance of mathematical Platonism in terms of the effects of this intimacy, to relate the exchange of meaning within mathematical languages to the exchange of commodities, to see in the notion of a "timeless, eternal, unchangeable" object the presence of a pure fetishized meaning, and so on; feasible, in other words, to see in the realist account of mathematics an ideological formation serving certain (techno-scientific) ends within twentieth-century capitalism.
>
> . . . Whether one sees realism as a mathematical adjunct of cap-

italism or as a theistic wish for eternity, the semiotic point is the same: what present-day mathematicians think they are doing—using mathematical language as a transparent medium for describing a world of pre-semiotic reality—is semiotically alienated from what they are, according to the present account, doing, namely, creating that reality through the very language which claims to describe it (p. 30).

2. Borba (199?), in a manuscript on the politics intrinsic to "academic" mathematics, argues that the use of mathematics in everyday life not only makes our choices seem more "rational," but also serves to end the discussion. Once we use mathematics to justify a decision, no one can question that discussion—after all, it is now "scientifically proved." Making another point, Frankenstein (1987) also contends that the shopping applications of school arithmetic curricula contribute to the appearance of "naturalness" in the way our social and economic structures are organized. Her adult students often find it ludicrous to think of restructuring society where food was free, for example, where eating was a civil right not a paid for commodity.

References

Bishop, A. J. (1990). Western mathematics: The secret weapon of cultural imperialism. *Race and Class* 32(2): 51–65.

Bloor, D. (1976). *Knowledge and social ideology*. London: Routledge and Kegan Paul.

Borba, M. C. (1990). Ethnomathematics and education. *For the Learning of Mathematics* 10(1): 39–43.

Carraher, T. N., Carraher, D. W., and Schliemann, A. D. (1985). Mathematics in the streets and in schools. *British Journal of Developmental Psychology* 3: 21–29.

Frankenstein, M. (1987). Critical mathematics education: An application of Paulo Freire's epistemology. In I. Shor (ed.). *Freire for the classroom* (pp. 180–210). Portsmouth, NH: Boynton/Cook.

Freire, P. (1973). *Education for critical consciousness*. New York: Seabury.

Gattegno, C. (1970). *What we owe children: The subordination of teaching to learning*. New York: Avon.

———. (1988). *The science of education: Part 2B: The awareness of mathematization*. New York: Educational Solutions.

Gerdes, P. (1985). *Marx demystifies calculus*. (Translated by B. Lumpkin). Vol. 16. Minneapolis: Marxist Educational Press. (Original published in

1983 as *Karl Marx: Arrancar o véu misterioso à matemática*. Maputo: Universidade Eduardo Mondlane.)

Ginsburg, H. P., Posner, J. K., and Russell, R. L. (1981). The development of knowledge concerning written arithmetic: A cross-cultural study. *International Journal of Psychology* 16: 13–34.

Ginsburg, H. P. (1982). The development of addition in the context of culture, social class, and race. In T. P. Carpenter, J. M. Moser, and T. A. Romberg (Ed.) *Addition and subtraction: A cognitive perspective* (pp. 191–210). Hillsdale, New Jersey: Lawrence Erlbaum.

Johnson, D. K. (1991). A pragmatic realist foundation for critical thinking. *Inquiry: Critical thinking across the disciplines* 7(3): 23–27

Lave, J. (1988). *Cognition in practice*. Cambridge, England: Cambridge.

Martin, B. (1988). Mathematics and social interests. *Search* 19(4): 209–214.

Marx, K. (1983). *Mathematical manuscripts*. London: New Park.

Powell, A. B. (1986). Marx and mathematics in Mozambique. *Science and Nature* (7/8): 119–123.

Rotman, B. (1988). Toward a semiotics of mathematics. *Semiotica* 72(1,2): 1–35.

Scribner, S. (1984). Studying working intelligence. In B. Rogoff and J. Lave (ed.). *Everyday cognition: Its development in social context* (pp. 9–40). Cambridge, MA: Harvard.

Struik, D. J. (1948). Marx and mathematics. *Science and Society* 12(1): 181–196.

Walkerdine, V. (1988). *The mastery of reason*. London: Routledge and Kegan Paul.

Chapter 6

The Myth of the Deprived Child: New Thoughts on Poor Children

Herbert P. Ginsburg

Editors's comment: Herbert Ginsburg, a cognitive psychologist and mathematics education researcher, evaluates psychological research, conducted from about 1970 to 1985, on the intellectual development and education of poor children. He documents that people of various cultural backgrounds—even from different socioeconomic classes—develop the requisite cognitive structures to do mathematics even before they attend school. This chapter first appeared in *The school achievement of minority children: New perspectives* (V. Neisser (ed.), pp 169–189, Hillside, NJ: Lawrence Erlbaum, 1986). The present version provides a brief update in a postscript.

Introduction

The aims of this chapter are to evaluate the past ten or fifteen years' psychological research on the intellectual development and education of poor children. Although much original work in this area was misguided, some insights have been gained into the intellectual functioning of the poor. In general, research suggests that poor children as a group do not suffer from massive intellectual deficiency. Their school failure—at least in the first several years of school—cannot be explained primarily in terms of cognitive developmental deficit. Hence, research on poor children and their education must take new directions. In the second part of the chapter, I offer speculations on the types of research we should undertake.

The Myth of the Deprived Child

In evaluating previous research on poor children, it is useful to begin by describing the political and social climate of the late 1960s and early 1970s, a period of considerable ferment and conflict. One major force at the time was a kind of liberal environmentalism. In 1964, Lyndon Johnson proclaimed the war on poverty. On the assumption that education is the gateway to middle-class prosperity, one of the major campaigns of that war was to be the education of poor children, especially blacks. The task for the government, then, was to provide adequate education for those lacking it. Doing this required both legal and psychological remedies. Legal battles over desegregation were fought and sometimes won. But the liberals decided early on that desegregation was not sufficient. Poor children, especially blacks, did not perform adequately in the public schools, even when segregation was no longer in place. To remedy this situation, the Johnson administration leaned heavily on the advice of social scientists, who recommended, among other things, the creation of Head Start and the funding of psychological and educational research. The government offered considerable support for social science generally, including educationally oriented research (e.g., Project Literacy), and work on poor children in particular. The general aim of the research was to be the understanding of the problems of poor children, so that appropriate remedies could be developed. The results of this work were of two types: a body of research on poor children's cognitive functioning, and various attempts at "compensatory education," programs designed primarily to remedy intellectual deficiencies in preschoolers and prepare them for schooling.

Clearly, the liberal environmentalist approach was well intentioned. And in the political climate of the 1980s, that is no small compliment. Yet, several of us were dissatisfied on a number of grounds with the liberal environmentalist position. In the early 1970s, I wrote *The Myth of the Deprived Child* (Ginsburg, 1972), the aim of which was to offer a critique and analysis of existing work on poor children's intellect and education. The book made a number of points, and from them, I have selected a few major themes that are useful as stepping points for an analysis of the situation today.

The book began by stating the obvious, namely, that poor children as a group were doing badly in public schools and that the educational system required drastic improvement. The question then became: what do we know about poor children's intellectual capacities, and how can that knowledge be used to improve education? The bulk of the book focused on an analysis of psychological knowledge concerning poor children's intellect. The argument was roughly as follows: Liberal environmentalists (e.g., Hunt, 1969) believe that poor children develop in a deprived environment that stunts their intellectual growth. The environment fails to provide sufficient stimulation or provides the wrong kind of stimulation. As a result, poor children suffer from cognitive deficits. Nativist theory also postulates a cognitive deficit, but offers a different explanation of its origins. Jensen (1969) proposes that lower class children, and blacks in particular, suffer from a specific cognitive deficit, an inability to engage in "conceptual learning," and this inability is a result of genetic inheritance. According to both the environmentalists and the nativists, the cognitive deficit (whatever its origins) prevents poor children from learning the conceptual material taught in school.

I argued that both views, and a good deal of the research stemming from them, were misguided, and proposed an alternative view of "cognitive difference." The reasoning behind the argument was something like this.

First, much of the empirical evidence supporting the deficit view could not be believed, largely on methodological grounds. As Labov (1972) pointed out at the time, many of these studies—like those of Bereiter and Englemann (1966), and Deutsch (1967)—employed rigid methodologies and were not based on an understanding of children in general or poor children in particular. It is easy to get poor children to do badly on some standard test; it is much harder to employ methods sensitive to their true competence. Anyone who had real contact with poor children, I felt, would realize that much of the psychological research was insensitive, narrow minded, and wrong.

Second, the cognitive deficit research seemed to ignore important cognitive universals. In the 1960s, cognitive theory was not well understood or accepted. I felt that, from a Piagetian perspective, the important point was not that poor children produced fewer Peabody correct responses than middle-class children, but that all children, poor- and middle-class alike, were probably capable of the concrete operations and even formal operations as well. The essence of the matter was commonalities in basic aspects of mental functioning, not individual differences. At the time, there was little empirical support for this position, aside from the work of Labov (1972) on black English. The main theoretical foundations were in Piaget's theory and Lenneberg's (1967).

Third, a third argument stressed possible cultural differences.

The general point was that poor children are not so much deficient as they are distinctive. Presumably, in response to the unique demands of their distinctive environments, they develop special kinds of adaptations—skills not possessed by middle-class children. Again, there was very little evidence to support this intuition. Labov's work on black English was again cited, as were anecdotes—perhaps apocryphal or even stereotypic—about black children's knowledge of arithmetic being expressed in numbers-running in Harlem.

Fourth, another theme was that poor children did not have to be instructed in basic intellectual skills; they were quite capable of learning on their own. A good deal of development takes place in a natural and spontaneous fashion. This position was partly a reaction against behaviorist theories of learning, some of which were still taken seriously at the time, and particularly work on cognitive socialization, like that of Hess and Shipman (1967). These researchers claimed essentially that (a) poor mothers did not know how to train their children in basic cognitive skills, and (b) by implication, if the cognitive skills were not trained, they would not develop. The counter argument was that many Piagetian-type skills develop independently, on their own, and that parents do not even know that the skills exist, let alone attempt to teach them. Hence, I stressed self-directed learning, and this position led to some recommendations concerning the implementation of "open schools," which were felt to provide a solution to the education of the poor.

Fifth, finally, I argued that it is necessary to examine what is meant by "success in school." Usually, school success is defined in terms of performance on standard achievement tests, which in turn conceptualize academic knowledge in limited, often trivial ways. Standard tests—and many educators and psychologists who use them—conceive of academic knowledge in terms of correct responses, quantitative traits (e.g., "verbal ability"), and the regurgitation of prepackaged knowledge. Yet we know from cognitive theory that knowledge is complex: it involves construction; it must be conceptualized in terms of process and strategy; and the least interesting aspect is the surface response. Thus, the conventional conception of school success was superficial, missing the heart of the matter, namely, the cognitive analysis of children's concepts and strategies in particular areas of academic work. In the absence of such cognitive analyses, it was not possible to discuss intelligently what children, poor or otherwise, did or did not know or need to know in the school context.

I believe that my assessment of the situation in the early 1970s was not too far off the mark. The empirical research of both environmentalists and nativists was, indeed, insensitive and unconvincing. Both the environmentalist and the nativist theories were basically wrong. The "cognitive difference" view offered some useful insights. Let us see what events in the subsequent ten or fifteen years have to say about these matters.

First, consider the social context. Over the past fifteen or twenty years, the political situation has changed drastically. Federal concern with problems of poverty has diminished steadily, so that the war on poverty seems light years away. Indeed, in the 1980s, the Reagan Administration seemed to be conducting a war on the poor, rather than on poverty, so that one positively yearns for the follies of the liberals. These comments on politics are not a digression; psychological research is heavily influenced by the political climate. The result of the political retreat from a concern with poverty is a correlated decrease in psychological research on the topic. With a few notable exceptions, the recent past has given us little research on poor children's intellect and education. An unanticipated benefit is that we do not have to contend with volumes of misguided research, but the general outcome is that psychological problems of poverty are being swept under the rug. With some notable exceptions, for example, the work of Yando, Seitz, and Zigler, 1979; Ogbu, 1978; and Feagans and Farran, 1982, research on poor children is no longer a popular topic in developmental psychology.

At the same time, some recent work does shed light on key aspects of the cognitive difference position and opens up important questions for future research and theory. Some of this work derives from the direct study of poor children; most of it stems from work in related areas.

1. Methodology. In the early 1970s, it seemed clear that much of the data purporting to demonstrate cognitive deficits in poor children were simply unbelievable. The research techniques employed were not sensitive enough to uncover the true extent of poor children's competence. Since that time, several developments have reinforced the basic point. Many cross-cultural researchers have become dissatisfied with standard methods. Traditional cross-cultural research relied on standard Western techniques, like translated intelligence tests, to investigate cognition in "primitive" peoples. The general, and perhaps predictable, finding was that non-Westerners lack whatever cognitive skills were under investigation. Cole and Scribner (1974), providing a strong critique of this approach, showed that standard, traditional techniques often yield absurd results concerning traditional people's competence. Western tasks are not always interpreted in the way intended; they may be misunderstood, with the result that standardization often precludes validity. One alternative is to make every effort to discover tasks that are relevant for individuals in the context of the local culture. One task may be suitable for tapping competence in one culture, whereas an objectively distinct task may be suitable for measuring the same skill in another culture. Although the tasks are objectively different (dissimilar instructions, materials, etc.), they may be subjectively equivalent in tapping the same cognitive processes. Conversely, the identical task may be subjectively inequivalent between cultures. The key for measuring competence is not necessarily objective identity, but subjective equivalence.

The implications for methodology are enormous. Standardization often makes no sense and defeats the purposes for which it was intended. Extending the argument, we may think of children as analogous to cultures. Each child, or at least each developmental level, has its own perspective (culture). For subjective equivalence to be achieved, objective identity often needs to be abandoned and tasks adapted to each perspective.

If cross-cultural work suggests sensitivity to individual cultures or to the cultures of individuals, recent developments in cognitive research also point to flexibility of method. A good deal of the Newell and Simon (1972) research on complex problem solving in adults employs the talking aloud method, in which individuals say "everything that comes into their head" as they are grappling with a difficult problem. The experimenter occasionally asks questions, but for the most part the data are essentially introspections. Several considerations lead these investigators to eschew simple quantifiable measures such as those usually obtained in the laboratory. One is that the investigators are interested in complex intellectual activities, which often cannot be expressed in simple ways. Another is that introspection can indeed be valuable at least for certain aspects of cognitive study. In any event, the use of the talking aloud procedure shows that serious psychologists are exploring flexible and nontraditional methods in the investigation of cognitive processes. (For further discussion of this point see Ginsburg, Kossan, Schwartz, & Swanson, 1983.)

A third example is even more directly pertinent to the question of poor children. Over the past ten years, the clinical interview technique has played a very important role in research on mathematical thinking. Most investigators have come to make a commonsensical distinction between competence and performance in intellectual functioning: it seems obvious that often children do not demonstrate in their performance the true extent of their knowledge. Young children in particular often know much more than they reveal. Often, too, standard tests fail to tap much more than the surface performance. Consequently, contemporary researchers often find it useful to employ the clinical interview method to assess competence. In this, they are indebted to Piaget, who recognized early on (1929) that the assessment of true competence often requires flexible questioning—questioning that is contingent on the child's response, that employs techniques like counter-suggestion, that deliberately manipulates phrasing, and that generally attempts to discover means for getting the child to perceive the problem in the manner intended. The clinical interview is deliberately nonstandardized; that is its strength for the purpose of assessing competence (Ginsburg, Kossan, Schwartz, and Swanson, 1983).

Following is an example showing how the clinical interview technique can be used to reveal unsuspected competence in a child who is performing poorly in arithmetic. Butch was in the third grade of an upstate New York elementary school. His teacher identified him as having severe problems in learning elementary school arithmetic. Both his grades and his achievement scores were low, in arithmetic as well as other school subjects, and he was a candidate for repeating third grade. He was not retarded or severely emotionally disturbed. Outside of the classroom, in the playground, he was lively and boisterous; his everyday behavior seemed to reveal at least average intelligence. Yet in the classroom he was quiet and obviously had considerable trouble learning arithmetic. Wanting to know why Butch was having problems in arithmetic, the teacher requested a diagnosis, which she hoped would produce an understanding of Butch and identify those factors responsible for his failure to learn.

In a clinical interview the following conversation took place. Asked what he was doing in school, Butch said he was working with fractions.

> Interviewer: Fractions? Can you show me what you are doing with fractions?
> Butch writes: $8\,\overline{)16}$
> Interviewer: OK. So what does that say?
> Butch: 8, 16.
> Interviewer: What do you do with it?
> Butch: You add it up and put the number up there.
> Interviewer: OK. What is the number?
> Butch writes: $\begin{array}{r} 23 \\ 8\,\overline{)16} \end{array}$

Several features of the interview are notable. First, the interviewer allowed Butch to determine the topic for discussion, so that she could explore the issues that concerned him. The aim was to let Butch determine the agenda, not to impose on him a preconceived plan of interviewing. Second, the interviewer tried to get Butch to explain in his own words what he was doing. The questions were designed to be open ended, like "What do you do with it?", so that Butch could answer them in a way that would reveal his mental processes. In short, the interview aimed at discovering the child's approach: what are his concerns and how does he operate? These central features of clinical interviewing distinguish it sharply from normative, standardized testing.

To this point, the interview shows that Butch is doing something unusual. He confuses fractions with division and cannot calculate the simple division. He seems to lack an understanding of school arithmetic and engages in highly irregular procedures. What is going on? The next excerpt clarifies some of these matters and illustrates the power of clinical interviewing.

> Interviewer: How did you do that?
> Butch: I went, 16, 17, 18, 19, 20, 21, 22, 23. I added from 16.

It was clear that Butch got the answer 23 just as he earlier had said he did: "You add them up." The interview also revealed that Butch did this adding by counting on from the larger number. Thus, in this brief excerpt, the interviewer learned that Butch used the word "fractions" but wrote a division problem, solved a written division problem by addition, and did addition by counting! The clinical interview showed that Butch was not wrong simply because of stupidity, poor conceptual ability or the like. Instead, there were clear reasons for his mistakes, and he possessed surprising skill in mental calculation. Further, the information was provided by Butch himself, prodded by clinical interview techniques.

We all know intuitively that clinical interviewing is a sensible way to proceed. Even hard-nosed experimentalists use it. Before starting an experiment, investigators often use the clinical method (calling it pilot work) to find out what to do in the experiment. Then, if the experiment does not work, the experimentalist uses the clinical method to find out what went wrong. The actual experiment may be something of a formality. And, of course, college professors use the clinical method. Even though they may find it convenient to use standardized methods like multiple choice tests to assess the performance of large numbers of freshmen, they employ something like the clinical interview in examining doctoral candidates, when things are really serious. What professor would conduct a doctorate exam in a multiple choice format? In any event, the clinical interview is increasingly popular in research on mathematical thinking and often reveals surprising competencies in children who perform poorly in school arithmetic.

These recent methodological developments strengthen the earlier arguments concerning the irrelevance of much research on cognitive deficits in poor children because it was based on standard methods of limited sensitivity. Moreover, these developments have influenced as least two recent studies of poor children's intellectual functioning. R. Yando, V. Seitz, and E. Zigler (1979) made serious attempts to employ flexible methods in their research, and so did Russell and I in our studies of preschool and kindergarten children's mathematical thinking (Ginsburg and Russell, 1981). In our research, we could not engage in extensive clinical interviewing, because the children were so young. Consequently, we spent many hours devising and revising experimental tasks so that children would understand them. Often, some minor change in wording or procedure made an immense difference, and it usually took many hours to discover these minor variations. For example, we began with what we thought was an easy task designed to measure the child's ability to determine the sum of two visible collections of objects. We presented a story in which the child was required to find the union of two static sets. For example: "Turtle has three nuts and owl has two nuts. That's three and two. How many do they have altogether?" When the task was presented in this manner, inner city children performed quite badly on the average. We revised the task, using problems that stressed the *active combining* of sets. Thus: "Puppet has two pennies. He's walking to the store and finds one more penny. How many pennies does he have altogether?" With this apparently minor change in the semantics of the problem, inner-city children's performance levels improved, and class and race differences did not achieve statistical significance.

In brief, flexible methodology is at the heart of research on intellectual competence. Much cognitive deficit research has not used this kind of methodology and hence is irrelevant. Recent research, using flexible methods, uncovers important areas of competence in poor children. It is to some of these competencies that we turn next.

2. *Universals.* Ten years ago, there was a bit of evidence suggesting that poor children are characterized by certain "cognitive univer-

sals," like the Piagetian concrete operations or basic syntactic processes. Recent research, most of it in the cross-cultural tradition, attests to the basic validity of this point of view. For example, a recent review of cross-cultural Piagetian research (Dasen and Heron, 1981) suggests that virtually all cultures examined seem to possess the capability for concrete operational thinking, although the evidence concerning formal operations is by no means clear. Similarly, the research of Cole and Scribner (1974) and their colleagues generally demonstrates that nonliterate West Africans show basic competencies in reasoning, memory, and the like, although these competencies may not be expressed in typically Western fashion. Thus, the Kalahari bushmen (Tulkin and Konnor, quoted by Flavell, 1977) demonstrate the capability for scientific thinking as they engage in the tracking of animals. At the same time, they would no doubt find it impossible to deal with such Piagetian formal operational tasks as the combining of chemicals.

Our own research (Ginsburg, Posner, and Russell, 1981a, 1981b, 1981c) examines the development of mathematical thinking in unschooled and schooled West Africans in two different cultures. One of these groups in the Ivory Coast, the Baoulé, is an animist, agricultural group, placing no particular emphasis on mathematics. Our research was conducted in areas where some Baoulé children attended schools, while others did not. The second group, the Dioula, are Muslims who have traditionally engaged in mercantile activities and are scattered throughout West Africa. The Dioula, although often illiterate, need to employ calculational processes in the course of commerce. Like the Baoulé, some Dioula subjects attended school and others did not. The two African groups provided a useful contrast in terms of hospitality to mathematical ideas and procedures. We used these groups to investigate the effects of schooling and culture on the development of informal mathematical skills.

One basic finding was that unschooled African children from both cultures possess fundamental informal concepts of mathematics, like more, equivalence, and adding. Posner (1982) found that Baoulé and Dioula children ranging from four or five years of age to nine or ten years of age perform about as well as Americans on the elementary concept of more. Young children in both African cultures ". . . possess the basic notion of inequality; by the age of 9–10, regardless of schooling or ethnic background, they display a high level of accuracy. Moreover their methods for determining the greater set are similar to those of American children . . . suggesting . . . a universal capacity" (Posner, 1982). Although they may acquire this concept at a later age than Americans, the Africans do acquire it, without the benefit of schooling or middle-class American culture. Posner also investigated elementary addition and found that both schooled and unschooled Dioula are extremely skilled in this area, and that schooled Baoulé are also adept. Only unschooled Baoulé, members of the agricultural society, did relatively badly, perhaps because their culture places little emphasis on counting. In any event, the important finding is that unschooled Dioula children are competent in elementary addition using counting and other effective strategies, and that even the unschooled Baoulé achieve some success in this area.

Ginsburg, Posner, and Russell (1981a) investigated more complex forms of addition in schooled and unschooled Dioula children and adults, and in American children and adults. In general, unschooled Dioula children eventually exhibit a high degree of competence in the solution of verbally presented addition problems. The young Dioula begin with elementary counting procedures, but older Dioula switch to the extensive utilization of regrouping methods (e.g., $23+42=20+40+3+2$), which are more efficient, particularly in the case of larger numbers. At first, the Dioula do not employ the strategies with great accuracy but with age become increasingly proficient in their use, learning to discriminate among different types of problems and to apply different strategies where appropriate. Moreover, the strategies employed by the Dioula are essentially the same as those observed in American children. Schooling and American culture are not *necessary* for the development of mental addition strategies. Other cross-cultural researchers like Saxe (Saxe and Posner, 1983) find similar results. Apparently, illiterate children growing up in what we would consider the abject poverty of traditional cultures nevertheless manage to develop some fundamental cognitive concepts and skills.

Poor children in our own culture develop similar skills. That was the hypothesis of ten years ago, and it is even more reasonable today. Basic cognitive skills should be no less prevalent in lower-class Americans than in unschooled Africans or middle-class Americans. Our research in the Washington, D.C. ghetto (Ginsburg and Russell, 1981) was designed to investigate the development of informal mathematical notions in lower- and middle-class children, both black and white, at the prekindergarten and kindergarten levels. Each child, seen individually, was given a large number of mathematical thinking tasks (17 in all), many derived from our work in Africa and from the work of investigators like Gelman. The tasks ranged from such informal skills as the perception of more, the understanding of addition operations, simple enumeration, and addition calculation, to such school-taught notions as the representation and writing of numbers.

In general, we found no social class differences and at most statistically insignificant trends favoring middle-class over lower-class children. In the vast majority of cases, children of both social classes demonstrated competence on the various tasks and used similar strategies for solving them. If these competencies and strategies were not evident at the preschool level, they emerged by kindergarten age in all groups. For example, middle- and lower-class children made effective use of counting strategies to solve addition calculation problems involving concrete objects. In general, the only large differences involved age: developmental changes from preschool to kindergarten far outweigh social class differences in this area. Furthermore, the research showed fewer racial than social class differences. Race has only trivial associations with early cognitive function. Jensen's (1969) notion that lower-class and black children exhibit weaknesses in abstract thought is wrong, at least with respect to early mathematical cognition. Many of the tasks employed in our study were prime examples of abstract thought (e.g., the understanding of addition operations) and yet were not associated with racial or social class differences. Our overall conclusion was that, at least at the age of four or five, poor children, black or white, possess fundamental competencies in early mathematical thinking; there is little evidence of pervasive cognitive deficit.

3. COGNITIVE DIFFERENCE. Ten years ago, a number of us hypothesized that poor children were characterized by cognitive *differences*, not deficits (e.g., Cole & Bruner, 1971). The major evidence was Labov's work, showing that black English is different on the surface, but employs the same basic syntax as standard English. The basic argument was that cognitive differences were an expression of distinctive adaptations to unique environments.

While the argument was largely conjectural, there is some recent cross-cultural evidence shedding light on the question of cognitive differences. We know in a general way that cultures develop distinctive techniques for dealing with distinctive problems. Thus, the Puluwat islanders develop clever methods for navigating without special instruments (Gladwin, 1970). Our own research in the Ivory Coast examined the development of mathematical thinking within the contexts of two very different cultures, one Moslem and mercantile and the other animist and agricultural. The Dioula traditionally engage in commerce and have wide experience in dealing with the money economy. By contrast, the Baoulé often do not. We encountered Dioula adults who engaged in complex forms of mental arithmetic, and even understood basic mathematical principles in a practical fashion (Petitto and Ginsburg, 1982). Research makes it clear that cultural groups sometimes develop distinctive patterns of cognition in response to local environmental demands. Unfortunately, we know little more than this; we have virtually no solid information concerning the distinctive cognitive activities of subgroups within our own culture. More about this below.

4. *Cognitive socialization.* My earlier critique stressed spontaneous development and downgraded the role of cognitive socialization. I argued that poor children can learn on their own, that cognitive socialization as carried on by the middle class is not necessary to instruct poor children in the basic intellectual skills that will later form the basis for school learning. In some ways the critique was accurate and in some ways a mistake. It was accurate in pointing out that children's learning—including that of poor children—is often spontaneous and does not always depend on adult instruction. Some aspects of cognition do indeed develop in a more or less self-directed fashion, without the necessity for parental involvement. Infants probably develop object permanence, action schemes, and perceptual skills on their own, without parental help or knowledge. Preschool children develop methods for addition and concepts of equivalence on their own, without explicit instruction (although parents or other agents of culture must of course directly or indirectly provide the basic counting numbers). We probably underestimate the extent to which children spontaneously develop basic concepts, skills, and sensible views of the world.

The critique was also accurate in pointing out that much of the early cognitive socialization research was badly conceived and executed. For example, the influential studies of Hess and Shipman (1967) gave an oversimplified view of cognitive socialization and studied it poorly. These investigators conceived of cognitive socialization as a one-way process in which parents shape children's intellectual development. Such theorizing does not do justice to the self-directed aspects of human development and to the complex interactions between parents and children. Also, Hess and Shipman examined the parent-child interaction of lower-class families in a laboratory setting that required mothers to instruct children in an artificial task. This situation may have been comfortable for middle-class mothers and children, but for their lower-class peers it may have been uninteresting, and even threatening and condescending. For this group, the laboratory task appears to be culturally biased and lacking in ecological validity. Little can be learned from research of this type.

At the same time, even though my critique may have been reasonable, I probably underestimated the extent to which cognitive socialization is important and necessary for the development of basic intellectual skills. Perhaps I was too much of a Piagetian. Piaget never really understood the role of the social-cultural environment. But Vygotsky did. And recent research in the Vygotsky tradition (e.g., Greenfield, 1984; and Rogoff & Gardner, 1984), illustrates the subtle and important ways in which parent-child interaction shapes early cognitive development. Children do not learn everything on their own; parents seem to play a major role in shaping certain key elements of children's cognition.

What does this mean for the understanding of poor children? Probably it is still correct to maintain that they acquire certain intellectual skills on their own, in a self-directed fashion. In these cases, cognitive socialization may be beside the point. At the same time, cognitive socialization may be crucial for the development of some intellectual and other skills that may later play a major role in schooling. Hence, we need to use sensitive methods and sophisticated theories to learn more about cognitive socialization in general and in poor children in particular.

5. Academic knowledge. Ten or fifteen years ago we seemed to know virtually nothing about academic knowledge and needed to know much more if we were going to say anything sensible about poor children's performance in school. Since then, we have made enormous strides in our understanding of schooled cognition. Now we can go far beyond our earlier intuition that achievement test scores do not tell the whole story. Now we have had more than a decade of serious research into such matters as reading (both decoding and comprehension), expository writing, and mathematical thinking. Having recently edited a book on the subject (Ginsburg, 1983), I am familiar with the latest research on mathematical thinking. We now know a good deal about early counting activities and their role in early calculation (Fuson & Hall, 1983); the mental numberline and subsequent concepts of base ten (Resnick, 1983); the role of systematic strategies in the generation of calculational errors (VanLehn, 1983); the semantics and syntactics of word problem solving (Riley, Greeno, & Heller, 1983); and the basic strategies of algebraic problem solving (Davis, 1983). Children's knowledge of academic mathematics involves complex cognitive activities, whose nature and extent we are just beginning to understand.

One benefit of the work on academic knowledge is that we are now in a position to perform sensible and informative studies of school learning—on "excellence" (Edmonds, 1986). When we relied solely on achievement tests as the dependent measure, this was not possible. In the absence of an adequate theory of academic knowledge, it was impossible to come to a sound understanding of what children really learn in school. Now, however, we can begin to understand these issues, and this should be of great benefit in dealing with poor children's education.

In particular, the new contributions to the theory of academic knowledge allow us to come to at least a preliminary understanding of school failure. The desire to ameliorate school failure was the motivating force behind the early studies of poor children's intellect; now we have research that sheds light on the nature of school failure and hence on the school performance of many poor children. Russell and Ginsburg (1984) conducted a study of cognitive factors underlying low achievement in school mathematics. The study included both middle- and lower-class children at the third and fourth grade levels. (By implication, the study is especially relevant to understanding poor children, because they are disproportionately represented in the ranks of school failures). Our general aim was to determine whether fourth-grade children who scored at least one year below the norm on standard mathematics achievement tests displayed unusual patterns of thinking in several different areas of mathematical cognition. We wished to know whether the low achievers displayed weaknesses in the areas of: (*a*) informal mathematical thinking, (*b*) abstract thought, (*c*) calculational aspects of arithmetic, and (*d*) basic concepts (e.g., knowledge of base ten). We tested each child individually, again using clinical interview techniques and large number of tasks developed over many years. Several findings are relevant for our concerns. First, the mathematics difficulty (MD) children possess fundamental informal concepts like the mental numberline and procedures like mental addition and estimation. They are capable of basic enumeration skills and even concepts of place value as applied to written numbers. They display insight into some structured tasks and can solve basic word problems. They make more errors, by definition, than do normal children, but the errors are the results of common "bugs" (Brown & Burton, 1978) or error strategies; MD children make errors for the same reasons as normally achieving children. Our general conclusion was that MD children are essentially normal with respect to basic mathematical thinking. They are capable of abstract thought and do not display cognitive processes of an unusual nature.

Of course, there are a number of cautions that must be raised in respect to this study. For one thing, we did not study all cognitive processes, so that our conclusions are necessarily limited to the particular measures employed. Second, we obtained some results that were incongruous and difficult to explain, for example, MD children have particular difficulty with elementary number facts. They had a harder time remembering that 2 + 2 is 4 than they did in performing some rather complicated mental addition strategies. This result is probably just the reverse of what Jensen would expect; in his view, poor children have special talent for rote memory and cannot handle conceptual tasks. The result contradicts Jensen's view, but I have no easy way of explaining it either.

Our results at present are inconclusive, and a good deal of research needs to be done. On the basis of clinical experience with children failing in school, I predict that research results would support the view that low-achieving children as a group do not suffer from serious cognitive deficiencies like inability to understand abstractions. Furthermore, in at least the first several grades of school, low-achieving children are not likely to display unusual patterns of academic cognition. They may get many wrong answers, but their basic understanding of school related work is not qualitatively different from that of normal achieving children. At the same time, there must eventually be a cumulative deficit that puts these children further and further behind. Thus, children who fail to learn simple addition and subtraction will be at a clear disadvantage in learning more complicated topics in arithmetic (like the long division algorithm). Thus, there must come a point where children who fail in school are really "out of it," and their academic cognition must eventually become deficient. But this does not reflect basic cognitive difficulties.

If the hypothesis is basically correct, that school failure does not originally derive from deficient cognition (but may eventually produce it), why do children exhibit school failure in the first place? The question leads us into areas like education, motivation, and style. With respect to education, it should be abundantly clear that many schools teach badly, and this is likely the major cause of children's academic failure. There is no evidence that under stimulating conditions poor children cannot learn quite well. Another factor is motivational: children prone to school failure may experience some form of distress that prevents them from exhibiting their capability or realizing their potential. And of course, once these children fall behind, the prophecy becomes self-fulfilling. Finally, there is the factor of style: some children's learning style or cognitive style may not mesh effectively with the teaching environment of the schools.

In brief, over recent years, we have made important advances in our understanding of appropriate methodology, in our understanding of basic and perhaps universal cognitive processes, in our knowledge of distinctive cognitive adaptations in response to unique environments, in our views of cognitive socialization, and in our conceptualization of academic knowledge. Most of these advances are indirectly relevant to the understanding of poor children, but importantly relevant nonetheless. In general, the findings support the hypothesis that poor children do not suffer from massive cognitive deficiencies. Poverty of intellect cannot explain their failure in school.

Needed Research

Cognitive developmental psychology is moving in new directions that can inform the study of poor children. Researchers are beginning to go "beyond the purely cognitive" (Schoenfeld, 1983) to propose new perspectives on issues of intellectual development. Some of these new ways of looking at mind may provide insights into poor children's intellectual growth and education.

1. Learning Potential. For the most part, cognitive psychologists have focused attention on a narrow aspect of mental life—the current cognitive structure of the individual. Traditional research has focused on such issues as the nature of concrete operations at ages X vs. Y, or the counting strategies of young children. So-called developmental studies typically examine process A (e.g., egocentric communication) in different groups of children at ages X, Y, and Z. Even the few studies employing as subjects the same children at different age levels typically examine existing cognitive structures in a static fashion and do not focus on the developmental process itself. To be sure, there are exceptions to the situation I have described; some developmentalists have focused on development and learning. For example, after many years' exploring cognitive structures (concrete operations, formal operations, etc.) the school of Piaget turned in the 1970s to the examination of issues of equilibration (as in Inhelder, Sinclair, & Bovet, 1974). Yet most cognitive developmental psychology is not directly concerned with the process of development, but with the characterization of differences in current structures at various age levels.

For the purposes of education—whether of poor children or anyone else—the perspective of this kind of developmental psychology is valuable but at the same time has shortcomings. As I argued previously, it is important to understand the structure of academic knowledge—to analyze its concepts and processes. It is valuable to determine, for example, that at the preschool level, poor children possess certain informal addition concepts and procedures (Ginsburg & Russell, 1981), or that mathematical errors are generated by common error strategies (Ginsburg, 1981). Yet, for purposes of education, it is even more important to know how one can build on the informal knowledge or eliminate the bugs; learning potential and development are the crux of the matter, not current cognitive structure. The focus is on *becoming* more than on *being*. Indeed, Papert (1980) even suggests that a focus on current cognitive structure may be counterproductive:

> The invention of the automobile and airplane did not come from a detailed study of how their predecessors, such as horse-drawn carriages, worked or did not work. Yet, this is the model for contemporary education research. . . . There are many studies concerning the poor notions of math or science students acquire from today's schooling. There is even a very prevalent "humanistic" argument that "good" pedagogy should take these poor ways of thinking as its starting point. . . .
>
> Nevertheless, I think that the strategy implies a commitment to preserving the traditional system. It is analogous to improving the axle of the horse-drawn cart. But the real question, one might say, is whether we can invent the "educational automobile." (p. 44)

Applying Papert's perspective to the case of poor children, we might argue that it is less important to know what informal knowledge poor children possess at age four or why third graders in the current schools make addition errors than it is to discover what poor children can do under more nearly ideal circumstances. Current cognitive structures, as they are shaped by the typical school environment, may be almost irrelevant to the issue of learning in more stimulating circumstances. Whether the poor child can or cannot count at age four or employs some error strategy may not be of great relevance for what he or she can accomplish in the atypical classroom.

Much of the literature on radical educational reform supports this point. Many years ago, educators like Kohl (1967), showed that unusual classrooms could produce atypically fine learning in poor children. More recently, Papert (1980) and his colleagues have shown that the LOGO computer environment can produce dramatic learning in physically handicapped children, some of whom are even judged to be retarded in ordinary classrooms. Edmonds (1986) shows that reading problems can be remedied even in children who might be considered learning disabled. It is clear that the existence theorem has been clearly proven: poor children (and various handicapped children as well) *can learn* under unusual conditions. The potential exists even if the effective educational environment is rare. Psychological theories of learning disabilities should be treated with great skepticism; they describe only what exists under current conditions, not what can occur.

Psychologists should focus more on the issue of learning potential and less on the description of cognitive structure conceived in static terms. For many years, the psychological study of learning was dull and irrelevant; perhaps there were good reasons for abandoning it. But now it seems to be time, as Brown and French (1979) have suggested, to return to this ancient but still central topic.

2. *NONCOGNITIVE FACTORS.* Motivation plays a fundamental role in education. We are all familiar with children who make great intellectual strides when they get "turned on," when there is interest in and passion for learning. The latest cultural phenomenon of this type is children who get hooked on computers and without the benfit of formal instruction—or despite it—become expert in their use. We are all familiar with other children who are frightened of learning, for example, fear of mathematics, or contemptuous of it, or afraid to exhibit to peers signs of intellectual interest. Ogbu's analysis (1986) focuses on the motivations, beliefs, and expectancies produced by the caste system in this country: some children do not learn in school because they perceive no social or economic benefits from doing so. Clinicians are familiar with children whose failure to learn is rooted in their neurotic character structure. For instance, the teacher's disapproval may be linked to the parent's, and subtraction in arithmetic may be seen as an instance of the "taking away" of love. Such aberrations may be more common than we think.

It is not enough to say, as Piaget seems to, that the cognitive structures are the source of their own motivation ("functional assimilation"). Certainly there is, can be, should be "intrinsic motivation," but many other forms of motivation are at the heart of education as well. Indeed, I would make two speculations. One is that most cases of learning problems or low achievement in the schools can be explained primarily on motivational grounds rather than in terms of fundamental cognitive deficit. Most children fail in school not because they are stupid (cognitively deficient, lacking in "formal operations", etc.) but because they are afraid, turned off, and the like. I think academic psychologists are out of touch with reality when they take so seriously the role of basic cognitive factors (i.e., intelligence, conceptual thought, Piagetian operations) in school failure. Of course, it's easier to measure cognitive variables than motivation and personality.

A second speculation is that understanding motivation may be at least as useful for educational practice—for remediation—as knowledge of cognitive structure or process. No doubt, as Brown, Palinesar, and Purcell (1986) have shown, focused intervention based on cognitive analysis (diagnosis) of disruptions in process can provide successful remediation. But it may also be true that motivating poor achievers in new ways, without paying much attention to their cognitive processes, may also dramatically improve their learning. The evidence for this is largely anecdotal: these are cases of children who read poorly until for some reason they "decide" to read, whereupon they learn very rapidly, without the benefit of tutorial help based on profound cognitive analysis.

Motivation is central to learning, just as is cognition. It is foolish to argue about which is more important. Both are vital. But we developmental and educational researchers have tended to slight the motivational.

3. Cognitive style. As I tried to illustrate in the foregoing, poor children probably do not suffer from fundamental cognitive deficits. Instead, there is evidence for the existence of universal basic cognitive processes. At the same time, poor children—or any children—may develop distinctive intellectual adaptations to the special demands of their environments. This is the cognitive difference view, usually put in opposition to the deficit theory.

An important research question for the future has to do with the nature and extent of such cognitive differences and their role in education. One way of conceptualizing the differences may be in terms of cognitive style. This concept has a long and checkered history, originating in psychoanalytic theory but eventually becoming entombed in psychometric practice. The basic idea seems to be that intellect, like other aspects of psychological functioning, has a personality, a style. Intellect can be impulsive, or defensive, or vivacious, or dull, just as our social behavior can be. In an informal study of letters of recommendation, I found that professional psychologists relied heavily on style concepts in evaluating their students and colleagues. Hardly anyone spoke of *g*; many described "independence of mind."

One research issue concerns the extent to which poor children exhibit distinctive cognitive styles that interfere with school work. Boykin (1979) has suggested that blacks exhibit a "verve" of intellect that may clash with the expectations of the middle-class school. (For a recent review, see Shade, 1982). The personality of intellect seems basic to cognitive function; we need to know much more about the role of cognitive style in poor children's intellect and education.

4. The individual in the social system. Intellect and personality are embedded in social life; they cannot be fully understood in isolation. We need a genuine ecological psychology that interprets behavior and cognition in the context of the larger social-political system. Johnny fails in school not solely or primarily because he is dumb, but because of the motivation linked to his implicit beliefs concerning his place in the class and caste system, because of the way in which he is treated by teachers whose choice of profession is itself influenced by the class system and by social expectations concerning sex roles, and because of political-economic factors beyond his control that place him in a jobless family with few material resources. Education is a social-political phenomenon as much as a psychological issue. The espousal of a narrowly psychological perspective is naive.

Although over the past several years psychologists have become increasingly aware of the need for an ecological psychology (Bronfenbrenner, 1979; Neisser, 1976; Ogbu,1986), a good deal of theoretical work needs to be done to make it a reality.

Conclusion

We have made progress in our understanding of poor children's intellect and education. The old myths of cognitive deficit are even less credible now than before. At the same time, we require more research and thinking about learning potential, motivation, cognitive style, and the role of social-political factors. To make progress, we need to supplement (or transform?) our cognitive notions with genuinely psychological and ecological considerations. This should lead to improved understanding of poor children, to reform of the educational system, and to the progress of psychology generally.

Postscript

From the vantage point of 1996, my roughly ten year old paper does not seem terribly dated in some respects. It spoke of the need to employ more flexible methods to examine poor children's intellectual development; to consider issues of cultural difference and distinctiveness as opposed to cognitive or cultural "deprivation" or "deficit"; to examine academic knowledge and motivation in detail; to focus on learning potential, on what is possible, rather than on the educational status quo; and to develop more genuinely ecological theories of development. Over the past ten years or so, researchers have in fact begun to conduct work along these lines (e.g., Brown & Campione, 1994; McLloyd, 1990; Moll, Amanti, Neff, & Gonzalez, 1992; Natriello, McDill, & Pallas, 1990), and in so doing have added greatly to our knowledge.

But in other respects, the paper seems to come from another world, a world more innocent than today's. The current landscape of poverty includes both an underclass whose suffering is almost unimaginable and a staggering array of ethnic groups struggling to find a home in the United States. The children of the underclass are not merely poor: they bear terrible burdens of violence, drugs, and an absence of social support. The children of ethnic groups bring a bewildering variety of differences to the educational system. Although Latinos are becoming one of the largest minorities in the U.S., the school system understands them poorly. And other groups, too numerous to mention—Vietnamese, Russians, Caribbeans—bring their styles and values to schools ill-prepared to appreciate them. Today the challenges for researchers and for educators—and for the body politic—are even more daunting than they were ten or twenty-five years ago.

References

Bereiter, C., and Englemann, S. (1966). *Teaching disadvantaged children in the preschool.* Englewood Cliffs, NJ: Prentice-Hall.

Boykin, A. W. (1979). "Psychological/behavioral verve." In A. W. Boykin, A. J. Franklin, and I. F. Yates (eds.). *Research directions of black psychologists.* New York: Russell Sage Foundation.

Bronfenbrenner, U. (1979). *The ecology of human development.* Cambridge, MA: Harvard University Press.

Brown, A. J., and French, L. A. (1979). The zone of potential development: Implications for intelligence testing in the year 2000. In R. J. Sternberg and D. K. Detterman (eds.). *Human intelligence perspectives on theory and measurement.* Norwood, NJ: Ablex.

Brown, A. L., and Campione, J. C. (1994). Guided discovery in a community of learners. In K. McGilly (ed.). *Classroom lessons: Integrating cognitive theory and classroom practice* (pp. 229–270). Cambridge, MA: MIT Press/Bradford Books.

Brown, A. L., Palinesar, A. S., and Purcell, L. (1986). Poor readers: Teach, don't label. In U. Neisser (ed.). *The school achievement of minority children* (pp. 105–143). Hillsdale, NJ: Lawrence Erlbaum Associates, Publishers.

Brown, J. S., and Burton, R. B. (1978). Diagnostic models for procedural bugs in basic mathematical skills. *Cognitive Science* 2: 155–192.

Cole, M., and Bruner, J. S. (1971). Cultural differences and inferences about psychological processes. *American Psychologist* 26: 866–76.

Cole, M., and Scribner, S. (1974). *Culture and thought.* New York: Wiley.

Dasen, P. R., and Heron, A. (1981). Cross-cultural tests of Piaget's theory. In H. C. Triandis and H. Heron (eds.). *Handbook of cross-cultural psychology.* Vol. 4. Boston: Allyn and Bacon.

Davis, R. B. (1983). Complex mathematical cognition. In H. P. Ginsburg (ed.). *The development of mathematical thinking.* New York: Academic Press.

Deutsch, M. (1967). (Ed.). *The disadvantaged child.* New York: Basic Books.

Edmonds, R. (1986). Characteristics of effective schools. In U. Neisser (ed.). *The school achievement of minority children* (pp. 93–104). Hillsdale, NJ: Lawrence Erlbaum Associates, Publishers.

Feagans, L., and Farran, D. C. (1982). (Eds.). *The language of children reared in poverty.* New York: Academic Press.

Flavell, J. H. (1977). *Cognitive development.* Englewood Cliffs, NJ: Prentice-Hall.

Fuson, K. C., and Hall, J. W. (1983). The acquisition of early number word meanings: a conceptual analysis and review. In H. P. Ginsburg (ed.). *The development of mathematical thinking.* New York: Academic Press.

Ginsburg, H. P. (1983). (Ed.). *The development of mathematical thinking.* New York: Academic Press.

———. (1972). *The myth of the deprived child: Poor children's intellect and education.* Englewood Cliffs, NJ: Prentice-Hall.

Ginsburg, H. P., Kossan, N. E., Schwartz, R., and Swanson, D. (1983). Protocol methods in research on mathematical thinking. In H. P. Ginsburg (ed.). *The development of mathematical thinking*. New York: Academic Press.

Ginsburg, H. P., Posner, J. K., and Russell, R. L. (1981a). The development of knowledge concerning written arithmetic: A cross-cultural study. *International Journal of Psychology* 16: 13–34.

———. (1981b). The development of mental addition as a function of schooling and culture. *Journal of Cross-cultural Psychology* 12: 163–179.

———. (1981c). Mathematics learning difficulties in African children: A clinical interview study. *The quarterly newsletter of the laboratory of comparative human development* 3: 8–11.

Ginsburg, H. P. and Russell, R. L. (1981). Social class and racial influences on early mathematical thinking. *Monographs of the society for research in child development* 46, serial no. 193.

Gladwin, T. (1970). *East is a big bird*. Cambridge, MA: Harvard University Press.

Greenfield, P. M. (1984). The role of scaffolded interaction in the development of everyday cognitive skills. In B. Rogoff and J. Lave (eds.). *Everyday cognition: Its development in social context*. Cambridge, MA: Harvard University Press.

Hess, R. D., and Shipman, V. (1967). Cognitive elements in maternal behavior. In J. P. Hill (ed.). *Minnesota Symposia on Child Psychology*. Vol. 1. Minneapolis: University of Minnesota Press, 57–81.

Hunt, J. McV. (1969). *The challenge of incompetence and poverty*. Urbana, IL: University of Illinois Press.

Inhelder, B., Sinclair, H., and Bovet, M. (1974). *Learning and the development of cognition*. Cambridge, MA: Harvard University Press.

Jensen, A. R. (1969). How much can we boost I.Q. and scholastic achievement? *Harvard Educational Review* 39: 1–123.

Kohl, H. (1967). *36 children*. New York: New American Library.

Labov, W. (1972). *Language in the inner city*. Philadelphia: University of Pennsylvania Press.

Lennenberg, E. H. (1967). *Biological foundations of language*. New York: Wiley.

McLloyd, V. (1990). The impact of economic hardship on Black families and children: Psychological distress, parenting, and socioemotional development. *Child Development* 61: 311–346.

Moll, L. C., Amanti, C., Neff, D., and Gonzalez, N. (1992). Funds of knowledge for teaching: Using a qualitative approach to connect homes and classrooms. *Theory into Practice* 31: 132–141.

Natriello, G., McDill, E. L., and Pallas, A. M. (1990). *Schooling disadvantaged children: Racing against catastrophe.* New York: Teachers College Press.

Neisser, U. (1976). *Cognition and reality.* San Francisco: W.H. Freeman.

Newell, A. and Simon, H. (1972). *Human problem solving.* Englewood Cliffs, NJ: Prentice-Hall.

Ogbu, J. (1986). The consequences of the American caste system. In U. Neisser (ed.). *The school achievement of minority children* (pp. 19–56). Hillsdale, NJ: Lawrence Erlbaum Associates, Publishers.

Ogbu, J. U. (1978). *Minority education and caste.* New York: Academic Press.

Papert, S. (1980). *Mindstorms: Children, computers, and powerful ideas.* New York: Basic Books.

Petitto, A. L., and Ginsburg, H. P. (1982). Mental arithmetic in Africa and America: Strategies, principles, and explanations. *International Journal of Psychology* 17: 81–102.

Piaget, J. (1929). *The child's conception of the world.* New York: Harcourt, Brace, and World.

Posner, J. K. (1982). The development of mathematical knowledge in two West African societies. *Child Development* 53: 200–208.

Resnick, L. B. (1983). A developmental theory of number understanding. In Ginsburg, H. P. (ed.). *The development of mathematical thinking.* New York: Academic Press.

Riley, M. S., Greeno, J. G., and Heller, J. I. (1983). Development of children's problem-solving ability in arithmetic. In Ginsburg, H. P. (ed.). *The development of mathematical thinking.* New York: Academic Press.

Rogoff, B. and Gardner, W. (1984). Developing cognitive skills in social interactions. In B. Rogoff and J. Lave (eds.). *Everyday cognition: Its development in social context.* Cambridge, MA: Harvard University Press.

Russell, R. L. and Ginsburg, H. P. (1984). Cognitive analysis of children's mathematics difficulties. *Cognition and instruction* 1: 217–244.

Saxe, G. and Posner, J. K. (1983). The development of numerical cognition: cross-cultural perspective. In H. P. Ginsburg (ed.). *The development of mathematical thinking.* New York: Academic Press.

Schoenfeld, A. H. (1983). Beyond the purely cognitive: metacognition and social cognition as driving forces in intellectual performance. *Cognitive Science* 7: 329–363.

Shade, B. J. (1982). Afro-American cognitive style. A variable in school success? *Review of Educational Research* 52: 219–244.

VanLehn, L. (1983). On the representation of procedures in repair theory. In Ginsburg, H. P. (ed.). *The development of mathematical thinking*. New York: Academic Press.

Yando, R., Seitz, V., and Zigler, E. (1979). *Intellectual and personality characteristics of children: Social class and ethnic group differences*. Hilldale, NJ: Lawrence Erlbaum Associates, Publishers.

Chapter 7

Mathematics and Social Interests

Brian Martin

Editors's comment: Brian Martin, originally a theoretical physicist and now working in science and technology studies, presents an overview of how mathematical knowledge is not neutral and discusses the ways in which mathematical knowledge is shaped by cultural influences. This chapter first appeared in *Search*, 19(4): 209–214 in 1988.

Mathematics is a product of society and it can both reflect and serve the interests of particular groups. The connection between mathematics and interest groups can be examined by looking at the social construction of mathematical knowledge and by looking at the social system in which mathematics is created and used.

Scientists have long believed that scientific knowledge is knowledge about objective reality. They commonly distinguish their enterprise from religious or political belief systems, seeing scientific truth as unbiased. This belief system has always had difficulties with certain applications of science such as nuclear weapons. The usual way in which the belief in the purity of science is maintained is by distinguishing between scientific knowledge and its applications. Scientific knowledge is held to be pure while its applications can be for good or evil. This is known as the use-abuse model.

This standard picture came under attack in the late 1960s and early 1970s. Radical critics argued that science is inevitably shaped by its social context. For example, funding of pesticide research by the chemical industry arguably influences not only what research topics are treated as important, but also what types of ecological models are considered relevant for understanding agricultural systems. Many critics argued that the key motive behind science is profit and social control (Rose & Rose 1976a, b; Arditti et al. 1980).

The political critics of science drew on and stimulated dramatic changes in the study of the history, philosophy, and sociology of science. Thomas Kuhn (1970) opened the door with his concept of paradigms, which are essentially frameworks of standard ideas and practices within which most scientific research proceeds. When a paradigm is overthrown in the course of a scientific revolution, the criteria for developing and assessing scientific knowledge change. The implication is that there is no overarching rational method to decide what is valid knowledge: scientific knowledge depends, on some level, on the vagaries of history and culture.

Sociologists studying scientific knowledge have developed and filled out this picture. They have examined not only the large-scale political and economic influences on scientific development but also the micro-processes by which scientists "negotiate" what is scientific knowledge (Barnes, 1974, 1977, 1982; Bloor, 1976; Latour and Woolgar, 1979, Mulkay, 1979; Knorr, et al. 1980).

Most of this analysis has been communicated using social science jargon in specialist journals and has had relatively little impact on practising scientists. The only philosopher of science taken note of by many scientists is Karl Popper, and even his ideas are used more as a "resource" in struggles over knowledge than as methodological aids (Mulkay & Gilbert 1981). Nowhere is this more true than in mathematics.

What does it mean to talk about the relationship between mathematics and social interests? It can refer to the impact of social factors—such as sources of funding, possible applications or prevalent beliefs in society—on the content and form of mathematical knowledge, such as on the choice of areas to study, the formulation of methods of proof and the choice of axioms. Alternatively, it can refer to the role mathematics plays in applications, from actuarial work to industrial engineering. Finally, it can refer to the social organization of the production of mathematics: the training of mathematicians, patterns of communication and authority in mathematical work, professionalisation, specialization and power relations.

"Interest" here refers to the stake of an individual or social group in particular types of actions or social arrangements. An interest can be small-scale, such as the personal advantage to a mathematician in publishing a paper to gain tenure, or large-scale, such as the strategic advantage to a military force in using an algorithm for tracking missiles. "Social interests" are those associated with major social groupings such as social classes, large organizations, occupational or ethnic groups.

My aim here is to survey some ideas bearing on mathematics and social interests. I approach the problem from two directions. The first is via the sociology of knowledge. Can sociological examination be applied to the creation and elaboration of mathematical knowledge? What does it mean to talk of the social shaping of mathematics? There are some provocative studies in this area, but in my view they do not lead by themselves to a comprehensive picture which can be used to evaluate the role of mathematical work in contemporary society.

The second path involves looking at the system of production and application of mathematical knowledge, and in particular at the use of expertise in modern society and at the relationship between mathematical theory and application.

Path One: Sociology of Knowledge

The sociology of knowledge attempts to explain the origin and evolution of knowledge using the same sorts of analysis which are applied to other phenomena, both natural and social. The dynamics of knowledge involve social, economic, political, religious, biological, and all sorts of other factors. Rather than assuming that the content and structure of knowledge is "given" by logic or the nature of reality—a transcendental explanation of knowledge—the sociology of knowledge looks for more mundane explanations.

David Bloor (1976) is a leading proponent of the "strong program in the sociology of science," which aims to investigate all knowledge using sociological methods. The key features of the strong program according to Bloor are that knowledge be explained in casual terms, that explanations be impartial and symmetrical with respect to the truth or falsity of the beliefs being explained, and that the theory be applied to itself.

Bloor adopts an approach to mathematics based on improving John Stuart Mill's view that all mathematics is ultimately based on physical models and human experiences, such as the manipulation of pebbles which can be seen as a motivation for arithmetic with natural numbers (Bloor 1976, Ch. 5). The traditional obstacles to Mill's view is F. L. G. Frege's point that mathematics seems to be "objective": mathematical reasoning has a compulsion about it which cannot always be attributed to a link with physical models. To extend Mill's theory, Bloor observes that Frege's definition of objectivity is equivalent to social convention: mathematicians have institutionalized a set of beliefs about the ways to proceed with the symbols they work with. These institutionalized beliefs are rather like rules in a game: they *must* be adhered to. Bloor's extension of Mill's perspective is that physical situations provide models for certain steps in mathematical reasoning (usually the more basic features) while mathematical convention gives an obligatory aspect to these steps and extensions of them. Mathematics thus deals not with physical reality but with social creations and conventions.

Bloor's reconstruction of Mill's position provides a powerful basis for the sociological investigation of mathematics. Since the "lawlike" features of mathematical reasoning are based on conventions, then it is natural to investigate how these conventions are created, sustained, and overturned.

Bloor investigates the history of mathematics to see what happened to alternative conceptions of mathematics, dealing with issues such as whether one is a number, Diophantine equations, and Pythagorean and Platonic numbers (Bloor 1976, ch. 6). His conclusion is that alternative concepts did exist, but that historians have relegated them to the historical rubbish bin of "non-mathematics." In this way only "genuine mathematics" remains part of the history of mathematics, which thus seems to be cumulative and without significant deviations or alternatives.

Bloor also examines the ways in which mathematical reasoning is socially "negotiated," namely, the practices through with mathematicians develop agreed-upon ways of using and interpreting the symbols and tools of their trade, including criticism, argumentation, reclassification and consensus (Bloor 1976, ch. 7). Bloor gives among other examples the case of the negotiation, over the years, of the proof of the formula $E + 2 = V + F$ relating the number of edges, vertices, and faces of a polygonal solid.

Bloor's program is a powerful one. It opens the foundations of mathematics to sociological examination by allowing the "objectivity" of mathematical reasoning to be seen as fundamentally social in nature. But Bloor does not extend his analysis to address the relation between mathematics and social interests. Even if it is accepted that the formula $E + 2 = V + F$ depends on somewhat arbitrary agreements among mathematicians rather than being inherent in the nature of polygonal solids (or the mathematical concepts of polygonal solids), that does not provide much insight into whether the social negotiation of the formula owes much or provides special benefits to particular groups in society.

At this stage it is worthwhile to spell out the different channels through which the form and content of mathematics can be shaped by society. Social interests can be connected with the choice of areas of mathematics to study, the interpretation of mathematics, and the development of mathematical frameworks.

The Choice of Mathematical Areas to Study

Differential funding or the availability of applications can affect the opening of branches of study and the prestige of different subjects. For example, the field of operations research grew out of military applications of mathematics during World War II and the strength of the field is maintained by continuing military interest.

Luke Hodgkin (1976) argues that the great surge in the "mathematics of computation," which encompasses numerical analysis and parts of computer science, is connected to the development of the needs of contemporary capitalism plus the availability of suitable technology for computing (such as transistors and now chips). He points out that the mathematics of computation is not a simple "reflection" of the economic system, as a simplistic Marxist account might suggest. Instead, the influence of the system of economic production is mediated through the social institutions of science, whose organization predated the great growth of computational mathematics.

Choice in mathematical research is also involved at the detailed level of application. Partial differential equations can be applied to many problems; the particular sets of equations which are selected out for formulation and solution can be influenced by applications, which in turn are linked to social interests.

The Interpretation of Mathematics

In many cases, especially in applied mathematics, mathematical constructions are chosen because they have desirable physical or social interpretations. An example here is Paul Forman's (1971) study of the effect of Weimar culture on the development of quantum theory. The most important strides in quantum theory occurred in Germany in the decade after World War I. Forman documents the intense antagonism to rationality which prevailed then in the Weimar Republic. Since causality was identified with rationality, physicists came under pressure to renounce their traditional allegiance to causality. Forman suggests that this pressure led the quantum physicists to search for, or at least latch on to, a mathematical formalism which could be interpreted as non-casual. In crude terms, the acausal Copenhagen interpretation and its associated mathematical framework were adopted because they looked good publicly.

Forman's study is quite relevant to mathematics, since theoretical physics constitutes the foremost application of mathematics. The case of quantum theory is intriguing because, in the decades since the establishment of the orthodox or Copenhagen interpretation, a number of alternative interpretations have been put forth. Some of these use the same mathematical formulations, but interpret their physical significance differently, while others use different mathematical formulations to achieve the same results.

> The statistical interpretation favored by Einstein uses the same mathematics (Ballentine 1970). . . .
>
> The hidden variable interpretation, a determinist approach, formulates the equations somewhat differently and, optionally, can give different results from the orthodox theory by addition of an extra parameter (Bohm 1952; Cushing 1994). . . .
>
> The splitting universe interpretation is a different interpretation of the same mathematics (DeWitt 1970). . . .
>
> The "realist" interpretation, which gets rid of the indeterminist element in quantum theory entirely, uses a different mathematical approach to achieve some of the same basic results (Landé 1965). . . .

The existence of these interpretations or reformulations of quantum theory adds support to Forman's analysis. At the least, the interpretation of the equations of quantum theory as supporting indeterminism was not *required* by the equations themselves. Furthermore, it seems possible that many of the achievements of the theory might have been accomplished using a somewhat different mathematical formulation, which could well have been *difficult* to interpret indeterministically.

So strong was the commitment to indeterminism that physicists accepted without question John von Neumann's proof in the 1930s that no hidden variable theory could be constructed. Although Bohm demonstrated such a theory in 1952, it was not until the 1960s that the flaw in von Neumann's proof was exposed (Pinch 1977).

In my experience, most physicists do not worry greatly about what quantum theory "means" but simply use mathematics in a pragmatic fashion. Indeed, one of the "crisis points" commonly experienced by physics students is when they give up their increasingly uncomfortable attempts to understand what the theory *really* means and instead just accept it, usually by sweeping their doubts under the carpet. Most historians and textbook writers have accommodated this process, as Bloor has argued about mathematics history, by exorcising alternative interpretations as unsuccessful, irrelevant or nonexistent.

The Development of Mathematical Frameworks

The choice of axioms, the types of theorems, the style of proofs and a host of other facets of mathematics can be shaped by factors such as views about the nature of social reality.

An example here is game theory, a mathematical theory which deals with conflict situations, originally developed to model economic systems (Martin 1978). Key concepts of the theory include the "players" in a game, each of which has a number of "choices," followed by "payoffs." The mathematical theory of games is built around determining the optimal strategies for making choices. The players, choices and payoffs are usually assumed to be fixed; competition is built in; payoffs tend to be quantifiable. Hence, game theory is especially suited for applications which assume and reinforce individualism and competition.

Game theory has been applied in many areas, such as international relations. What often happens in practice is that the values of the modelers are incorporated into the game theoretic formulation, which usually ensures that the game gives results which legitimate those very same values. Game theory in this situation provides a "mystifying filter": values are built into an ostensibly value-free mathematical framework, which thus provides "scientific" justification for the decision desired. Arguably, game theory has become popular because its mathematical framework makes it easy to use in this way.

The above-mentioned studies and others (Thomas 1972; Ogura 1974; Bos & Mehrtens 1977; MacKenzie 1978; Mehrtens 1987; for a comprehensive survey and analysis see Restivo 1983) show how the social context, such as economics or belief systems, can influence the areas of mathematics that are opened up and made fashionable, the types of theories that are developed, and the particular mathematical formalisms that are formulated and used. These are examples of the impact of social factors on mathematical knowledge, but they hardly establish that all mathematics is influenced in these sorts of ways. To establish this would require many studies in the line of Bloor's strong program, in an attempt to whittle down the areas of apparent autonomy of mathematical knowledge. Only if the range of sociological studies was very broad could the burden of proof be put on those who claim that there are areas of mathematics free of such formative influences.

Even if the strong program could be so developed, what would it say about mathematics and social interests? The existence of influences on the creation and adoption of mathematical knowledge does not automatically mean that knowledge preferentially serves particular groups in society.

The studies in the sociology of knowledge *initiate* the case that mathematics is connected with social interests, by refuting the view that mathematical knowledge always springs antiseptically from the nature of logic, from physical reality or from mathematicians' heads. The limits of sociological examination of mathematics remain to be tested. Some such as Bloor (1981) think the prospects are good while others disagree (Laudan 1981). In any case, since most of the sociology of knowledge studies deal with influences on the origin and development of mathematical knowledge in earlier eras, they only partially address concerns about the uses of present-day mathematics. To pursue the case further, I turn to the second path.

Path Two: The Mathematics-Society System

This approach to looking at mathematics enters not at the level of mathematical knowledge but at the level of the social systems in which that knowledge is created and applied. The social system of science refers to patterns of employment, funding, communication, training, authority, decision making, and so forth. The aim here is to look at the way systems of production and application of mathematics relate to social interests. To do this, I select out some salient features of the social systems associated with mathematical expertise.

Sources of Patronage

Most of the money for mathematics research—which is largely for salaries, but also for offices, libraries, computing and travel—comes from governments and large corporations. The source of funding inevitable has an influence on the areas of mathematics studied and the types of mathematical applications undertaken. As argued by Hodgkin (1976), much of the stimulus for work in computational mathematics also comes from actual or potential military applications.

At the detailed level of application, the formulation of mathematical problems is strongly influenced by funding and opportunities for application. In manufacturing industry, mathematical problems grow out of the need to cut costs, improve technologies, or control labor. A mathematical model for the rapid cooling of a metal bar without cracking is tied to an immediate problem. The mathematics of light transmission in optical fibres is driven by interest in application in telecommunications. The number of examples is endless.

What happens in many cases is that a practical problem, such as modeling air pollution dispersion or the trajectories of missiles, leads to a more esoteric mathematical project in numerical analysis or differential equations. The applications, and thus the funding, in these cases have an indirect influence on the type of mathematical problems studied and thought to be "interesting." That particular types of parabolic partial differential equations become whole fields of study in themselves is not due simply to some abstract mathematical significance of these equations, but to their significance in practical applications, even if at several stages removed.

Professionalization

Today, most mathematicians—taking a mathematician to be a person who creates or applies mathematical knowledge at a high level—are full-time professionals, working for universities, corporations or governments. There are few amateurs, nor do many mathematicians work for trade unions, as farmers, in churches, or as freelancers. Mathematics, like the rest of science, has been professionalized and bureaucratized. The social organization of mathematics influences the ways that ambitious mathematicians can pursue fame and fortune (Collins & Restivo 1983)

Mathematicians have a vested interest in their salaries, their conditions of work, their occupational status, and their self-image as professionals. Their preferences for types and styles of mathematics are influenced by these factors.

Judith Grabiner (1974) argues that there have been "revolutions in thought which changed mathematicians' views about the nature of mathematical truth, and about what could or should be proved." Grabiner examines one particular revolution, the switch from the 1700s when the main aim of mathematicians was to obtain results to the 1800s when mathematical rigor became very important. Of the various reasons for this which Grabiner canvasses, one is worth noting here. Only since the beginning of the 1800s have the majority of mathematicians made their living by teaching. Rather than just obtaining mathematical results for applications or to impress patrons, teachers need to provide a systematic basis for the subject, to aid students but also to establish a suitable basis for demarcating the profession and excluding self-taught competitors from jobs. This is an example of how the social organization of the profession of mathematics can affect views about the nature of mathematical truth.

Gert Schubring (1981) has argued that in the professionalization of mathematics in Prussia in the early 1800s, the "meta-conception" of pure mathematics played an important role. By defining "mathematics" as separate from externally defined objectives, the mathematicians oriented the discipline to internal values that they could control. To do this, support from the state had to be available first. Given state patronage for academic positions, the mathematicians could proceed to establish a discipline by establishing training which channelled students into the new professional orientation, reducing the number of self-taught mathematicians obtaining jobs in the field and socializing students into the meta-conception of pure mathematics. This account meshes nicely with that of Grabiner.

This process continues today. Especially in universities, the home grounds of pure mathematics, mathematicians stake their claims to autonomy and resources on their exclusive rights, as experts, to judge research in mathematics. This is no different from the claims of many other disciplines and professions (Larson 1977). The point is that if mathematicians emphasized application as their primary value, their claims to status and social resources would be dependent on the value of the application. The conception of "pure" mathematics enables an exclusive claim to control over the discipline to be made.

Herbert Mehrtens (1987, p. 160) develops the thesis that "a scientific discipline exchanges its knowledge products plus political loyalty in return for material resources plus social legitimacy." He shows how German mathematicians in the 1930s were able to accommodate the imperatives of the Nazis, especially by providing useful tools to the state. The adaptability of the German mathematics community grew out of its social differentiation, specifically the different functions of teaching, pure research, and applied research. Mehrtens' study provides an excellent model for analyzing the interactive dynamics of the two factors of patronage and the structure of the profession.

Male Domination

Most mathematicians are men, and mathematics like the rest of natural science is seen as masculine: a subject for those who are rational, emotionally detached, instrumental, and competitive. Mathematicians are commonly thought, especially by themselves, to have an innate aptitude for mathematics, and claims continue to be made that males are biologically more capable of mathematical thought than females. The teaching of pure mathematics as concepts and techniques separated from human concerns, plus the male-dominated atmosphere of most mathematics research groups, make a career in mathematics less attractive for those more oriented to immediate human concerns, especially women.

Male domination of mathematics is linked with male domination of the dominant social institutions with which professional mathematical work is tied, most notably the state and the economic system, through state and corporate funding and through professional and personal contacts (Bowling & Martin 1985).

The high status of mathematics as a discipline may be attributed in part to its image as a masculine area. Mathematical models gain added credibility through the image of mathematics as rational and objective—characteristics associated with masculinity—as opposed to models of reality that are seen as subjective and value-laden.

Specialization

There are various ways in which mathematicians shape and use their expert knowledge to promote their interests vis-à-vis other social groups. If mathematical knowledge was too easy to understand by others—both non-mathematicians and other mathematicians—the claims by mathematicians for social resources and privilege would be harder to sustain. Specialization enables enclaves of expertise to be established, preventing scrutiny by outsiders. In applications work, specialization ensures that only particular groups are served. In all cases, specialization plus devices such as jargon prevent ready oversight by anybody other than other specialists. Since hiring professionals to understand specialist bodies of knowledge can be afforded on a large scale only by governments and large corporations, specialization serves their interests more than those of the disabled or the unemployed, for example.

The role of these factors is particularly obvious in mathematical modeling. A mathematical model may be a set of equations, which is thought to correspond to certain aspects of reality. For example, most of theoretical physics, such as elementary theory for projectiles or springs, can be considered to consist of mathematical models. In most parts of physics, the models are considered well established, and physicists work by manipulating or adapting the existing models. But in other areas the choice of models is open. Various parts of reality may be chosen as significant, and various mathematical tools may be brought to bear in the modeling process.

Many people who have been involved in mathematical modeling will realize the great opportunities for building the values of the modeler into the model. I have seen this process at work in a variety of areas, including mathematical ecology, game theory, stratospheric chemistry and dynamics, voting theory, wind power, and econometrics.

A good example is the systems of difference equations used in the early 1970s to determine the "limits to growth." The choice of equations and parameters more or less ensured that global instability would result (Cole et al., 1973). When different assumptions were used by different modelers, different results—for example, that promotion of global social equality would prevent global breakdown—were obtained, nicely compatible with the values of the modelers. Another example is the values built into global energy projections developed at the International Institute for Applied Systems Analysis (Keepin & Wynne 1984).

Mathematical models are socially significant in two principal ways: as practical applications of mathematics and as legitimations of policies or practices. Most models are closely tied to practical applications, such as in industry. The narrow specialization involved in the modeling ensures that few other than those developing or funding the application would be interested in or capable of using the model. This sort of applied mathematics is closely linked to the social interests making the specific application. Whether the application is telecommunications satellites, anti-personnel weapons or solar house design, one may judge the mathematics by the same criteria used to judge that application. It is not adequate to say that the killer is guilty while the murder weapon is innocent, for in these sorts of applications the mathematical "weapon" is especially tailored for its job. Certainly applied mathematicians cannot escape responsibility for their work by referring to "neutral tools," whether this refers to their mathematical constructions or to themselves.

Models serving as legitimations are involved in a more complicated dynamic. In many cases such as limits-to-growth studies the models do no more than mathematicise a conclusion which would be obvious without the model. But the models are seen as important precisely because they are mathematical, thus drawing on the image of mathematics as objective. A mathematics-based claim also has the advantage of being the work of professionals. Anyone can make a claim, but if a *scientist* does so, relying on the allegedly objective tools of mathematics, that is much more influential. Although exercises in mathematical modeling are often shot through with biases, for public consumption this often is overlooked; the modelers draw on an aura of objectivity which is sustained by the more esoteric researches of pure mathematicians.

What then of pure mathematics? There are two major ways in which a link to social interests can be made. First is *potential* applications. These are not always easy to assess, but a good guess often can be obtained by looking at actual applications in the same or related specialities. If any new application turns up, it is likely to be in the same areas and to be used by the same groups.

It is a debatable point whether mathematics should ever be evaluated separately from applications. Arguably, the study of nature is the primary motivation for the development of and importance of mathematics, and the "correctness" of pure mathematics should be judged by its ultimate applicability to the physical world (Kline 1959, 1980). The primary reason for the ascension of pure mathematics, namely, mathematics which is isolated from application, is the social system of modern science.

This system—including funding, professionalization, male domination and specialization—in which claims to sole authority over areas of knowledge are used to claim resources, is the second way that pure mathematics is connected with social interests. Even if some bit of pure mathematical research turns out to have no application, it is still usually the case that social resources have been expended to support professional workers who are mostly male and who produce intellectual results of interest only to a handful of others like themselves. Furthermore, the work of pure mathematicians, and indeed their very existence, helps legitimate the claims of mathematics to objectivity.

Conclusions

The question, "What is the link between mathematics and social interests?", is usually answered in advance by assumptions about what *mathematics* really is. If mathematics is taken to be that body of mathematical knowledge which sits above or outside of human interests, then by definition social interests can only be involved in the practice of mathematics, not in *mathematics* This Platonic-like conception sees mathematics as value-free, but is itself a value-laden conception: it serves to deflect attention from the many links between mathematics and society.

Most people would agree that nuclear weapons have not been constructed to serve all people equally; particular social interests are involved in designing, building, testing, and deploying nuclear weapons. But what of the uranium, plutonium, iron, and other atoms contained in nuclear weapons? Are these atoms "value-laden?" A reasonable stance in my view is that the atoms in themselves are not linked to any particular groups—except the plutonium atoms which were manufactured by humans—but that the connection enters through the humanly constructed configuration of atoms. The idea of a value-free atom in isolation is all very well, but that is not what we encounter in human constructions.

Elements of mathematical knowledge can be likened to atoms, except that all mathematical concepts have been created by humans. In isolation, the mathematical concepts of an integral or a ring seem not to be associated with the interests of particular groups in society. But mathematical concepts do not exist in isolation. They are organized together for particular purposes, very narrowly for detailed applications, more generally for teaching. The more specialized and advanced ideas are mostly restricted to a small segment of the population, which claims social resources and status due to its expertise.

The belief that mathematics is a body of truth independent of society is deeply embedded in education and research. This situation, by hiding the social role of mathematics behind a screen of objectivity, serves those groups which preferentially benefit from the present social system of mathematics. Exposing the links between mathematics and social interests should not be seen as a threat to "mathematics" but rather as a threat to the groups that reap without scrutiny the greatest material and ideological benefits from an allegedly value-free mathematics.

References

Arditti, R., Brennan, P., and Cavrak, S. (Eds.). (1980). *Science and liberation.* Boston: South End Press.

Ballentine, L. E. (1970). The statistical interpretation of quantum mechanics. *Review of Modern Physics* 42: 358–381.

Barnes, B. (1974). *Scientific knowledge and sociological theory.* London: Routledge and Kegan Paul.

———. (1977). *Interests and the Growth of Knowledge.* London: Routledge and Kegan Paul.

———. (1982). *T. S. Kuhn and Social Science.* London: Macmillan.

Bloor, D. (1976). *Knowledge and Social Imagery.* London: Routledge and Kegan Paul.

———. (1981). The strengths of the strong programme. *Philosophy of the Social Sciences* 11: 199–213.

Bohm, D. (1952). A suggested interpretation of quantum theory in terms of "hidden" variables. *Physical Review* 85: 166–193.

Bos, H. J. M., and Mehrtens, H. (1977). The interactions of mathematics and society in history: some exploratory remarks. *Historia Mathematica* 4: 7–30.

Bowling, J., and Martin, B. (1985). Science: a masculine disorder? *Science and Public Policy* 12: 308–316.

Cole, H. S. D. et al. (1973). *Thinking about the Future.* London: Chatto and Windus.

Collins, R., and Restivo, S. (1983). Robber barons and politicians in mathematics: a conflict model of science. *Canadian Journal of Sociology* 8: 199–227.

Cushing, J. T. (1994). *Quantum mechanics: Historical contingency and the Copenhagen hegemony.* Chicago: The University of Chicago Press.

DeWitt, B. S. (1970). Quantum mechanics and reality. *Physics Today* 23: 30–35.

Forman, P. (1971). Weimar culture, casualty, and quantum theory, 1918–1927: adaptation of German physicists and mathematicians to a hostile intellectual environment. *Historical Studies in the Physical Sciences* 3: 1–115.

Grabiner, J. V. (1974). Is mathematical truth time-dependent? *American Mathematical Monthly* 81: 354–365.

Hodgkin, L. (1976). Politics and physical sciences. *Radical Science Journal* 4: 29–60.

Keepin, B., and Wynne, B. (1984). Technical analysis of IIASA energy scenarios. *Nature* 312: 691–695.

Kline, M. (1959). *Mathematics and the Physical World.* New York: Crowell.

———. (1980). *Mathematics: The Loss of Certainty.* New York: Oxford University Press.

Knorr, K. D., Krohn, R., and Whitley, R. (Eds). (1980). *The Social Process of Scientific Investigation.* Dordrecht: D. Reidel.

Kuhn, T. S. (1970). *The Structure of Scientific Revolutions.* Chicago: The University of Chicago Press.

Landé, A. (1965). *New Foundations of Quantum Mechanics.* Cambridge, MA: Cambridge University Press.

Larson, M. S. (1977). *The Rise of Professionalism: A Sociological Analysis.* Berkeley: University of California Press.

Latour, B., and Woolgar, S. (1979). *Laboratory Life: The Social Construction of Scientific Facts.* London: Sage.

Laudan, L. (1981). The pseudo-science of science? *Philosophy of the Social Sciences* 11: 173–198.

MacKenzie, D. (1978). Statistical theory and social interests. *Social Studies of Science* 8: 35–83.

Martin, B. (1978). The selective usefulness of game theory. *Social Studies of Science* 8: 85–110.

Mehrtens, H. (1987). The social system of mathematics and National Socialism: a survey. *Sociological Inquiry* 57: 159–182.

Mulkay, M. (1979). *Science and the Sociology of Knowledge.* London: Allen and Unwin.

Mulkay, M., and Gilbert, N. (1981). Putting philosophy to work: Karl Popper's influence on scientific practice. *Philosophy of the Social Sciences* 11: 389–407.

Ogura, K. (1974). Arithmetic in a class society: notes on arithmetic in the European Renaissance. In *Science and Society in Modern Japan: Selected His-*

torical Sources (pp. 19–23). Edited by S. Nakayama, D. L. Swain, and E. Yagi. Tokyo: University of Tokyo Press.

Pinch, T. J. (1977). What does a proof do if it does not prove? In *The Social Production of Scientific Knowledge* (pp. 171–215). Edited by E. Mendelsohn, P. Weingart, and R. Whitley. Dordrecht: D. Reidel.

Restivo, S. (1983). *The Social Relations of Physics, Mysticism, and Mathematics.* Dordrecht: D. Reidel.

Rose, H., and Rose, S. (Eds.). (1976a). *The Political Economy of Science.* London: Macmillan.

———. (Eds.). (1976b). *The Radicalisation of Science.* London: Macmillan.

Schubring, G. (1981). The conception of pure mathematics as an instrument in the professionalization of mathematics. In *Social History of Nineteenth Century Mathematics* (pp. 111–134). Edited by H. Mehrtens, H. Bos, and I. Schneider. Boston: Birkhauser.

Thomas, M. (1972). The faith of the mathematician. In *Counter Course: A Handbook for Course Criticism.* Edited by T. Pateman. Harmondsworth: Penguin. pp. 187–201.

Chapter 8

Marx and Mathematics

Dirk J. Struik

Editors' comment: In our conception of ethnomathematics, "ethno" not only refers to a specific ethnic, national, or racial group, gender, or even professional group, but also to a cultural group defined by a philosophical and ideological perspective. The social and intellectual relations of individuals to nature or the world and to such mind-dependent, cultural objects as productive forces influence products of the mind that are labeled mathematical ideas. In this chapter, Dirk J. Struik, an eminent mathematician and historian of Mathematics, indicates how a particular perspective—dialectical materialism—decisively influenced Marx's theoretical ideas on the foundation of the calculus. This chapter first appeared in *Science and Society* 12(1): 181–196, in 1948. At the end of the present version, the author includes a brief update in a postscript.

Marx received his early training in mathematics at the Gymnasium of Trier (Treves), the Rhineland city where he was born. At his graduation, in 1835, his knowledge of mathematics was considered adequate. This means that he started his career with some knowledge of elementary arithmetic, algebra to the quadratic equations, and plane and solid geometry. He also may have had trigonometry, and a little higher algebra, analytical geometry and calculus.

There are no indications that he showed any interest in mathematics during the turbulent years before and after 1848, in which he and Engels developed their outlook on the world. The first token that Marx had returned to his study of mathematics is from the period in which he settled in London and was working on his great scientific projects. In a letter to Engels of January 11, 1858,[1] he wrote:

> During the elaboration of the economic principles I have been so damned delayed by computational errors that out of despair I undertook again a quick scanning of the algebra. Arithmetic was always alien to me. Via the algebraic detour, however, I catch up quickly.

From this period until his death in 1883, Marx showed continued interest in the study of mathematics, often returning to it as a diversion during his many days of illness.

His study of algebra was followed by that of analytical geometry and the calculus. In a letter to Engels of July 6, 1863, he reported progress:

> In my spare time I do differential and integral calculus. Apropos, I have plenty of books on it and I will send you one if you like to tackle that field. I consider it almost necessary for your military studies. It is also a much easier part of mathematics (as far as the purely technical side is concerned) than for instance the higher parts of algebra. Aside from knowledge of the common algebraic and trigonometric stuff no preparatory study is needed except general acquaintance with the conic sections.[2]

It seems therefore that Marx found algebra easier than arithmetic and the calculus easier than algebra. But he was not so much interested in the technique of the calculus. He was irresistibly drawn to the age-old question of the foundation of the calculus, the more so, since in the books which he consulted on this subject, the calculus was treated in a most unsatisfactory and occasionally in a controversial way. Marx, like so many dialectical thinkers before and after him, found unending fascination in the different definitions of the derivative and the differential, as is shown by a large amount of manuscript material which was found among his papers.

In the years after 1870, Marx even tried to develop his own views. Engels reports on this phase in the preface to the second volume of *Capital:*

> After 1870 an intermission set in again, mainly due to sickness. The content of the many notebooks with abstracts of this period consists of agronomy, American and especially Russian agrarian relations, money, market and banking systems, and finally natural science, geology and physiology, and especially independent mathematical papers.[3]

Marx, in the later days of his life, cast some of his reflections concerning the differential calculus into a readable form and dispatched the manuscript to Engels. A letter of August 18, 1881 shows that Engels had studied them:

> Yesterday I found at last the courage to study your mathematical manuscripts even without reference to textbooks, and I was glad to see that I did not need them I compliment you on your work. The matter is so perfectly clear (*sonnenklar*) that we cannot be amazed enough how the mathematicians insist upon mystifying it.[4]

Engels continues to present Marx' viewpoint in his own words and to compare it with Hegel's views, with which both he and Marx were thoroughly familiar. He ends with the words:

> The matter has taken such a hold of me that it not only turns around in my head the whole day, but that also last week in a dream I gave a fellow my shirt buttons to differentiate and this fellow ran away with them [*und dieser mir damit durchbrannte*].[5]

Marx, who at that time was preoccupied with his wife's sickness—she died in December of the same year—did not, it seems, return to the subject in his subsequent correspondence. When, however, Engels reported to Marx (November 21, 1882) on an exchange of letters between him and their friend Sam Moore on the subject of Marx' mathematical theories, Marx made a prompt reply the next day. We return to this correspondence later in this article.

Marx died before he could add anything more to his ideas. Engels later thought of publishing Marx' mathematical manuscripts together with his own on the dialectics of nature. In the preface to the second edition of the *Anti-Dühring* (1885), he mentions his own studies in mathematics and the natural sciences, and adds that he had to discontinue them after the death of Marx. He concludes: "there will perhaps later be an opportunity to collect and to publish the obtained results, together with the posthumous, and very important, manuscripts of Marx."[6]

Engels did not find the time to accomplish this work, and the papers of Marx and Engels dealing with the exact sciences remained in the archives. The German Social Democrats, who inherited the papers of Marx and Engels, were unable to appreciate the dialectics of mathematics, physics, and chemistry. Understanding had to wait until the Russians began to show the fundamental importance of Marx' and Engels' philosophical work. Lenin's *Materialism and Empirio-criticism* (1908) was a trail blazer, but it did not become known outside of strictly Russian circles until it was published in German, long after the revolution of 1917. Later the Russians published Engels' *Dialectics of Nature,* first in Russian, then (1927) in the original German.

Both Lenin's and Engels' books are now available in English, Lenin's in a translation of 1927, Engels' in a translation of 1940.

Still later some of the most characteristic of Marx' mathematical manuscripts were published, but only in a Russian translation.[7] Our study is based on the papers published by the Russians. It is to be hoped that all of his mathematical note books will eventually be published not only in Russian, but also in the original German.

The extent of Marx' interest in mathematics is shown by the fact that the Marx-Engels-Lenin Institute in Moscow has obtained, since 1925, photographic copies of about 900 pages of Marx' mathematics manuscripts, all of which have been deciphered and put in order.[8] They consist essentially of abstracts of textbooks, studied by Marx, often with notes, of comprehensive accounts of special subjects, and of independent investigations, expressing different stages in Marx' studies, from preliminary sketches to finished manuscripts probably prepared for the benefit of Engels. Only a few pages, hardly twenty-four are devoted to computational work.

By far the most voluminous of these manuscripts deal with algebra, which Marx studied from Lacroix', Maclaurin's and perhaps from other texts. Most of this algebra deals with the solution of equations of higher degree, but Marx also showed an interest in series, notably divergent series. There are also abstracts dealing with analytic geometry, notably from a book by Hymers.

Other manuscripts contain Marx' reflections on the differential calculus. There are again plenty of abstracts and comprehensive accounts based on the textbooks of Lacroix, Boucharlat, and Hind, supplemented by those of Hall and Hemming, all popular school texts from the early decades of the nineteenth century. This work deals mainly with the conception of function and of series, of limit and of derivative, the series of Taylor and Maclaurin, and the determination of maxima and minima. Marx showed particular interest in Lagrange's famous use of the Taylor series for the "algebraic" foundation of the calculus, and compared the different definitions of the derivative and the differential in the various texts. Marx, in one of his own notes, reproduces the derivation of the binomial theorem from Taylor's theorem, and remarks that "Lagrange, on the contrary, derives Taylor's theorem from the binomial theorem," a fact which he often repeats and to which he devotes some thought. One of his manuscript papers is entitled "A somewhat modified development of Taylor's theorem on purely algebraic base according to Lagrange,"[9] others have such significant headings as: "Taylor's theorem—is based on the translation from the algebraic language of the binomial theorem into the differential way of expression," and "Maclaurin's theorem is also only translation from the algebraic language of the binomial theorem into the differential language." Two notebooks, probably dating from a later period in Marx' life, contain examples of the method of differentiation which Marx eventually preferred, as well as a paper on the differential and a historical sketch of the methods of differentiation used by Newton, Leibniz, D'Alembert, and Lagrange. These notebooks present the position which Marx seems to have placed before Engels. They also contain a long paper on the integral calculus, which contains a critical analysis of Newton's "Analysis per aecquationes numero terminorum infinitas." Their published contents form the subject of the present article.

Marx studied the calculus from textbooks which were all written under the direct influence of the great mathematicians of the late seventeenth and the eighteenth centuries, notably of Newton, Leibniz, Euler, D'Alembert, and Lagrange. He was not so much interested in the technique of differentiation and integration as in the basic principles on which the calculus is built, that is, in the way the notions of derivative and differential are introduced. He soon found out that a considerable difference of opinion existed among the leading authors concerning these basic principles, a difference of opinion often accompanied by confusion. This confusion only increased in the school textbooks written by the minor authors.[10] Different answers were given on such questions as whether the derivative is based on the differential or vice versa, whether the differential is small and constant, small and tending to zero, or absolutely zero, and so forth. Marx felt the challenge offered by a problem which had attracted some of the keenest minds of the past and which dealt with the very heart of the dialectical process, namely, the nature of change. Not finding any satisfying answer in the books, he tried to reach an answer for himself in his own typical way: by going to the sources, comparing the results, and forging beyond them into new regions. It may perhaps strike the reader that among the sources studied by Marx there seems to be no reference to Augustin Cauchy—at any rate as far as we can judge from the published material. Cauchy's work, which underlies the exposition of the foundation of the calculus in our present day textbooks, could have been available to Marx.[11] The reason that Marx took no notice of Cauchy may be that Cauchy's ideas only slowly penetrated into textbooks, so that they might have escaped Marx, who did not move among professional mathematicians.[12] A more likely reason is that Cauchy's way of defining the derivative was essentially that of D'Alembert, so that Marx did not consider his method a new one.

Whatever Marx' reasons were to ignore Cauchy's work, his feeling of dissatisfaction with the way the calculus was introduced was shared by some of the leading younger professional mathematicians of his day. In the same year (1858) in which Marx resumed his study of mathematics, Richard Dedekind at Zürich felt similar dissatisfaction, in his case while teaching the calculus. Writing in 1872, he first stated that in his lessons he had recourse to geometrical evidence to explain the notion of a limit; then he went on:

> But that this form of introduction into the differential calculus can make no claim to being scientific, no one will deny. For myself this feeling of dissatisfaction was so overpowering that I made the fixed resolve to keep meditating on the question till I should find a purely arithmetic and perfectly rigorous foundations for the principles of infinitesimal calculus.[13]

This led Dedekind to a new axiomatic approach to the conception of continuum and irrational number, which was one of the great pioneering efforts in what we call the "arithmetization of mathematics." Some years later one of the other pioneers of the new methods of rigor in mathematics, Paul Du Bois Reymond, exclaimed:

> What mathematician would deny that—especially in its published form—the conception of limit and its closest associates, the conception of the limitless, the infinitely large and the infinitely small, the irrational, etc., still lack rigor? The teacher in write and word is used to hurry quickly through this questionable entrance to analysis, in order to roam the more comfortable on the well blazed roads of the calculus.[14]

It was not until the last decades of the nineteenth century, under the influence of Dedekind and Du Bois Reymond, as well of Weierstrass and Cantor that the thorough overhauling of the principles of the calculus took place, which underlies modern methods, and has shown that Cauchy's approach can lead to full rigor. This work appeared too late to influence Marx and Engels.[15] The result is that Marx' reflections on the foundations of the calculus must be appreciated as a criticism of eighteenth-century methods. We feel, however, that his work, developed contemporaneously with but independently of the leading mathematicians of the second half of the nineteenth century, even now contributes to the understanding of the meaning of the calculus.

We should never forget, of course, that Marx never published his material, and that there is not even an indication that he intended publication, even though Engels seems to have played with the idea. Marx worked on mathematics in spare hours, for relaxation, often in hours of sickness, guided by some books which he happened to have in his library, such as Boucharlat's, which introduced the principles of differentiation in an unsatisfactory way. He looked for elucidation in the sources quoted in Boucharlat and similar books, which led him to Newton, Leibnitz, D'Alembert, and Lagrange. His notes were in the first place intended for his own clarification, after reading those classics in attempts to understand the often obscure texts. Struck by the unsatisfactory formulations in these books, he tried in a characteristic way to straighten out the difficulties for himself.

The difficulties which Marx tried to overcome are at present as real as in his time, even if our formal apparatus is more carefully elaborated and practically foolproof. These difficulties are as old as Zeno of Elea and as young as the latest philosophical or physiological attempt to understand how rest can pass into motion, and how motion can lead to rest. This is the reason why Marx studied so carefully the conception of the derivative of a function and the related conception of the differential. He found that there are three main methods by which these conceptions have been developed. Marx classified them, called them the mystical, the rational and the algebraic method (connected with the names of Newton-Leibnitz, D'Alembert, and Lagrange respectively), and then opposed to them his own mode of understanding the derivative, the differential, and the calculus in general. Let us explain the difficulty by differentiating the function $y = x^3$ in the different ways criticized by Marx.

(1) *Newton-Leibnitz.* ("The mystical differential calculus")[16]—x changes into $x + \dot{x}t$ in Newton, into $x + dx$ in Leibnitz; we follow Leibnitz. Then y changes into $y_1 = y + dy$ and $y_1 = y + dy = (x + dx)^3 = x^3 + 3x^2dx + 3x(dx)^2 + (dx)^3$. Since $(dx)^2$ and $(dx)^3$ are infinitesimal as compared with $3x^2dx$, they may be dropped, and we obtain the correct formula

$$dy = 3x^2dx.$$

This is highly mysterious, and the mystery does not disappear if we first divide dy by dx

$$dy/dx = 3x^2 + 3xdx + (dx)^2$$

and then let $h = dx$ be zero. It is true that we obtain the right formula

$$\frac{0}{0} = \frac{dy}{dx} = 3x^2,$$

but as Marx remarks:

> the nullification of h is not permitted before the first derived function, here $3x^2$, has been liberated from the factor h by division, hence $(y_1 - y)/h = 3x^2 + 3xh + h^2$. Only then can we annul (*aufheben*) the finite difference. The differential coefficient $dy/dx = 3x^2$ must therefore also originally be developed before we can obtain the differential $dy = 3x^2dx$.

In other words, we knew in advance what the answer must be, and build up some reasoning to make it plausible. It was this loose way in which Newton and Leibnitz usually founded the calculus which led Bishop Berkeley to his famous criticism in *The Analyst* of 1734. Here he asked whether the dx are zero or not zero, called them "Ghosts of departed quantities" and concluded that no mathematician who believed these absurdities could reasonably object to the miraculous tenets of religion. It has not been the only case in which foundation difficulties in science have been exploited for idealist and obscurantist reasons.

Mathematicians felt the difficulty and tried to cope with it by suggesting more exact ways of founding the calculus.[17] The most important contributions were those of D'Alembert and Lagrange.

(2) D'Alembert ("The rational differential calculus").[18] In Marx' words: D'Alembert starts directly from the starting point of Newton and Leibnitz

$$x_1 = x + dx$$

but he makes immediately the fundamental correction $x_1 = x + \Delta x$, that means, Δx becomes an undetermined, but prima facie finite increment, which he calls h. The transformation of this h or Δs into dx (he used the Leibnitz notation, like all Frenchmen) is only found as the last result of the development or at least just before closing hour (*knapp vor Torschluss*), while it appears as starting point with the mystics and the initiators of the calculus:

$$\frac{f(x+h)-f(x)}{h} = \frac{y_1 \times y}{x_1 \times x} \frac{x_1^3 - x^3}{h} = \frac{(x+h)^3 - x^3}{h} = 3x^2 + 3xh + h^2$$

Now, by placing $h=0$, the expression $[f(\mathrm{x}+\mathrm{h})-f(x)]/h$ changes into $\frac{dy}{dx}$:

$$\frac{0}{0} = \frac{dy}{dx} = 3x^2 = f'(x)$$

The way in which D'Alembert differentiates is very much akin to Cauchy's method. We write at present with Cauchy

$$\frac{dy}{dx} = \lim_{h \to 0} \frac{f(x+h)-f(x)}{h}$$

Marx' objection to this method is that though it is formally correct, the derivative $f'(x)$ is already present in $3x^2 + 3xh + h^2$, that is, before differentiation. It is simply the first term of a sum, $3x^2 + 2xh + h^2$, and D'Alembert's method only consists in devising a way in which to get rid of the member (or members) of the sum which follows $3x^{\cdot}$ Marx calls this "Loswicklung" (separation); while the correct method should be *Entwicklung* (development):

> The derivation therefore is the same as in Leibnitz and Newton, but the ready-made derivative is in strictly algebraic way *separated* from its further context.[19] There is no *development* but a separation of the $f'(x)$,[20] here $3x^2$ from its factor h and the members which appear next to it in the other members marching on in rank and file. What has really developed is the left hand symbolic side, namely, dx, dy and their ratio, the symbolic differential coefficient dy/dx or $0/0$ (rather in the other way $0/0=dy/dx$), which in its turn again provoked some metaphysical shudders, though the symbol was mathematically derived. D'Alembert had, by stripping the differential calculus from its mystical garb, made an enormous step ahead.

Marx' evaluation of D'Alembert's work as "an enormous step ahead" still stands. This is the more remarkable, since even modern historians of mathematics have a way of glossing over it. Marx next proceeds to Lagrange.

(3) Lagrange ("The purely algebraic differential calculus").

$$y = f(x) = x^3$$

$$y_1 = (x + h)^3 = x^3 + 3x^2h + 3xh^2 + h^3$$

Lagrange simply defines the coefficient of h as the derivative: $\frac{dy}{dx} = f'(x) = 3x^2$, or more generally by Taylor's theorem for a general $f(x)$:

$$y_1 = f(x + h) = y \text{ (or } fx) + \frac{dy}{dx}h + \frac{d^2y}{dx^2}\frac{h^2}{2} + \ldots$$

> Marx then paraphrases Lagrange's method in the words: In the first method (1), as well as in the rational one (2), the required real coefficient is fabricated ready made by the binomial theorem and can be found already as second term of the series expansion, hence in the term which necessarily contains h^1. The whole further differential procedure, be it as in (1) or be it as in (2), is therefore luxury. Let us therefore shed the useless ballast. We know once and for all from the binomial expansion that the first real coefficient is the factor of h, the second one that of h^2, etc. These real differential coefficients are nothing but *the derived functions of the original function* in x, expanded binomially in succession. . . . The whole real problem reduced itself to the finding of methods (algebraic ones) of expanding all kinds of functions of x + h into integral ascending powers of h, which in many cases cannot be effected without great prolixity of operations.[21] Up to now there appears nothing in Lagrange, but what can be found directly from D'Alembert's method (since this also includes the whole development of the mystics).

The objection which Marx raised against the classical writers was that all four had the derivative already prepared before the process of differentiation really begins. Marx wanted a method which actually followed the process of variation of the variable and in this process itself defined the derivative as $0/0$, in which case it can be endowed with a new symbol dy/dx. The derivative, he claimed, should be derived by a process of differentiation, not be produced from the beginning by the binomial theorem.

> Whether we start falsely from $x + dx$ or correctly from $x + h$, if we substitute this undetermined binomial into the given algebraic function of x, we change it into a binomial of a definite degree, e.g. $(x +$

> $h)^2$ instead of x^3, and this in a binomial in which in one case dx, in the other case h, figures as its last member. Hence it also figures in the expansion only as a factor, with which the functions, derived from the binomial, are externally affected.[22]

This lack of internal development can be avoided in the method which Marx suggests, say for $y = x^3$:

$$y = f(x) = x^3$$

$$f(x_1) - f(x) = y_1 - y = x_1^3 - x^3 = (x_1 - x)(x_1^2 + xx_1 + x^2)$$

$$\frac{f(x_1) - f(x)}{x_1 - x} = \frac{y_1 - y}{x_1 - x} = x_1^2 + xx_1 + x^2.$$

When $x = x_1$, or $x - x_1 = 0$, we obtain:

$$\frac{0}{0} = \frac{dy}{dx} = x^2 + xx + x^2 = 3x^2.$$

In this method, writes Marx, we obtain first a *preliminary derivative*, namely, $x_1^2 + xx_1 + x^2$, and this passes by $x = x_1$ into the *definite derivative*. This passing form s_1 to x does away with any "infinitesimal" approximation, it shows that the derivative is actually 0/0, obtained when $x - x_1$ is actually zero:

Here we see in striking form:

> *Firstly*; to obtain the derivative we must place $x = x_1$, hence $x - x_1 = 0$ in *the strict mathematical sense*, without a trace of only infinitesimal approximation. *Secondly*: Though the fact that x_1 has been placed $= x$ hence $x - x_1 = 0$, nothing symbolic enters into the "derivative." The quantity x_1, originally introduced by the variation of x, does not disappear, it is only *reduced* to its minimal boundary x. It remains an element introduced as new into the original function, which by its combination partly with itself, partly with the x of the original function produces at the end the "derivative," that is the preliminary "derivative" reduced to its minimum value. . . . The transcendental or symbolic accident ($0/0 = dy/dx = 3x^2$) occurs only on the left hand side, but it has already lost its terror, as it appears now only as the expression of a process that already has shown its real content on the right hand side of the equation.[23]

At the moment that $x = x_1$ the quotient $\Delta y/\Delta x$ becomes 0/0. Since in this expression 0/0 every trace of its origin and of its meaning has disappeared it is replaced by the symbol dy/dx, in which the finite differences Δy and Δx appear in symbolical form as liquidated (*aufgehobene*) or vanished (*verschwundene*) differences. At this moment algebra disappears and the differential calculus, which operates with the symbols dy/dx, begins.

In order to understand Marx' intentions better, we translate here part of the letter which Engels wrote him August 18, 1881, after he had read Marx' manuscript:

> When we say that in $y = f(x)$ the x and y are variables, then this is, as long as we do not move on, a contention without all further consequences, and x and y still are, pro tempore, constants in fact. Only when they really change, that is *inside the function*, they become variables in fact. Only in that case is it possible for the relation—not of both quantities as such, but of their variability—which still is hidden in the original equation, to reveal itself. The first derivative $\Delta y/\Delta x$ shows this relation as it occurs in the course of the real change, this is in every *given* change; the final derivative dy/dx shows it in its generality, pure. hence we can come form dy/dx to every $\Delta y/\Delta x$, while this itself ($\Delta y/\Delta x$) only covers the special case. However, to pass from the special case to the general relationship the special case has to be liquidated as such (*als solcher aufgehoben werden*). Hence, after the function has passed through the process from x to x' with all its consequences, x' can be quietly allowed to become x again, it is no longer the old x_1 which was only variable in name, it has passed through *real change*, and the *result* of the change remains, even if we liquidate it again *itself* (*auch wenn wir sie selbst wieder aufheben*).
>
> We see here at last clearly, what many mathematicians have claimed for a long time, without being able to present rational reasons for it, that the *derivative* is the original, the differentials dx and dy are derived.

The difference between Marx' method and D'Alembert's method (and also that of Cauchy) should not be misunderstood and rejected as trivial or insignificant ($x - x' = h$ versus $x' = x + h$). Marx, as I see it, was perfectly satisfied that D'Alembert's method is formally correct. However, he wanted to come to an understanding of the process of differentiation itself. Is the derivative obtained by letting x (and y) pass through a sequence of constant values, or is it necessary to let x (and y) really change? Thus understood, we see the old "paradox" of Zeno emerging: can the motion of a point be obtained by following a sequence of positions of this point at rest? Zeno showed that a sequence of such positions will never produce motion; he also showed by a similar reasoning that Achilles will never reach the tortoise. D'Alembert's methods, Marx claimed, represents a mode of thought which does not do justice to the actual event which happens when a function is differentiated. What happens is a real change, and this is better understood when we first write $\Delta y / \Delta x$ as a function of x and an entirely new x', and then let $x = x'$. Moreover, $h = x' - x$ does not only *approach* zero, *h becomes* zero. Emphasis is placed on the fact that the derivative only appears when both δy and δx are absolutely zero. This never became clear with the "mystics" Leibnitz-Newton, and appeared as an accidental thing in D'Alembert-Lagrange[24]. It is so little understood that in some popular texts, such as Hogben's *Mathematics for the Million*, the impression is given that the process of differentiation is only approximately true. But even in our modern textbooks, though they use a formal apparatus which is unimpeachable, some of the thought behind the apparatus is not fully clarified.

Let us take, as an example, the textbook *Pure Mathematics* of G. H. Hardy, who is one of our greatest living mathematicians. The derivative is explained in the Cauchy-D'Alembert way:

$$\phi'(x) = \lim_{h \to 0} \frac{\phi(x+h) - \phi(x)}{h}$$

which means that $\{\phi(x + h) - \phi(x)\}/h$ tends to a limit when h tends to zero. What does this mean? We are told that $\phi(y)$ tends to the limit l as y tends to zero, if, when any positive number δ, however small, is assigned, we can choose $y_0(\delta)$ so that $|\phi(y) - l| < \delta$ when $0 < y \leq y_0(\delta)$.[25]

This definition is exact, in the sense that we have a correct and subtle criterium to test any limit. But $\phi(y)$ always hovers near the limit, since we are told that y "tends" to zero. Similarly, $\phi'(x)$ is defined by means of an h which "tends" to zero. The question is, is the event $h = 0$ ever reached? Marx not only affirms it, he stresses it. The usual modern textbook definition does not take this question seriously, because it is satisfied with a pragmatic criterium which allows us to recognize a limit when it appears.[26]

The result is that much teaching of the elements of calculus proceeds as follows—and I confess to it myself in my own teaching. First, it is shown that a limit can be approached as closely as we like, but never reached. Then the derivative is defined with the aid of this conception of limit. And then suddenly we begin to work with this derivative, which could never be reached (as we have before demonstrated) as if it actually had been reached. The case $h = 0$, $x' = x$, though present in the formal apparatus, is somehow obscured in the reasoning. An exception is found in the work of Moritz Pasch, who in his very careful analysis of the derivative develops a formal apparatus in which there is full room for the case $h = 0$.[27]

Marx therefore belonged to that school of thinkers who insist on utmost clarity of thought in interpreting a formal apparatus. His position contrasts sharply to that of those mathematicians or mathematical physicists who believe that the formal apparatus is the only thing that matters. Marx' position was that of the materialist, who insists that significant mathematics must reflect operations in the real world.

It is interesting to notice that the differences between Marx' and D'Alembert's formal apparatus diminish when we consider more complicated functions. For the case $y = \sin x$ the derivative, in the D'Alembert way of differentiation, is still obtained by separation (*Loswicklung*), but by $y = \log x$ the derivative can only be obtained from $\Delta y/\Delta x$ by letting h pass through a real change.

As soon as dy/dx is established as the result of a real change, it becomes itself the subject of a calculus, the differential calculus. Marx, in a manuscript on the meaning of the differential, derived as one of the first formulas of this calculus, that the derivative of $y = uz$, u and z functions of x, is given by

$$\frac{dy}{dx} = z\frac{du}{dx} + u\frac{dz}{dx}$$

When $uz = f(x)$, then dy/dx can be written $f'(x)$ and "the $f'(x)$ stands opposed to dy/dx as its own symbolic expression, as its double or symbolic equivalent."

> The symbolic differential coefficient has become *an independent starting point,* whose real equivalent has first to be found. The initiative has been moved form the right hand pole, the algebraic one (in $dy/dx = f'(x)$) to the left hand one, the symbolic one. With this, however, the differential calculus appears also as a specific kind of computation, operating already independently on its own territory. Its starting points du/dx, dz/dx are mathematical quantities which belong exclusively to this calculus and characterize it. And this reversal (*Umschlag*) of the method resulted here form the algebraic differentiation of uz. The algebraic method changes automatically into its opposite, the differential method.

Now, by removing in the equation (a), $\frac{dy}{dx} = z\frac{du}{dx} + u\frac{dz}{dx}$ the com- common denominator dx, we obtain (b), $d(uz) = dy = udz + zdu$ in which every trace of its origin from (a) has been removed.

> It (b) is therefore valid in the case that u and z depend on x as well as in the case that they only depend on each other without any relation to x. It is from the beginning a symbolic equation and can serve from the beginning as a symbolic operational equation.

The differential is therefore a symbolic form—we would say an operational form—$dy = f'(x)dx$ appears as just another form of $dy/dx = f'(x)$ and is always convertible into the differential form. Modern mathematicians will have no fault to find with this method, and V. Glivenko has specially shown[28] how Hadamard, the French mathematician, had stressed the operational character of the differential. Marx does not mention, however, the now common interpretation that dy should be $f'(x)\Delta x$, obtained by arbitrarily placing $dx = \Delta x$. This way of representing dy, which dates back to Cauchy, may have escaped Marx (he criticizes Boucharlat for his introduction of the differential, but Boucharlat's methods is an antiquated one). We believe however that Marx would in any case have objected to this equation $dx = \Delta x$, which established an identity between two conceptions with an entirely different operational meaning. The interpretation of dy by Cauchy, which has found its way in all our texts, is mechanical and can only be justified by the use to which the formula $dy = f'(x)dx$ can be put as an approximation to certain changes of a constant x into an equally constant $x + \Delta x$.[29] And the fact that this difference between dx and Δx, dy and Δy can be neatly represented in a figure would not have impressed Marx and Engels, whose interest was in the arithmetical-algebraic relationship of the symbols of the calculus with the real process of change. This may be shown from the following correspondence between Marx and Engels after Sam Moore had written his opinion on the manuscript material of Marx:

> Enclosed first a mathematics attempt by Moore. The result that "the algebraic method is only the differential method disguised" refers of course only to his own method of geometrical construction and is there also relatively correct. I have written to him that you do not care about the way in which the matter is represented in the geometrical construction, the application to the equation of the curves is indeed sufficient (*reiche ja hin*). Moreover, the fundamental difference between you and the old method is that you make x change into x',

> hence make them really vary, while the other one departs form $x + h$, which is always only the sum of two quantities, but never the variation of a quantity. Your x therefore, even when it has passed through x' and has again become x, is yet another than before; while x remains constant during the period when h is first added to x and later again subtracted. However, every graphical representation of the variation is necessarily the representation of the *past* process, of the *result*, hence of a quantity which became constant, the line x; its complement is represented as $x + h$, two segments of a line. From this already follows that a graphical representation of how x becomes x' and x' again becomes x is impossible (Engels to Marx, Nov. 21, 1882).[30]

Marx' answer followed the next day:

> Sam, as you have seen immediately, criticizes the analytical method which I have used by simply pushing it aside, and instead keeps himself busy with the geometrical application, to which I did not devote one word.
>
> I could in the same way get rid of (*konnte damit abspeisen*) the development of the proper so-called differential method—beginning with the mystical method of Newton and Leibnitz, then continuing with the rationalist method of D'Alembert and Euler, and finishing with the strictly algebraic method of Lagrange (which however always starts fro the same original principle as Newton-Leibnitz)—I could get rid of this whole historical development of analysis by saying that *practically* nothing essential has changed in the geometrical application of the differential calculus, that is, in the geometrical representation (*Versinnlichung*).[31]

This last remark of Marx shows affinity with that of Dedekind, who also endeavored to build up the calculus independent of the geometrical representation of the derivative. We can consider this as one of the characteristics of Marx' analysis, in which it agreed with our modern approach. Another important feature was his insistence on the operational character of the differential and on his search for the exact moment where the calculus springs form the underlying algebra as a new doctrine. "Infinitesimals" do not appear in Marx' work at all. In his insistence on the origin of the derivative in a real change of the variable he takes a decisive step in overcoming the ancient paradox of Zeno—by stressing the task of the scientist in not denying the contradictions in the real world but to establish the best mode in which they can exist side by side.[32] Here his position is directly opposite to that taken by Du Bois Reymond, who thought that the increments *dx, dy* have to be taken as being at rest, invariable[33], or of the modern mathematician Tarski, who denies the existence of variable quantities altogether.[34] Marx' position in this respect will be appreciated by most mathematicians.

We believe that this survey of Marx' opinions on the origin of the calculus demonstrates that publication of his other mathematical manuscripts is also desirable.

Postscript

After this paper was published in 1948, the collected mathematical works of Marx were published in Moscow, the original German text together with a Russian translation and introduction: *K. Marks, Matemati*ˇ*ceskie rukopisi* (1968, 640 pp). A partial English translation exits: *The mathematical manuscripts of Karl Marx* (London, New Park, 1983). See the review by D. J. Struik in the *Archives Internationales d'histoire des sciences* 97 (1976) 343; H. Kennedy, "Karl Marx and the foundations of differential calculus," *Historia Mathemtica* 4 (1977): 303–318, and P. Gerdes, *Marx demystifies calculus* (translated from Portuguese by Beatrice Lumpkin), MEP Publications, Minneapolis, 1983, with extensive bibliography.

Notes

1. Marx-Engels, *Gesamtausgabe* (Berlin: 1930), Abt. III, Bd. II, p. 273.

2. Ibid., III, p. 149.

3. Marx, Karl, *Capital* (Chicago: 1919), II, p. 10.

4. *Marx-Engels Gesamtausgabe*, Abt. III, Bd. IV, p. 513.

5. Ibid., p. 514.

6. Engels, Friedrich, *Anti-Dühring* (New York: 1939), p. 17.

7. Marx, Karl, *Marksizm i Estestvoznanie* (Moscow: Partisdat, 1933). The Russian translation of the manuscripts occupies p. 5–61; it is followed by articles by E. Kolman, S. Ianovskaia, D. J. Struik, H. J. Muller, and others. The original German text of the manuscript has not, as far as I know, been pub-

lished, though there seems to have been plans; see *Unter dem Banner des Marxismus* (1935), no. 9, p. 104, n. 1. I received in 1935 a typewritten copy of the original German text of the published mathematical manuscripts from the Marx-Engels Institute in Moscow, and the quotations in the present article are translated from this text.

8. The information concerning the general character of the 900 pages of Marx' mathematical manuscripts is taken from S. Ianovskaia, "O Matematicheskich Rukopisiakh K. Marksa," p. 136–180. See also E. Colman, *Science at the Cross Roads* (London: 1931), p. 233–235.

9. "Nach Lagrange somewhat modified Entwicklung des Taylorschen Theorems auf bloss algebräischer Grundlage."

10. A good survey of the various theories is given by F. Cajori, "Grafting of the Theory of Limits on the Calculus of Leibniz," *Am. Math. Monthly* 30 (1923): p. 223–34.

11. A. Cauchy, *Résumé des leçons données a l'Ecole Royale Polytechnique sur le calcul infinitésimal* (Paris: 1823).

12. The preface to the sixth edition of Boucharlat's book (1856), which Marx consulted, though mentioning in detail the work of Newton, Leibniz, D'Alembert, and Lagrange, is silent about Cauchy. One of the first widely used textbooks which explicitly used Cauchy's methods was C. Jordan, *Cours d'analyse*, which appeared in 1882.

13. R. Dedekind, *Stetigkeit und Irrationalzahlen* (1872). Translated in "Essays on the Theory of Numbers" (Chicago: 1901), p. 1 f.

14. P. Du Bois Reymond, *Die allgemeine Funktionentheorie*, I (1882), p. 2. The author was the brother of the physiologist Emil, who framed the slogan of agnosticism: "Ignorabimus."

15. It is even doubtful if any pertinent information on the work of the great German mathematicians of the second half of the nineteenth century reached Marx and Engels. The England of their days was an excellent place to study capitalism, as well as physics and chemistry and biology, but it was backward in mathematics, except in some specialized branches of geometry and algebra.

16. Leibnitz issued his first publication on the calculus in 1684, Newton his in 1693.

17. See e.g., F. Cajori, *A History of the Conceptions of Limit and Fluxion in Great Britain from Newton to Woodhouse* (Chicago, and London: 1919).

18. D'Alembert on "Différentiel" in Diderot's *Encyclopédie* (1754).

19. "losgewickelt von ihrem sonstigen Zusammenhang."

20. "Es ist keine *Entwicklung*, sondern cine *Loswicklung des* $f(x)$."

21. We know that often it cannot be done at all, but this requires an extension of the functional conception beyond Lagrange's horizon.

22. *"nur als Faktor, womit die durch das Binom abgeleiteten Funktionen äusserlich behaftet sind."*

23. *"Das transzendentale oder symbolische Unglück ereignet sich nur auf der linken Seite, hat aber seine Schrecken bereits verloren, da es nun als Ausdruck eines Prozesses erscheint, der seinen wirklichen Gehalt bereits suf der rechten Seite der Gleichung bewährt hat."*

24. More information in F. Cajori, "The History of Zeno's Arguments on Motion," vi, *Am. Math. Monthly* 22 (1915): p. 143–149.

25. G. H. Hardy, *Pure Mathematics* (Cambridge, England: Cambridge University Press, 6th ed., 1933), esp. p. 116, 198. This definition of limit is valid when y tends to zero by positive values. In a similar way, a definition of limit can be reached when y tends to zero by negative values.

26. See e.g. F. Cajori, *Am. Math. Monthly*, 22 (1915), p. 149, concerning variables reaching their limits: "In modern theory it is not particularly a question of argument, but rather of assumption. The variable reaches its limit if we will that it shall; it does not reach its limit, if we will that it shall not." Such a reasoning seems to lead to the conclusion that it depends on our will whether Achilles will reach or will not reach the tortoise.

27. M. Pasch, "Der Begriff des Differentials," in *Mathematik am Ursprung* (Leipzig: 1927), p. 46–73, esp. 61, 68.

28. V. Glivenko, "Der Differentialbegriff bei Marx und Hadamard," *Unter dem Banner des Marxismus* (1935) no. 9, p. 102–110; Russian text in *Pod Znamenem Marksizma* (1934) no. 5. See J. Hadamard, *Cours d'analyse*, I (Paris: 1927), p. 2 and 6.

29. Compare C. De la Vallée Poussin, *Cours d'analyse infinitésimale*, I (Paris: Louvain, 1923), p. 52: "For the substitution of dx for Δx in the equation $df(x) = f'(x)\Delta x$ there is no necessity, but it is hallowed by custom and this custom is justified."

30. The words between quoation marks are in English in the letter—see *Marx-Engels Gesamtausgabe*, Abt. III, Bd. II, p. 571.

31. *Marx-Engels Gesamtausgabe*, Abt. III, Bd. IV, p. 572. Compare Marx, *Capital*, Part I, ch. 3, Section 2: "The Metamorphosis of Commodities," (Engl. translation, ed. 1889, p. 76).

32. *Marx-Engels Gesamtausgabe*, Abt. III, Bd. IV, p. 572. Compare Marx, *Capital*, Part I, ch. 3, Section 2: "The Metamorphosis of Commodities," (Engl. translation, ed. 1889, p. 76).

33. Du Bois Reymond, Die allgemeine Funktionentheorie, I, (1882), p. 141, states his dislike for the conception of dx as a "quantité évanouissante,"

since he disapproves (*geht mir entschieden wider den Mann*) quantities which begin to move only when we look at the formulas: "As long as the book is closed, profound rest previals. As soon as I open it, the race to zero begins of all quantities provided with the *d*." Marx, without coming to Du Bois Reymond's conclusion, might have shared his criticism, since he wanted to express not only a change on paper, but a change in reality.

34. A. Tarski, *Introduction to Logic* (New York: 1941), p. 4.

Section IV

Reconsidering What Counts as Mathematical Knowledge

Arthur B. Powell and Marilyn Frankenstein

In a French mathematics education study, a seven-year-old was asked the following question: "You have ten red pencils in your left pocket and ten blue pencils in your right pocket. How old are you?" When he answered: "twenty years old," it was not because he didn't know that he was *really* seven, or because he did not understand anything about numbers. Rather it was, as Pulchalska and Semadeni (1987) conclude, because the unwritten "social contract" between mathematics students and teachers stipulates that "when you solve a mathematical problem . . . you use the numbers given in the story. . . . Perhaps the most important single reason why students give illogical answers to problems with irrelevant questions or irrelevant data is that those students believe mathematics does not make any sense" (p. 15).

As this situation described by Pulchalska and Semadeni reveals, we can observe the split between "everyday" mathematical knowledge and "school" mathematics in many different contexts. Earlier, we noted that D'Ambrosio (1985/reprinted here as chapter 1) traces the historical development of this split to the social stratification of Egyptian and Greek societies. In a contemporary context, Frankenstein (1989) finds that working-class adult students in the United States are often surprised to learn that the decimal point is the same point used to write amounts of money. Similarly, Spradbery (1976) worked with sixteen-year-old students in England who

> had failed consistently to master anything but the most elementary aspects of school Mathematics. . . . They had received, and remained unhelped by, considerable "remedial" teaching and, finally, they left

> school "hating everyfink what goes on in maffs." Yet in their spare time some of these same young people kept and raced pigeons. . . . Weighing, measuring, timing, using map scales, buying, selling, interpreting timetables, devising schedules, calculating probabilities and averages . . . were a natural part of their stock of commonsense knowledge (p. 237).

Besides social and class divisions, Harris (1987/reprinted here as chapter 10) shows that sexism also underpins the dichotomy between "school" mathematics and one's stock of commonsense knowledge and perverts what counts as mathematical knowledge. For example, a problem about preventing the lagging in a right-angled cylindrical pipe from inappropriately bunching up and stretching out, is labeled engineering and considered to be "mathematics," whereas the identical domestic problem of designing the heel of a sock is called "knitting" and not considered to have mathematical content. We, instead, classify both the engineering and domestic problems as examples of ethnomathematics. Further, Gerdes (1988/reprinted here as chapter 11) and his students raise questions about how imperialism might have been responsible for arresting the development of various "abstract" geometrical and algebraic ideas inherent in traditional Mozambican material culture.

The mathematical knowledge embedded in the activity of adults handling money, students racing pigeons, and women knitting socks is not fragmented from the knowledge of each of these activities; rather, it is created and recreated in praxis. However, the academically enforced disjuncture between "practical" and "abstract" mathematical knowledge contributes to students feeling that they do not understand or know any mathematics. Further, Joseph (1987/reprinted here as chapter 3) considers that this disjuncture fuels the intellectual elitism that regards mathematical discovery as following only "from a rigorous application of a form of deductive axiomatic logic." Moreover, this elitism, combined with racism, considers nonintuitive, nonempirical logic a unique product of European, Greek mathematics. This Eurocentric view dismisses Egyptian and Mesopotamian mathematics as merely the "application of certain rules or procedures . . . [not] "proofs" of results which have universal application" (pp. 22–23). Joseph disputes this biased definition of proof, arguing that

> the word "proof" has different meanings, depending on its context and the state of development of the subject. . . . To suggest that because existing documentary evidence does not exhibit the deduc-

> tive axiomatic logical inference characteristic of much of modern mathematics, these cultures did not have a concept of proof, would be misleading. Generalizations about the area of a circle and the volume of a truncated pyramid are found in Egyptian mathematics. . . . As Gillings [1972, pp. 145–6] has argued, Egyptian "proofs are rigorous without being symbolic, so that typical values of a variable are used and generalization to any other value is immediate" (pp. 23–24).

Henderson (1990) argues that formal, academic mathematics "masquerades" as real mathematics. He believes "no formal definition can capture all aspects of our experience [of a mathematical entity]" and further, a more "real" concept of proof is "something which is sufficient to convince a reasonable skeptic" (p. 3). Henderson (1996) informs his geometry students that "[a] proof as we normally conceive of it is not the goal of mathematics—it's a tool—a means to an end. The goal is understanding. Without understanding we will never be satisfied—with understanding we want to expand that understanding and to communicate it to others." Further, Hersh (1991) argues that although "rigorous" proofs are infallible *in principal,* many of them are fallible *in practice*:

> The difference maybe negligible for computations of a few lines. It's not so negligible for computations of a few dozen pages. In fact, a simple probabilistic argument shows that if the probability of error in each line has magnitude "epsilon," then the probability of error in sufficiently many lines has magnitude "one minus epsilon." In other words, a sufficiently long computation is practically sure to contain errors (pp. 25–26).

He goes on to state that "additional grounds for confidence in our conclusions . . . [such as] examples and special cases, analogies with known results, an expected symmetry or an unexpected elegance . . . " are what convinces us that our conclusions "must be true" (p. 26).

An illustration of points that Henderson and Hersh make and one that also challenges the "Greek deductive" model of proof from within academic mathematics is the computer demonstration of topology's famous four-color conjecture. Gardner (1980) discusses how this proof, "buried in printouts that resulted from 1200 hours of computer time" forces us to rethink what counts as a proof and what distinctions we make between empirical science and theoretical mathematics. He summarizes Tymoczko (1979) who makes the point that this computer proof is "a program for attacking the [four-color] problem by computer along with the results of an 'experiment' performed on the computer . . . [blurring] the distinction between mathematics and natural science and [lending] credibility to the opinions of those contemporary philosophers of science such as Hilary Putnam who see mathematics as a 'quasi-empirical' activity" (p. 14).

The Eurocentric bias that denies the rigor of Egyptian mathematics also considers to be "childlike" and "primitive" the mathematical knowledge of traditional, non-literate cultures. Ascher and Ascher (1986) argue that "there is not one instance of a study or restudy that upon close examination supports the myth of the childlike primitive" (p. 131). They provide examples that not only support this point, but also reveal how false assumptions about the mathematical knowledge of others and lack of respect for the logic of others intersects with racism when one considers what counts as mathematical knowledge. For instance, they discuss a well-known anecdote about a trade between an African sheep herder and an explorer. The herder agrees to accept two sticks of tobacco for one sheep but becomes confused and upset when given four sticks of tobacco for two sheep. The story is supposed to show that the herder cannot comprehend that 2 + 2 = 4.[1] An alternative interpretation, respecting the herder's knowledge, "raises the issue of the difference between a mathematical concept and its application. . . . Sheep are not standardized units." So it is logical that a second, different animal would not also be worth two sticks of tobacco. "[T]he applicability of even the simplest of mathematical models becomes a question of cultural categorization" (p. 128). For another example, the Aschers quote Lévy-Brühl, an anthropologist who felt that the occasional substitution of 3, 7, or 9 for each other in Veddic religion, rituals, and legends was "an absurdity in logical thought . . . quite natural to prelogical mentality, for the latter, preoccupied with the mystic participation, does not regard these numbers in abstract relation to other numbers, or with respect to the arithmetical laws in which they originate" (1910, p. 211). They counter his conclusion with information from a recent field study of the Kédang who also use this kind of number substitution

> When used in symbolic contexts, odd numbers are associated with life and even numbers with death. Substitutions within these classes are possible if circumstances require it. If, for example, a ceremonial period of four days is stipulated but cannot be met, 2 days will do but 3 would be a serious infringement. . . . The formation of these equivalence classes is an example of an abstract idea about numbers (p. 130).

An example that shatters the notion of a dichotomy between concrete and abstract thought and demonstrates the subjective, culturally determined nature of "abstract" categories is provided by Glick (as cited in Rose 1988, p. 291) in recounting the frustrations of researchers working with a group of people whom "academic anthropologists" would label "primitive." The investigators had twenty objects, five each from four categories: food, clothing, tools, and cooking utensils. When asked to sort the objects, most of the people produced ten groups of two, basing their sorting on practical connections among the objects (for example, "the knife goes with the orange because it cuts it").

> [the people] at times volunteered "that a wise man would do things in the way this was done." When an exasperated experimenter asked finally, "How would a fool do it?" he was given back groupings of the type . . . initially expected—four neat piles with foods in one, tools in another.

Walkerdine (1990/reprinted here as chapter 9) cites an encounter which illuminates the creation of the categories "concrete" and "abstract." She describes two observations: a mother and her sons arguing about buying drinks they could not afford; and, a father and his son making a game out of calculating change—"what if I bought . . .?" She contrasts the concrete material necessity in the conversation between the mother and sons with the imaginary constructions in the dialogue between the father and son. She asks intriguing questions about these exchanges:

> What is the effect of relative poverty and wealth on the way in which certain problems can be presented as "abstract" versus "concrete," or, as I would prefer to put it, problems of practical and material necessity versus problems of "symbolic control"? And what is the relationship between the classic concrete/abstract distinction and the one between a life in which it is materially necessary to calculate for survival and a life in which calculation can become a relatively theoretical exercise? Might calculation as a theoretical exercise have become the basis of a form of reasoning among imperial powers which depended for the accumulation of their capital on the exploitation of the newly discovered colonies? Do theoretical concepts come with wealth and what, if so, does this mean for economic and psychological theories of development and underdevelopment? (p. 52)

She goes on to argue that to describe the interaction between the father and son as "abstracted" from everyday practices is misleading because the imaginary calculation "exists as a discursive relation in a new set of practices, namely, those of school mathematics, with its own modes of regulation and subjection" (p. 54). Rather than dichotomizing concrete versus abstract, Walkerdine speaks for viewing the different conversations as shifting from "one discursive practice to another."

The kind of linear thinking that puts knowledge into separate categories of "practical" and "abstract" not only limits our conceptions of what are mathematical ideas in everyday life, but also in academic settings. For example, Bigum (1990) theorizes that the Eurocentric conception of "rational thought," based on Newtonian determinism, results in linear numeracy being considered the underlying foundation of academic mathematical knowledge. This limits the concept of mathematical applications to the sphere of "linear predictability and control: the mathematics of *reason*. . . . In our rational constructions, Newton's system of the world is long forgotten, what remains is the illusion of controlling, knowing, and accessing the world: history collapses" (pp. 4–5). Diop (1991) argues that the discoveries and conceptions of modern physics have changed our concept of "reason" and "called into existence a new logico-mathematical formalism that will raise, for the first time in the history of the sciences, "doubt," "uncertainty" at the level of logical value" (p. 363). Bigum (1990) suggests a more appropriate foundation for an academic mathematics that can apply to comprehending our post-modern, quantum existence. It would be rooted in the unpredictability of chaotic nonlinear mathematics that "has proven that even simple questions can have answers so complicated that they contain more information than man's [*sic*] entire logical system" (Ford, cited in Bigum, p. 7). Moreover, a more appropriate foundation would also be grounded in the self-referential logic of Kurt Gödel's Incompleteness Theorem that asserts the impossibility of completely describing or deciding the veracity or falsity of all questions within a given system, which "reflects, perhaps, the unfinished character of mathematical logic" (Diop, 1991, p. 365).

Note

1. Zaslavsky (1973) relates that this tale was first told by Sir Frances Galton after he visited Africa (p. 289). Galton, who coined the term "eugenics" in 1883, considered measurement "the primary criterion of a scientific study." In

essence, he tried to "standardize" anything that might possibly be measured, including prayer, beauty, and boredom—the latter by "counting the number of [a person's] fidgets." He further believed that nearly everything he could measure was inheritable. When his cousin Charles Darwin pointed out that "men did not differ much in intellect, only in zeal and hard work," Galton countered that "the aptitude for work is heritable like every other faculty." So, it is not surprising that Galton could not see a more sophisticated reason for the sheepherder's confusion. Moreover, an important note for the politics of knowledge is that Galton was considered a leading intellect of his time and his "scholarship" had significant influence on the development of mental measurements such as the IQ test (Gould, 1981, pp. 75–77).

References

Ascher, M., and Ascher, R. (1986). Ethnomathematics. *History of Science* 24: 125–144.

Bigum, C. (1990). "Postmodern curriculum: The mathematical basis; or, who was Kurt Gödel anyway?" In R. Noss, A. Brown, P. Drake, P. Dowling, M. Harris, C. Hoyles, and S. Mellin-Olsen (Ed.). *Proceedings of the First International Conference: Political Dimensions of Mathematics Education: Action & Critique*. Institute of Education: University of London.

D'Ambrosio, U. (1985). Ethnomathematics and its place in the history and pedagogy of mathematics. *For the Learning of Mathematics* 5(1): 44–48.

Diop, C. A. (1991). *Civilization or Barbarism: An authentic anthropology*. New York: Lawrence Hill.

Gardner, M. (1980). The coloring of unusual maps leads into uncharted territory. *Scientific American* (February): 14–21.

Gillings, R. J. (1972). *Mathematics in the time of the pharaohs*. Cambridge, MA: MIT Press.

Gould, S. J. (1981). *The Mismeasure of Man*. New York: W. W. Norton & Co.

Harris, M. (1987). An example of traditional women's work as a mathematics resource. *For the Learning of Mathematics* 7(3): 26–28.

Henderson, D. W. (1990). The masquerade of formal mathematics and how it damages the human spirit. In R. Noss, A. Brown, P. Drake, P. Dowling, M. Harris, C. Hoyles, and S. Mellin-Olsen (Ed.). *Proceedings of the First International Conference: Political Dimensions of Mathematics Education: Action & Critique*. (pp. 115–118) Institute of Education: University of London.

Henderson, D. W. (1996). *Experiencing geometry on plane and sphere*. Upper Saddle River, NJ: Prentice Hall.

Hersh, R. (1991). What proof? Unpublished manuscript.

Joseph, G. G. (1987). Foundations of eurocentrism in mathematics. *Race & Class*, 28(3): 13–28.

Pulchaska, E., and Semadini, Z. (1987). "Children's reaction to verbal arithmetic problems with missing, surplus or contradictory data." *For the Learning of Mathematics*, 7(3): 9–16.

Rose, M. (1988). Narrowing the mind and page: Remedial writers and cognitive reductionism. *College Composition and Communication* 39(3): 267–302.

Spradbery, J. (1976). "Conservative pupils? Pupil resistance to curriculum innovation in mathematics." In G. Whitty and M. Young (Ed.). *Explorations in the Politics of School Knowledge* (pp. 236–243). Driffield: Nafferton.

Tymoczko, T. (1979). The Four-Color problem and it's philosophical significance. *The Journal of Philosophy*, 76(2): 57–83.

Walkerdine, V. (1990). Difference, cognition, and mathematics education. *For the Learning of Mathematics* 10(3): 51–56.

Zaslavsky, C. (1973). *Africa counts: Number and patterns in African culture*. Boston: Prindle, Weber and Schmidt.

Chapter 9

Difference, Cognition, and Mathematics Education*

Valerie Walkerdine

Editors's comment: Through her argument against pre-social models of cognition, Valerie Walkerdine, a cognitive psychologist and mathematics education researcher, questions distinctions between practical and theoretical knowledge as it applies to the learning of mathematics. In our analysis, this lays the groundwork for reconsidering what counts as mathematical knowledge. This chapter first appeared in *For the Learning of Mathematics* 10(3): 51–56, in 1990.

It seems apt, when our conference takes place in the so-called Third World, with participation of academics from the old European imperial powers and the newer ones north of the border, to discuss how understandings of difference might affect our analyses of cognition and mathematics education. To explore these issues, I want to discuss some implications of my research on cognitive development, class, and gender, the analysis of mathematics education as discursive practice, and the approaches to practices of mathematics in formal and informal settings as in, for example, the work of Lave (1988), Scribner (1984), and Carraher (1988).

My aim is to speak generally, to open up a debate by examining some basic questions concerning how and why the issues that are raised are characterized in a particular way. In "The Mastery of Reason" (1988), I set out some ways in which I felt that post-structuralist theories might help us to understand the issues of context and transfer. I argued in that volume that there were some major problems

* An invited talk given at the International Group for the Psychology of Mathematics Education, Mexico City, July 1990.

with the way in which context and transfer were theorized, stemming from a view of context as something grafted onto a single model of cognitive development. I suggested that the problem lay within the theory itself and that simply adding on context was not enough. I proposed a theory of practices in which, instead of a unitary, fixed model of the human subject possessing skills in contexts, linked to models of learning and transfer, we might understand subjectivity itself as located in practices, examining the discursive and signifying methods through which a person becomes "subjected" in each practice. It is that analysis which I want to draw on here in suggesting that we might approach the issue of context and transfer differently in order to propose a theoretical account which deals adequately with the social and historical. To do this, I suggest, we need to abandon our view of the pre-given subject with skills and pre-social models of human cognition altogether. Such a view is, of course, not new, but I do want to suggest one theoretical pathway which might aid in this endeavor.

Let me begin by examining certain assumptions about childhood which are made in so-called First World theory. I vividly remember attempting to write some notes for a review of Lave's book sitting outside a cafe in the fashionable Coyoacan district of Mexico City. Of course there were the ubiquitous small children selling things: here artisanias, elsewhere Chiclets, and in some corners of the Third World, themselves. It would be easy to imagine a research project which aimed to examine the advanced calculating skills of such children of tender years and to compare this with school performance. But what would it mean and what would we be doing it for? I also remember that, just as I was watching a small boy trying to sell toys, a little boy of about the same age, but clearly from an entirely different class, cycled past on his tricycle with his mother. It was this child who embodied the classical bourgeois idea of "child." Of course, one could argue that these two children represented difference, or rather cultural differences and strengths, one in informal mathematics perhaps, the other in formal. But to approach the problem of difference in this way would be socially and politically a problem—for the two little boys are not "equal but different." They differ also in the type and extent of their exploitation and oppression. What I want to do here is to sketch out what might be the beginnings of a way of understanding psychologically and socially how difference is lived. When we concentrate solely on the cognitive aspects of performance we fail to engage with certain central aspects of the way in which oppression is experienced. That is, as in the above example, the child selling on the street is earning money which is likely to be a central plank in his family's survival. He has to learn to calculate then as if (to use an English phrase) "his life depended on it." Meanwhile we might point to the way in which such calculation is "low level" (a very common complaint in research on girls and mathematics, for example, see Walkerdine et al (1989) for a discussion) compared to the so-called "higher order reasoning" which the middle class boy might be able to perform. We might add that the first child is deprived and that this explains his poor performance at higher level tasks. But what I want to do here is to question this very line of argument. What is higher level and how can we make sense of an argument like this outside certain historical and social questions about the nature of mathematics and mathematics education itself? My claim is that if we begin to address these questions we set up our psychological arguments in a completely different way.

Let me cite another example. I remember sitting in a seaside cafe watching a mother and her sons negotiating the buying of drinks. The boys wanted cokes and the mother argued that these were too expensive and that they should have "a warming cup of tea." By contrast, I watched a father and son sitting in a cafe in a park in central London making a game out of calculating change: "What if I bought . . .'?", and so on. There was no economic necessity at stake here. Now, although it might look at first sight as though these two examples were similar, I would argue that there are some important differences. What does it mean that the father and son in the park are constructing imaginary problems as opposed to the material problems faced by the mother trying to regulate her sons' consumption of expensive commodities? What is the effect of relative poverty and wealth on the way in which certain problems can be presented as "abstract" versus "concrete," or, as I would prefer to put it, problems of practical and material necessity versus problems of "symbolic control?" And what is the relationship between the classic concrete/abstract distinction and the one between a life in which it is materially necessary to calculate for survival and a life in which calculation can become a relatively theoretical exercise? Might calculation as a theoretical exercise have become the basis of a form of reasoning among imperial powers which depended for the accumulation of their capital on the exploitation of the newly discovered colonies? Do theoretical concepts come with wealth and what, if so, does this mean for economic and psychological theories of development and underdevelopment?

How do our ideas of "real mathematics" and of mathematical "truth" become incorporated into the "truth" about the human subject which is used in the regulation of the social. The "truth" of reason and reasoning, of the world as a book written in the language of mathematics, become important aspects of historically specific regimes of truth. Carraher (1989) discusses this issue, but Foucault's idea of "truth" is useful because it allows us to link that mathematical "truth" with the "truths" in forms of management and government which aim to regulate the subject. So, for example, when Carraher tells us that Brazilian street children did not solve a problem by one-to-one correspondence we are left little option but to pathologize them since we have no other (socially and historically specific) theories on offer.

Historically, European accounting practices, for example, shifted in complexity when they introduced double entry bookkeeping. The money economy introduced in relation to mercantile capitalism a system of positive and negative integers. The transformations of "gold," into a commodity is accomplished by means of a number of transformations of signification in which the same signifier, "gold", becomes a different sign (Walkerdine, 1986, 1988). Many people chart the fact that shifts within psychology and mathematics are related to particular social and political events and practices. How do these issues get caught up in fantasy, the kind of fantasy which other classes and groups have of each other? Samantha, a white middle class English four-year-old is asking her mother why the window cleaner will get paid for cleaning the windows of her home. She is puzzling over the exchange relation between work and money and goods. Tizard and Hughes (1984) cite this example as showing the "power of the puzzling mind" of the four-year-old. Here they use a generic concept, and yet they point out that many of the working-class four-year-olds in their study did not puzzle over the exchange relation. It was made painfully clear to the children: they were often told, for example, that they could not have certain things because money was scarce, and indeed that money was earned through labor. Why then do Tizard and Hughes assume that such a specific issue, which clearly relates to poverty and wealth, reflects a "general state of mind?" Are there indeed any general states of mind at all?

When poor children fantasize they may often dream of wealth, of fantasies of plenty in which they have, as one nursery school child put it in her domestic play, "chicken and bacon and steak." Or, like the seven-year-olds I cited in "The Mastery of Reason," they may find a shopping game that they are playing very pleasurable because, while they are supposed to be counting change and subtracting small amounts of money from ten pence, they actually laugh at the disjunction between prices and goods (a yacht for two pence, for example) and pretend to be wealthy shoppers, put on middle class accents and generally have a good time. However, they do not get better at maths. The fantasies of poverty are wealth, the fantasies of mathematicians, according to Brian Rotman (1980), are of an ordered world. He calls mathematics "Reason's Dream," a dream in which "things once proved stay proved forever," a dream of order outside the confines of time and space. And the dream of developmental psychologists? Certainly for a specific configuration of developmental psychology and education that I shall go on to describe, there is a dream of a classroom as a natural environment with pain and oppression left outside the classroom door.

Mathematics and Discursive Practices

Theories of cognitive development, at least those stemming from the work of Piaget, have their roots in theories of evolution. They offer us generic theories of the development of a "species being" in interaction with an object world, ontogeny recapitulating phylogeny. Indeed, the development of the human species, at home with a mother, is often taken to be at the highest point of the evolutionary scale. In this view, there is little room for history or for the social, except a social which is grafted on or which regulates rates of development according to a fixed sequence. This fixed sequence takes us from pre-logical to logico-mathematical reasoning which is at first concrete and then abstract. The assumed pinnacle of abstract reasoning is rarely if ever questioned. And yet, of course, it is precisely this which various groups are routinely accused of not being able to reach: girls, working-class children, blacks, third-world children, and so forth. And what I am putting forward is the germ of a suggestion that this simple sequence is itself a historical product of a certain world view produced out of European models of mind at a moment in the development of European capitalism dependent on the colonization and domination of the Other, held to be different and inferior. It was the European aristocratic and bourgeois male who was to become the model of a rationality founded upon a life-style in which economic necessity was not an issue and in which the domination of the Other was to become to a certain extent justified by a reading of difference as inferiority. That the position of those Others, the working classes and the colonized peoples, for example, was produced by their oppression and exploitation, their poverty, their appalling working conditions, letting a few of them slowly into education in order that European and colonial administration might eventually be undertaken by members of the dominated groups themselves, is a feature which is rarely brought into question when attempting to understand the production of psychological aspects of development.

Here I am referring to the way in which certain colonized peoples and members of the European working class were "educated out" for entry into the middle class, but that this meant effectively that regulators did not need to be applied by colonizers or the upper and middle classes because oppressed peoples could be taught to regulate the less educated members of their own groups. This made the whole thing more complex, and sedimented the idea that "the normal" was something that such peoples could aspire to and was something which they were not. Such concepts of normality and pathology could then become central planks of recognition and self-regulation that people took into themselves.

In order to set out some of the ways in which we might begin to understand this, I want to outline very briefly the place of theories of cognitive development in British mathematics education and then go on to examine the idea, as set out in "The Mastery of Reason," that we might understand mathematics education in terms of discursive practices. The idea of child development as a central plank of the early education of children in Europe generally, and in Britain in particular, has a long history, especially in relation to debates about childhood as a natural state associated with the idea of an education according to nature. The idea that reasoning is a natural phenomenon was to become the centerpiece of the new "scientific pedagogy," using psychology, promoted from the end of the nineteenth century onward (Walkerdine, 1984). The promotion of reasoning can be understood as part of what Foucault (1980) has described as the new modes of government, based as they were on the necessity for the production of scientific knowledge in the population, with particular emphasis on the new urban proletariat. Child study societies were set up in England around the turn of the century and many people followed Darwin's example of monitoring the development of his infant son. The idea of mapping development was taken to mean that education could be scientifically controlled according to an idea of stages of development. There was an overwhelming emphasis on the idea of the norm and normality, through which the regulation of the population was to be assured (Rose, 1985). In the early twentieth century, following the work of Itard and Seguin in France with Victor, the wild boy of Aveyron, which implied that humanity could be taught, Maria Montessori applied their methods to the education of children from Italian slums. It is here, then, that the idea that from the feral child to the child of the Other (working class, colonized) is just a short step, begins to take shape. Normal development can be monitored, humanness can be trained. The Other can be regulated by attempting to render him/her normal and by monitoring the pathology of development to try to put it right. The idea of development is, of course, presented as though it were a matter of "nature"; but this nature is very particular indeed. Many authors have noted that the model for reasoning normality is that of a white, bourgeois male (e.g. Sayers, 1982; Le Doeff, 1979). It should be noted, therefore, that the idea of normal development carries with it a very oppressive model of the natural, in which the idea of a normal course of development is used as a regulative device. While such ideas were incorporated into pedagogic experiments in the early decades of the twentieth century, it was not until the postwar period that they really began to come into their own in state education. The climate was ready in the 1930s, but the war intervened and it was not until the expansive years of the 1960s that "childcentredness," as it became known, became incorporated in a large way in curriculum development. Mathematics education was rather slower than other curriculum areas to take up these ideas, but there were two reports in the 1950s and 1960s, one by the Mathematical Association (1956) and one by the Schools Council (1965) which advocated the "new pedagogy." In doing so, the central idea of mathematics as reason became enshrined within the curriculum. This was widely interpreted to mean that logico-mathematical principles could be used to code all activities, and this became translated into a kind of commonsense understanding in which everything became potentially mathematics. There was an inherent confusion because it was assumed that children were unable to recognize that mathematics is everywhere. In this analysis, representation was grafted on to an unproblematic base of action. In "The Mastery of Reason," I challenged these central notions, arguing that "mathematical" signs are produced within specific practices and that these practices are always discursive.

Let me give some examples. I analyzed the way in which so-called mathematical signifiers, such as "more" and "less," were produced within domestic settings in the homes of a sample of thirty four-year-old girls and their mothers (Tizard and Hughes, 1984; Walkerdine and Lucey, 1989). While it is commonly assumed in early education that "less" is more complex than "more" and that the two form a pair, a contrastive opposition, describing the comparison of quantities, analysis of the mother-daughter exchanges revealed that, although there were plenty of examples of the comparison of quantities, these were not described using the word pair "more/less," and that while "less" was not used at all, "more" was used frequently, but in the context of the regulation by the mothers of their daughters' consumption of commodities. For example, a mother might tell her daughter that she could have no more of a particularly expensive commodity or that she could not have more food until she had eaten what was on her plate. The contrastive pair here was not "more/less" but something more like "more/no more." It will come as no surprise that such terms were used more frequently by mothers in working-class families, so that such little girls would be more likely to understand their mothers as more regulative and to have very strong negative associations with the terms "more." (In Walkerdine (1990), I cite an example from my own history: my mother's use of the phrase "much wants more.")

What then will such children make of the use of the term "more" to describe the comparison of quantities in early mathematics? I argue that this may be the same signifier as in the practices of the home, but it is not the same sign. This difference is crucial for it suggests a more complex issue than existing practice might suggest. I argue further that such signifiers are made to signify when united with a signified within a particular practice, from which they take their meaning. Such practices are discursively regulated with the participants positioned in particular ways. The idea of the production of mathematical signs within practices renders them at once both socially and historically specific and links them to the nonrational noncognitive axis by the use of Lacan's (1977) transformation of Freud's theory of unconscious chains of association into chains of signifiers.

I further analyzed the ways in which domestic practices are discursively different from or similar to school mathematical ones. Although the analysis of "mathematics is everywhere" stresses the similarity between practices, such an analysis glosses over important differences. This discourse stresses the idea of transfer and the sense that all experiences can be analyzed logico-mathematically. My analysis stresses why and how practices are made to signify and suggests that the relation between family and school practices is far more complex than is suggested by the notion of doing mathematical examples in familiar contexts. I examined examples of mothers and daughters cooking together and asked when and how cooking could be said to have become mathematics. Certain quite specific discursive transformations took place when cooking became mathematics. In every case, the discourse moved away from the product of the task, something to be cooked, towards a mathematical string, with a particular linguistic form, in which all external reference was removed from the string itself (as for example in the string 2 + 3 = 5). I argued that cooking could not be said to *be* mathematics, only to act as a foil for it, until this transformation had occurred. This concentration on the mathematical string for its own sake, moving away from a product, is typical of the mathematical tasks which I observed in early education. Indeed, the analysis of the shopping game to which I referred earlier, makes it clear that one of the problems for the group of seven-year-olds was that this game was represented as shopping but that the regulation of the game was quite distinct from that of shopping practices. For example, each child had to choose a card with an item to be bought and an amount of money less than ten pence. They had to work out what change from ten pence they would get if they bought the item using plastic coins and to record the calculation on paper. As I have explained, the group found the disjunction between the game prices and "real" prices the basis for considerable humor and fantasy. They also had a fresh ten pence piece each time so that their money never decreased as it would have in real shopping, and their end-product was a calculation written on a piece of paper and not a number of purchases. In other words there was absolutely no exchange. Now, this is the issue that I referred to right at the beginning of the paper. The calculation has apparently become abstracted from its insertion in everyday practices. Yet to use the term abstracted is misleading, for the new calculation exists as a discursive relation in a new set of practices, namely, those of school mathematics, with its own modes of regulation and subjection. The child moves from the position of a shopper to that of a student, for example. What I am trying to establish is that this move is not best described as a shift from concrete to abstract but as a move from one discursive practice to another. Second, what comes to be valorized as a higher order activity might have everything to do with attempts to regulate and control through reason in a social order which finds its norm in a bourgeois subject who does not need to calculate to survive. Third, the new discursive practice of school mathematics has its own mode of regulation and subjectification. By this I mean that each child becomes positioned as a subject in a new way. That way may be similar to or different from the patterns of subjectification in other practices, but evidence suggests that for oppressed groups the patterns are substantially different. This may have important affective consequences. All of this suggests that the idea of children and adults possessing different skills in different contexts can be shown in a new light. Scribner and Carraher's subjects, for example, are not bourgeois subjects: they are oppressed and exploited groups—working-class men in the U.S.A. and children from the Brazilian lumpen proletariat.

I should like to end, then, by attempting to exemplify the ways in which oppressed subjects may live the different positioning from practice to practice. This disturbs the cozy picture of the rational unitary subject (Henriques et al. 1984), the "natural child" of developmental psychology, and substitutes an account which is specific to time and place and against which Reason's Dream looks like one more colonial fantasy.

Splitting the Difference

How do children manage the transition from one practice to another? Although it is common in psychological accounts, especially from the 1960s and 1970s, to suggest that it is good mothering that prepares children for success in school (see Walkerdine and Lucey, 1989 for a review) such accounts are problematic in that they imply that the problems experienced by children from oppressed groups are the result of inadequate mothering. Such accounts deny the complexity of the pain of moving from subjectification in one practice to another which appears to have a completely different set of rules and expectations. How might children from oppressed groups cope with and defend themselves against the pain? Althusser (1971) in his famous Ideological State Apparatuses paper used Lacan's theory of the mirror stage to argue that schools interpellated children as subjects, creating imaginary identities for them. Lacan used the idea of the mirror to suggest that the child's first view of itself as whole and unitary was the first ideological illusion. Now, while the identity created by the school may well be a fiction, it has powerful effects. While Lacan may be quite correct in asserting the illusory nature of the idea of a coherent identity, it is undoubtedly the case that subjects from oppressed groups experience more keenly a disabling sense of fragmentation (Mama, 1987).

The title of this section refers to the psychoanalytic term "splitting," which is one of the mechanisms of defense against extreme anxiety. While Freud (1951) and Melanie Klein (1975) use this term in rather different ways, both refer to the way in which the unconscious defends itself (see Walkerdine, 1985; Hollway, 1984; and Urwin, 1987 for further discussion). Although on the surface some children may appear to be dealing with the transition from one practice to another in a detached manner, it is precisely this detachment which psychoanalytic accounts suggest is a key to extreme distress. Super-rationality may be a defense against extreme anxiety. One of the six-year-old girls in a study which I conducted presented in class the appearance of extreme stupidity. She could not follow a simple instruction and was extavagantly vague. It later became clear that her vagueness was her best defense, the way in which she routinely cut off from the fact that her mother was being systematically physically abused by her father. Her violent feelings only emerged in an incident in which she had broken the heads off some dolls in the Wendy House.

Patsy is a working-class girl who at the age of four was part of the Tizard and Hughes (1984) study. We (Walkerdine and Lucey, 1989) saw her at ten. At four she was, like many other girls in the study, having difficulty coming to terms with being a "big girl." However, if her mother positioned her as a "clever girl" she was willing and able to carry out certain tasks. The positioning as her mother's clever girl was important. She also scored high on an IQ test. However, at ten she was certainly not positioned by her teacher as "clever". Rather the teacher categorized her as "nowhere near as bright as the rest (of the class)." She said that Patsy resorted to infantile behavior and that basically she had no saving graces. How come this "clever" little girl became stupid and infantile at ten?

It is a shocking fact that three other working class girls in the sample who gained high IQ scores at four were also regarded as stupid at ten and they all, like Patsy, positioned themselves as victims. While many girls mentioned the violence of others, especially boys, these girls saw themselves as the target of that violence. Using the psychoanalytic discourse, which I have discussed, it is possible to see this as a defensive response to unendurable pain. What if Patsy and these other girls felt frightened in an alien world that they did not understand and which did not understand them? They could not easily unleash their anger against those who they needed desperately to call them "clever," to make them feel safe and at home. To project their violent emotions into others and to present themselves as victims as reminiscent of some of the symptoms displayed by colonized peoples as described by the psychiatrist Franz Fanon (1967) when talking about the Algerian War. Sometimes to learn to split is to learn to survive and to long to be loved in an alien world in which it is all too easy to be rejected. Another defense, of course, is to do the rejecting first, so as to make the pain of failure more bearable. Sociologists have tended to describe such strategies as anti-school resistance (e.g. Willis, 1977).

For all of these children, crossing the boundaries from one practice to another cannot be easy. In Walkerdine et al (1989) we pointed out that no girls cross the boundary from home to school as an easy transition from dependency to autonomy. When girls enter school they are classified, categorized. The readings of their behaviors and performance are highly gender-specific. We presented ample evidence to support the view that even when girls displayed the characteristics valorized in boys this did not mean they were judged as being successful. Often, precisely these designations rendered them pathological when viewed in relation to femininity. We argued that it was necessary to understand how highperforming girls came to be designated as "only hardworking" when poorly achieving boys could be understood as "bright" even though they presented little evidence of high attainment. Poorly achieving girls in the study, quite simply, were *never* designated bright (see for example, Walkerdine et al, 1989, page 102).

In other words, we presented a whole system of subjectifications through which girls are judged. That these subjectifications have little empirical foundation in relation to the girls' performance further points to the importance of the sense that some fiction is being created to account for what it is necessary to prove time and time again: the inferiority of the Other. The Other constantly threatens the dominant group and no end of fantasies and fictions are employed to position the oppressed subject as Other, pathological. We argued further that since Reason has to be understood as the possession of "man," there will always be a push to prove Otherness as "lack." It is indeed the paranoias of the powerful that are at stake here: the fear that the oppressed might be able to take away their position of dominance. It is our contention that this dominance has to be assured by a number of social and psychic strategies for constituting the oppressed groups as Other and so pathologizing them.

Such issues bring us back full circle to the pathologization of difference. It is my contention that any psychological approach to the issue of difference and mathematical performance must deal with the complex psychic issues raised above. The fantasies of the colonizer are inscribed in the regulation of colonial subjects (Bhabha, 1984); they become the "truths" through which development and performance are understood. Those fantasies and the attempts at regulation are inscribed in the very history of the insertion of theories of reason and reasoning into mathematics education, and wherever we find the Other, the working class, the peasant, the black, the girl, there we find claims of the proof of abnormality, of irrationality. My argument finally is then that in order to address these issues properly we need to construct accounts which move away from the stagewise progressions of most First World developmental models to an understanding of development as specific to social and historical circumstances. Only then, I suggest, will we be able to engage with oppression as something other than individual pathology.

References

Althusser, L. (1971). *Lenin and philosophy, and other essays*. Monthly Review Press.

Bhabha, H. (1984). The other question: the stereotype and colonial discourse. *Screen*.

Carraher, T. N. (1988). Street mathematics and school mathematics. *Proceedings of the Twelfth International Conference on the Psychology of Mathematics Education*. Hungary.

Foucault, M. (1980). *Discipline and punish*. Harmondsworth: Penguin Books.

Freud, S. (1951). The splitting of the ego in the process *of* defence. *The Complete Psychological Works of Sigmund Freud.* Vol 23. London: Hogarth Press.

Hollway, W. (1984). "Gender identity in adult social relations." In J. Henriques, et al. *Changing the subject*. London: Methuen.

Lacan, J. (1977). *Ecrits: a selection*. London: Tavistock.

Lave, J. (1988). *Cognition in practice.* Cambridge, MA: The University Press; Le Doeff, M. (1978). Operative philosophy. *Ideology and Consciousness;* Klein, M. (1975). *Love, guilt and reparation, and other works*. London: The Hogarth Press.

Mama, A. (1987). Subjectivity, race and gender. Ph.D. diss. University of London, Birkbeck College.

Mathematical Association. (1956). *The teaching of mathematics in primary schools.* London: G. Bell and Sons.

Rose, N. (1985). *The psychological complex*. London: Methuen.

Rotman, S. (1980). *Mathematics: an essay in semiotics*. Bristol, Eng.: University of Bristol, mimeo.

Sayers, J. (1982). *Biological politics*. London: Methuen.

Scribner, S. (1984). Studying working intelligence. In B. Rogoff and J. Lave (Eds.). *Everyday cognition; its developmental and social context*. Cambridge, MA: Harvard University Press.

Schools Council. (1965). *Mathematics in primary schools*.

Tizard, B. and Hughes, M. (1984). *Young children learning*. London: Fontana Books.

Urwin, C. (1987). "Splitting the difference." Paper presented to the British Psychological Society Developmental Section.

Walkerdine, V. (1984). Developmental psychology and the child-centred pedagogy. In J. Henriques, et al., *Changing the Subject*.

Walkerdine, V. (1985). On the regulation of speaking and silence. In Steedman, Urwin, and Walkerdine (Eds.). *Language, gender and childhood*. London: Routledge and Kegan Paul.

Walkerdine, V. (1988). *The mastery of reason*. London: Routledge.

Walkerdine, V. *et al* (1989). *Counting girls out*. London: Virago Press.

Walkerdine, V. and Lucey, H. (1989). *Democracy in the kitchen*. London: Virago Press.

Walkerdine, V. (1990). Post-structuralism and mathematics education. In: R. Noss (Ed.). *Political dimensions of mathematics education*. London: University of London, Institute of Education.

Chapter 10

An Example of Traditional Women's Work as a Mathematics Resource

Mary Harris

Editors's comment: Mary Harris presents a specific cultural activity usually considered women's work and reconsiders it as mathematical activity. This chapter first appeared in *For the Learning of Mathematics* 7(3): 26–28, in 1987.

"Ex Africa semper aliquid novi" Pliny is supposed to have written: "There is always something new from Africa."

Part of the newness of Paulus Gerdes' work in Mozambique (Gerdes, 1986a) is that he offers "non-standard problems," easily solved by many illiterate Mozambican artisans, to members of the international mathematics education community—who cannot (at first) do them. They have trouble in constructing angles of 90, 60, and 45 degrees and regular hexagons out of strips of paper, problems which are no trouble at all to people for whom the intellectual and practical art of weaving is a necessary part of life.

Recently I have been offering to experienced teachers and teachers in training, some of the "non-standard problems" that are easily solved by any woman brought up to make her or her family's clothes. Many of the male teachers are so unfamiliar with the construction and even shape and size of their own garments that they cannot at first perceive that all you need to make a sweater (apart from the technology and tools) is an understanding of ratio and all you need to make a shirt is an understanding of right angles and parallel lines, the idea of area, some symmetry, some optimization and the ability to work from two-dimensional plans to three-dimensional forms.

What makes the problems non-standard is the viewpoint of those who set the standards. Gerdes' work, and the work of others in the field of ethnomathematics offer a rather threatening confrontation to the traditional standard setters. Gerdes is up against a number of factors that until recently have tried to determine the education, or previous lack of it, in his country. The freshness of his work is his illustration of the mathematics that already exists in Mozambican culture and how he is setting about "defrosting" it.

It is interesting to take Gerdes' analysis and his energy and commitment and to apply them worldwide and in the different context of women's culture. There will be those who will maintain that "women" is too wide a term to allow of a single culture, but the "set of women who make and use textiles in home-making" would certainly seem to fit the definition of culture used by Wilder (1981) in his *Mathematics as a Cultural System*.

In mathematical activity, women lose out in two ways. Until very recently, in most histories of mathematics, women mathematicians barely got a mention. Throughout Western history, as Alic (1986) shows, the work of women mathematicians and scientists has been "ignored, robbed of credit and forgotten." Better that they should stay at home and do their needlework, a harmless, practical and nonintellectual activity. Many women do their needlework in factories; indeed, in the textiles industry, women are at the "cutting edge" (Chapkis and Enloe 1983). Here their lowly status is rationalized by all the usual tales: they "are said to have patience, a tolerance for monotony, nimble fingers, attention to detail, little physical strength, no mechanical aptitude" and so on. In a world where there is little pre-disposition to take women's intellectual work seriously, the potential for giving them any credit for thought in their practical work, at home or in the factory, is severely limited.

The social distinction between the practical mathematics considered appropriate for an artisan class and the theoretical mathematics of élites is older than Plato (D'Ambrosio 1985). Yet "practical, scientific, aesthetic, and philosophical interests have all shaped mathematics" (Kline 1953). Nowadays practical mathematics is nearly always seen in terms of the application of theoretical mathematics learned in the formal context of school. It often still carries the stigma of being particularly suitable for Low Attainers. Women doing practical mathematics really are at the bottom of the pile.

In fact a single activity can, by its nature, generate more mathematics than the application of the theory to a particular case. A recent conference of about 240 mathematics teachers offers an example. The teachers were given an activity posed not as a mathematical one but in the form of an internal memo to his visualization team of the managing director of an expensive design company, asking for some more than usual ingenuity of design for a pack for a single sports ball, for a company he wished to impress. The teachers worked in groups and produced an impressive array of packs together with a list of mathematics they had found themselves involved in, which ran to three pages. Most significant though was the behaviour of the groups. Those who perceived the problem as just a mathematical one and only attempted it theoretically, found themselves confined to the very limited range of mathematics they had chosen. Those who set about the task in a practical way, however, continually came up with further problems both of construction and design which in turn generated more problems and more mathematics. In short, it was the theoreticians who limited themselves by their "theory and applications" approach. The practitioners found themselves involved not only with wider, but also with deeper mathematics as it emerged from the task itself. Of course, the practitioners were mathematics teachers and therefore knew what to look for, but there is no doubt at all that they surprised themselves.

Perhaps the role of school mathematics teachers in such circumstances should not be to teach some theory and then look for application but to analyze and elucidate the mathematics that grows out of the students' activity. Gerdes uses the term "frozen" for mathematics hidden in cultural artifacts, and it is a good one, but there is the danger that a pre-perception of what is there to be frozen could limit what is defrosted. We need a term that implies hatching or germination of undefined potential, as well as defrosting, because much of the mathematical thinking that has become frozen in an artifact has been put there by someone who has not been reared on North American textbooks or standard Western mathematics education with all its attitudes and prejudices. Traditional North American and Western mathematics has ignored it until now (Islamic mathematics has always revelled in it), though weavers have always known that weaving is an intellectual activity. A glance at any piece of traditional weaving reveals a huge range of obvious geometry to anyone who chooses to notice. How could it have got there without mathematical thinking? Why do people choose *not* to notice?

A student studying symmetry in a Western school might be congratulated by his teacher on producing the design in figure 10–1, after an investigation starting with an isosceles triangle and involving translation and rotation. The same teacher might draw the attention of the rest of the class to such a pattern and suggest the class analyze

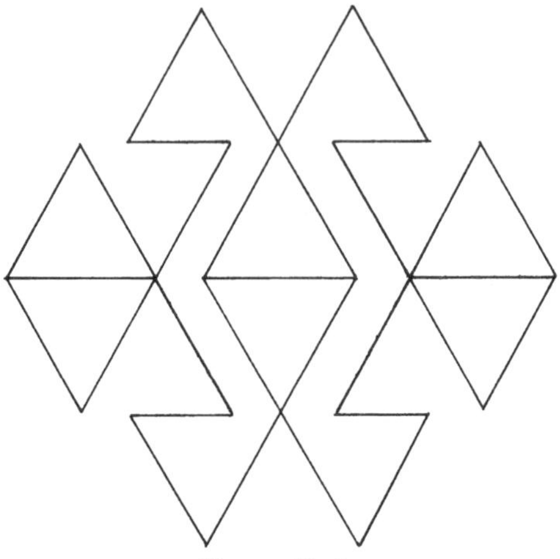

Figure 10–1.

together some of the things that have happened to the original triangle; thus, synthesis and analysis could take place over one activity, one student's thinking could be revealed to others, and further thought developed.

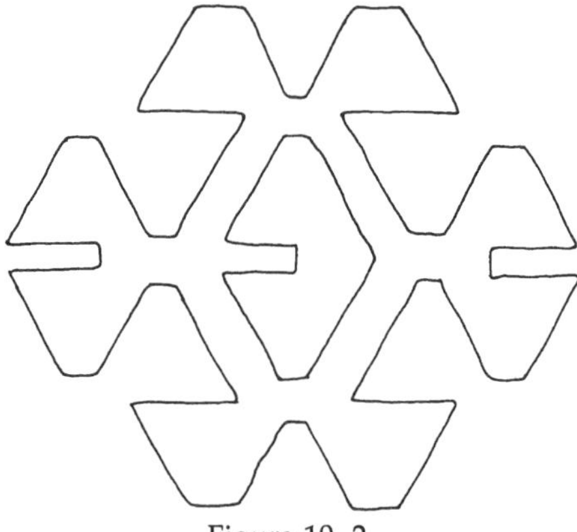

Figure 10–2.

Figure 10–2 is a sketch of a motif worked on a Turkish flat woven rug, a kilim in the possession of the writer. The motif is one of many on the rug and it is sketched in such a way that the lines of a background grid would represent generally the warp and weft threads. Unlike figure 10–1 where a student can "plant" a triangle anywhere on squared paper and work outwards from it, the weaver has to work under several constraints both of design and construction. The motif has to be placed exactly in relation to other motifs, in relation to the symmetrical border of the rug and the symmetrical center panel within a wide border, so that the whole rug itself is symmetrical. Construction constraints mean that the weaver works from right to left and left to right on a grid already defined and confined by the number of warp threads. The length of the rug is determined by the design, so that the weaver finishes it when the whole thing is symmetrical.

For reasons which need to be closely examined, figure 10–1 seems to count as mathematics, figure 10–2 does not. Most of the reasons suggested to the writer so far, do not stand up to much examination. A summary of them is that figure 10–2 is simply not taken seriously as mathematics because first the weaver has had no schooling and is illiterate and second, she is a girl. It has even been stated by more than one mathematics educator that the weaver "is not thinking mathematically," to which the immediate response must be: "How do you know? Have you asked her what her thinking was?" It seems that the whole question as to whether figures 10–1, and 10–2, neither, or both are mathematical is hedged about with not much more than attitude. The weaver is simply assumed not be capable of thinking mathematically.

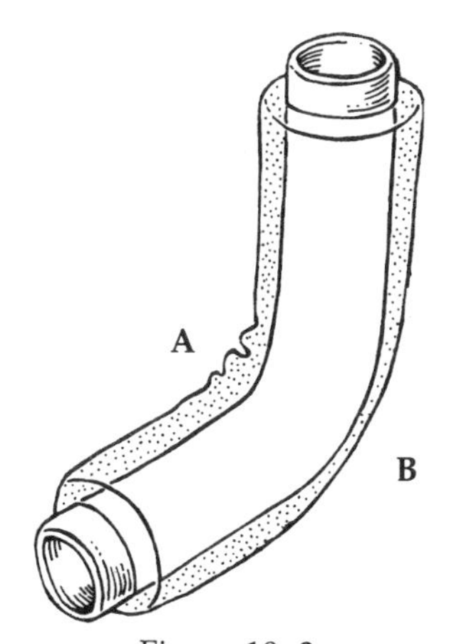

Figure 10–3.

Take another example, this time from industry, the problem of lagging a right-angled cylindrical pipe in a factory (figure 10–3). The problem is to prevent the lagging from bunching up at the inside of the angle at A and thus developing a hot spot and stretching out at B, thus developing a cold spot. The factory is a chemical works and the lagging problem is compounded by the fact that the right angle has to open out on occasion to about 180 degrees. So as well as being inefficient in crucial places, the lagging will soon wear out. In industry this is just the sort of problem to which mathematical thinking would be brought.

Why is it that this industrial problem is considered to be inherently mathematical whereas the identical domestic problem, that of the design of the heel of a sock, is not? Dare it be suggested that the reason is that socks are traditionally knitted by Granny—and nobody expects her to be mathematical. What, dear old Gran?

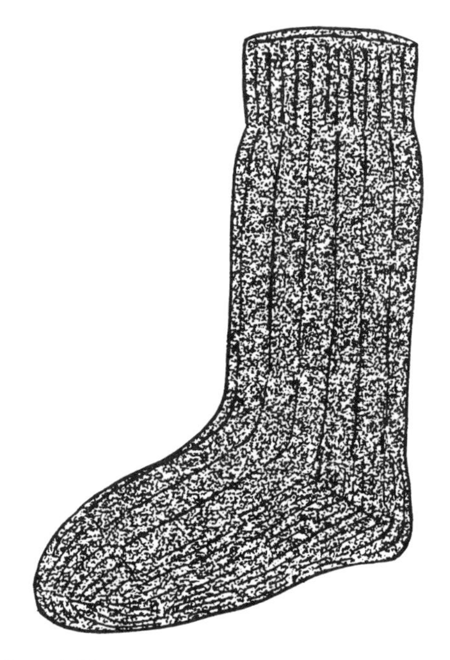

Turn heel as follows: —

1st row - K.12 [14, 16], sl.1, K.1, psso., K.1, turn.

2nd row - P.6 [6, 8], P.2 tog., P.1, turn.

3rd row - K. to last 6 [8, 8] sts., sl.1, K.1, psso, K.1, turn.

4th row - P. to last 6 [8, 8] sts., P.2 tog., P.1, turn.

5th row - K. to last 4 [6, 6] sts., sl.1, K.1, psso, K.1, turn.

6th row - P. to last 4 [6, 6] sts., P.2 tog., P.1, turn.

Figure 10–4.

Or is there really some mathematical reason like the problem not being capable of general solution? But the way experienced knitters actually work *is* to a general solution. Sock instructions are tediously written out row by row as in figure 10–4, but only the most in experienced or unskilled worker would actually use them like this. The normal procedure for an experienced person is to follow the pattern only for as long as it is necessary to get the feel of the way the thing is shaping. Women who make socks for families of children often work to their own general solution (there are several), modifying them for a particular child who has grown.

Why, when we take children to look at the environment so that we can draw out the mathematics, do we always show them the *man*-made environment: cranes, bridges, paving slabs, brick patterns? Why do we *choose* not to look at the closer environment, the one we carry with us? How many men who wear ties are aware of the interesting geometry that is going on literally under their noses?

So what is it about the sock problem that ensures that it will not be taken seriously? Why is it non-standard? Is it simply that it is not normal to couch it in mathematical jargon? Very well then, here it is formally: "Derive a general expression for the heel of a sock."

And as to whether you still wonder if it is mathematics or not, to quote Gerdes again, "Please answer for yourself."

References

Alic, Margaret. (1986). *Hypatia's heritage: A history of women in science from antiquity to the late nineteenth century*. London: The Women's Press.

Chapkis, Wendy and Enloe, Cynthia. (1983). *Of common cloth: Women in the global textiles industry*. Amsterdam and Washington DC: Transnational Institute.

D'Ambrosio, Ubiritan. (1985). "Ethnomathematics and its place in the history and pedagogy of mathematics." *For the Learning of Mathematics* 5, no. 1 (February).

Gerdes, Paulus. (1986a). "How to recognise hidden geometrical thinking: A contribution to the development of anthropological mathematics." *For the Learning of Mathematics* 6, no. 2 (June).

Gerdes, Paulus. (1986b). "On culture: Mathematics and curriculum development in Mozambique. "In Marit Johnsen Hoines and Stieg Mellin-Olsen (Eds.). *Mathematics and culture*. Report of a seminar held at

Bergen College of Education, 26–28 September 1985. Rådal, Norway: Caspar Forlag.

Kline, Morris. (1953). *Mathematics in western culture*. Oxford: Oxford University Press.

Wilder, R. (1981). *Mathematics as a cultural system*. Oxford: Pergamon Press.

Chapter 11

On Culture, Geometrical Thinking and Mathematics Education

Paulus Gerdes

Editors's comment: Paulus Gerdes, a Mozambican mathematician and mathematics educator, has been a leading researcher in uncovering mathematical ideas embedded in African cultural practices and artifacts and in presenting these findings to the mathematical community. In this chapter, he demonstrates an alternative construction of certain "Euclidean" geometrical ideas by reconsidering the mathematics inherent in traditional Mozambican culture. Consequently, this chapter raises questions concerning the origins of these geometrical ideas. Finally, he demonstrates how ethnomathematical research into Mozambican material culture can create an empowering curriculum for Mozambican students. This chapter first appeared in *Educational Studies in Mathematics*, Dordrecht 19(3): 137–162, in 1988. It has also been published in Bishop, A. (Ed.). *Mathematics education and culture*. Dordrecht: Kluwer, 1988, 137–162; and in Gerdes, P. *Ethnomathematics and education in Africa*. Stockholm: University of Stockholm Institute of International Education, 1995, 30–52.

"Colonization is the greatest destroyer of culture that humanity has ever known. . . . Long-suppressed manifestations of culture have to regain their place . . ."

—Samora Machel, 1978.

"Education must give us a Mozambican personality which, without subservience of any kind and steeped in our own realities, will be able, in contact

---

This article is dedicated to Samora Machel, President of Mozambique, who died on the 19th October 1986, the day I finished this chapter.

with the outside world, to assimilate critically the ideas and experiences of other peoples, also passing on to them the fruits of our thought and practice."

—Samora Machel, 1970.

Some Social and Cultural Aspects of Mathematics Education in Third World Countries

In most formerly colonized countries, post-independence education did not succeed in appeasing the hunger for knowledge of its people's masses.

Although there had occurred a dramatic explosion in the student population in many African nations over the last twenty five years, the mean illiteracy rate for Africa was still 66 percent in 1980. Overcrowded classrooms, shortage of qualified teachers, and lack of teaching materials contributed towards low levels of attainment. In the case of mathematics education, this tendency has been reinforced by a hasty *curriculum transplantation* from highly industrialized nations to Third World countries.[1] With the transplantation of curricula their *perspective* was also copied: "(primary) mathematics is seen only as a stepping stone towards secondary mathematics, which in turn is seen as a preparation for university education."[2] Mathematics education is therefore structured in the interests of a social elite. To the majority of children, mathematics looks rather useless. Maths anxiety is widespread; especially for sons and daughters of peasants and laborers, mathematics enjoys little popularity. Mathematics education serves the selection of elites: "Mathematics is universally recognized as the most effective education filter," as El Tom underlines.[3] Ubiratan D'Ambrosio, president of the Interamerican Committee on Mathematics Education agrees: ". . . mathematics has been used as a *barrier to social access*, reinforcing the power structure which prevails in the societies (of the Third World). No other subject in school serves so well this purpose of reinforcement of power structure as does mathematics. And the main tool for this negative aspect of mathematics education is evaluation."[4]

In their study of the mathematics learning difficulties of the Kpelle (Liberia), Gay and Cole (1967) concluded, that there do not exist any inherent difficulties: what happened in the classroom, was that the contents did not make any sense from the point of view of

Kpelle-culture; moreover the methods used were primarily based on rote memory and harsh discipline.[5] Experiments showed that Kpelle illiterate adults performed better than North American adults, when solving problems, like the estimation of number of cups of rice in a container, that belong to their "indigenous mathematics."[6] Serious doubts about the effectiveness of school mathematics teaching are also raised by Latin American researchers. Eduardo Luna (Dominican Republic) (1983) posed the question if it is possible, that the practical mathematical knowledge that children acquired outside the school is "repressed" and "confused" in the school.[7] Not only possible, but this happens frequently, as shown by the Brasilians Carraher and Schliemann (1982): children, who knew before they went to school, how to solve creatively arithmetical problems which they encountered in daily life, for example, at the marketplace, could, later in the school, not solve the same problem, that is, not solve them with the methods taught in the arithmetic class.[8] D'Ambrosio concludes that "*'learned' matheracy* eliminates the so-called *'spontaneous' matheracy*,"[9] that is, "An individual who manages perfectly well numbers, operations, geometric forms, and notions, when facing a completely new and formal approach to the same facts and needs creates a *psychological blockade* which grows as a barrier between the different modes of numerical and geometrical thought".[10] What happens in the school, is that "the former, let us say, spontaneous, abilities (are) *downgraded, repressed,* and *forgotten,* while the learned ones (are not being) assimilated, either as a consequence of a learning blockage, or of an early dropout."[11] For this reason, "the early stages of mathematics education (offer) a very efficient way of instilling the *sense of failure,* of *dependency* in the children."[12] How can this psychological blockade be avoided?

How can this "totally inappropriate education, leading to misunderstanding and sociocultural and psychological alienation"[13] be avoided? How can this "pushing aside" and "wiping out" of *spontaneous, natural, informal, indigenous, folk, implicit, non-standard,* and / or *hidden (ethno)mathematics* be avoided?[14]

Gay and Cole (1967) became convinced that it is necessary to investigate first the "indigenous mathematics," in order to be able to *build effective bridges* from this "indigenous mathematics" to the new mathematics to be introduced in the school: ". . . the teacher should begin with materials of the indigenous culture, leading the child to use them in a creative way,"[15] and from there advance to the new school mathematics. The Tanzanian curriculum specialist Mmari stresses, that: ". . . there are traditional mathematics methods still being used in

Tanzania. . . . A good teacher can utilize this situation to underline the universal truths of the mathematical concepts."[16] And how could the good teacher achieve this? Jacobsen answers: "The (African) people that are building the houses are not using mathematics; they're *doing* it traditionally . . . if we can bring out the scientific structure of why it's done, then you can teach science that way."[17] For D'Ambrosio, (1984) it becomes necessary ". . . to generate ways of understanding, and methods for the incorporation and compatibilization of known and current popular practices into the curriculum. In other words, in the case of mathematics, recognition and incorporation of ethnomathematics into the curriculum."[18] ". . . this. . . . requires the development of quite difficult anthropological research methods relating to mathematics; . . . *anthropological mathematics*. . . constitutes an essential research theme in Third World countries . . . as the *underlying ground upon which we can develop curriculum in a relevant way.*"[19]

Towards a Cultural-Mathematical Reaffirmation

D'Ambrosio stressed the need for incorporation of ethnomathematics into the curriculum in order to avoid a psychological blockade. In former colonized countries as Mozambique, there exists also a related cultural blockade to be eliminated. "Colonization—in the words of Samora Machel, first President of Mozambique—is the greatest destroyer of culture that humanity has ever known. African society and its culture were crushed, and when they survived they were coopted so that they could be more easily emptied of their content. This was done in two distinct ways. One was the utilization of institutions in order to support colonial exploitation. . . . The other was the 'folklorizing' of culture, its reduction to more or less picturesque habits and customs, to impose in their place the values of colonialism." "Colonial education appears in this context as a process of denying the national character, alienating the Mozambican from his country and his origin and, in exacerbating his dependence on abroad, forcing him to be ashamed of his people and his culture."[20] In the specific case of mathematics, this science was presented as an exclusively white men's creation and ability; the mathematical capacities of the colonized peoples were negated or reduced to rote memorization; the African and American-Indian mathematical traditions became ignored or despised.

A *cultural rebirth* is indispensable, as Samora Machel (1978) underlines: ". . . Long-suppressed manifestations of culture (have to) regain their place."[21] In this cultural rebirth, in this combat of racial and colonial prejudice, a *cultural-mathematical-reaffirmation* plays a part: it is necessary to encourage an understanding that the peoples of the Third World have been capable of developing mathematics in the past, and therefore regaining *cultural* confidence[22]—will be able to assimilate and develop the mathematics we need; mathematics does not come from outside the African, Asian, and American-Indian cultures.

We may conclude that the incorporation of mathematical traditions into the curriculum will—probably—contribute not only to the elimination of individual and social psychological blockade, but also of the related cultural blockade. Now, this raises an important question: *which mathematical traditions*? In order to be able to incorporate popular (mathematical) practices, it is first of all necessary to *recognize their mathematical character.* In this sense, D'Ambrosio (1985) speaks about the need to broaden our understanding of what mathematics is. [23] Ascher and Ascher (1981) remark in this connection "Because of the provincial view of the professional mathematicians, most definitions of mathematics exclude or minimize the *implicit* and *informal*; . . . involvement with concepts of number, spatial configuration, and logic, that is, implicit or explicit mathematics, is *panhuman*."[24]

Broadening our understanding of what mathematics is, is necessary, but not sufficient. A related problem is how to reconstruct mathematical traditions, when probably many of them have been—as a consequence of slavery, of colonialism. . .—wiped out. Few or almost none (as in the case of Mozambique) written sources can be consulted. Maybe for number systems and some aspects of geometrical thinking, oral history may constitute an alternative. What other sources can be used? *What methodology* ?

We developed a complementary methodology that enables one to uncover in traditional, material culture some hidden moments of geometrical thinking. It can be characterized as follows. We looked to the geometrical forms and patterns of traditional objects like baskets, mats, pots, houses, fishtraps, ad so forth and posed the question: *why* do these material products possess the form they have? In order to answer this question, we learned the usual production techniques and tried to vary the forms. It came out that the form of these objects is almost never arbitrary, but generally represents many practical advantages and is, quite a lot of times, the only possible or optimal solution of a production problem. The traditional form reflects accumulated experience and wisdom. It constitutes not only biological and physical knowledge about the materials that are used, but also mathematical knowledge, knowledge about the properties and relations of circles, angles, rectangles, squares, regular pentagons and hexagons, cones, pyramids, cylinders, and so forth.

Applying this method, we discovered quite a lot of "hidden" or "frozen" mathematics.[25] The artisan, who imitates a known production technique, is, generally, doing some mathematics. But the artisans who discovered the techniques, did and *invented* quite a lot of mathematics, were *thinking* mathematically. When pupils are stimulated to *reinvent* such a production technique, they may be encouraged to do and learn mathematics. Hereto they can be stimulated only if the *teachers* themselves are conscious of hidden mathematics, are convinced of the cultural, educational, and scientific value of rediscovering and *exploring* hidden mathematics, are aware of the potential of "unfreezing" this "frozen mathematics." Now we shall present some of our experiences in this necessary "cultural conscientialization" of future mathematics teachers.

Examples of "Cultural Conscientialization" of Future Mathematics Teachers

Study of Alternate Axiomatic Constructions of Euclidean Geometry in Teacher Training

Many alternate axiomatic constructions for euclidean geometry have been devised. In Alexandrov's construction,[26] Euclid's famous fifth postulate is substituted by the "rectangle, axiom":

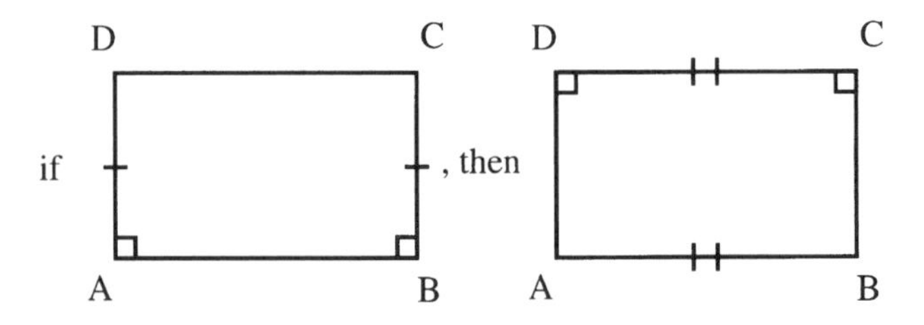

that is, if AD = BC and A and B are right angles, then AB = DC and C and D are also right angles. In one of the classroom sessions of an introductory geometry course, we posed the following provocative question to our future mathematics teachers—many of whom are sons and daughters of peasants—: "Which 'rectangle axiom' do the Mozambican peasants use in their daily life?" The students' first reactions were rather sceptical in the sense of "Oh, they don't know anything about geometry." Counterquestions followed: "Do the peasants use rectangles in their daily life?". "Do they construct rectangles?" Students from different parts of the country were asked to explain to their colleagues how their parents construct, for example, the rectangular bases of their houses. Essentially, two construction techniques are common:

(*a*) In the first case, one starts by laying down on the floor two long bamboo sticks of equal length.

Then these first two sticks are combined with two other sticks also of equal length, but normally shorter than the first ones.

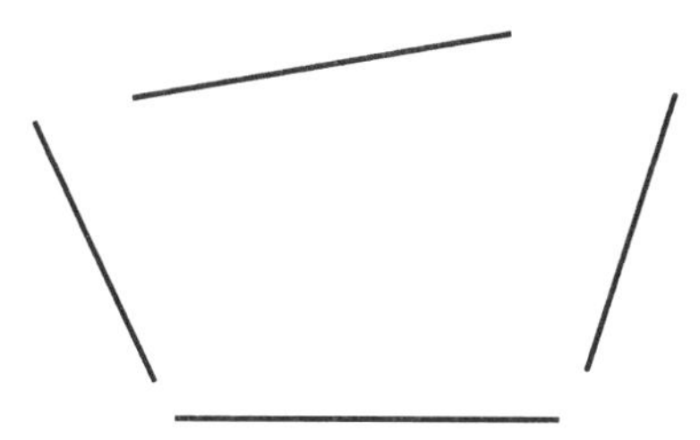

Now the sticks are moved to form a closure of a quadrilateral.

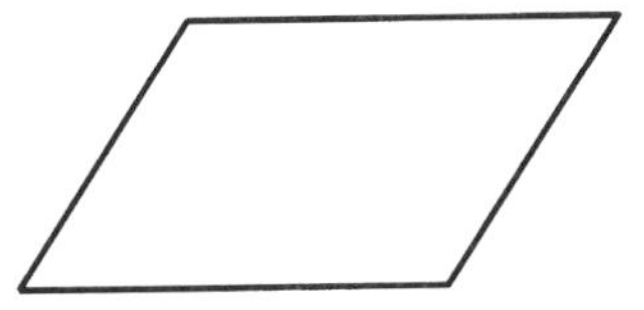

One further adjusts the figure until the diagonals—measured with a rope—become equally long. Then, where the sticks are now lying on the floor, lines are drawn and the building of the house can start.

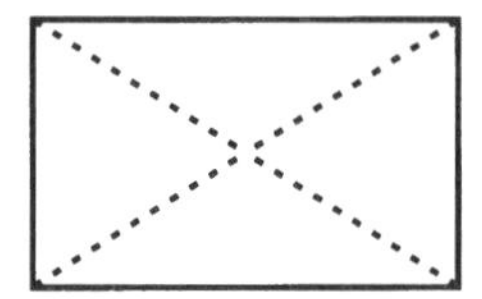

(*b*) In the second case, one starts with two ropes of equal length, that are tied together at their midpoints.

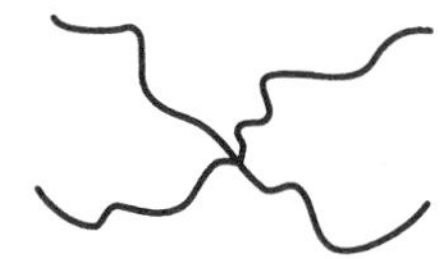

A bamboo stick, whose length is equal to that of the desired breadth of the house, is laid down on the floor, and at its end-points pins are hit into the ground. An endpoint of each of the ropes is tied to one of the pins.

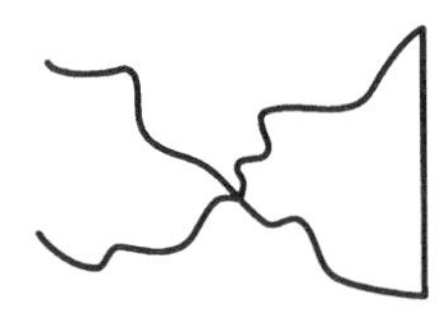

Then the ropes are stretched and at the remaining two endpoints of the ropes, new pins are hit into the ground. These four pins determine the four vertices of the house to be built.

"Is it possible to formulate the geometrical knowledge, implicit in these construction techniques, into terms of an axiom?" "Which 'rectangle axiom' do they suggest?" Now the students arrive at the following two alternate "rectangle axioms":

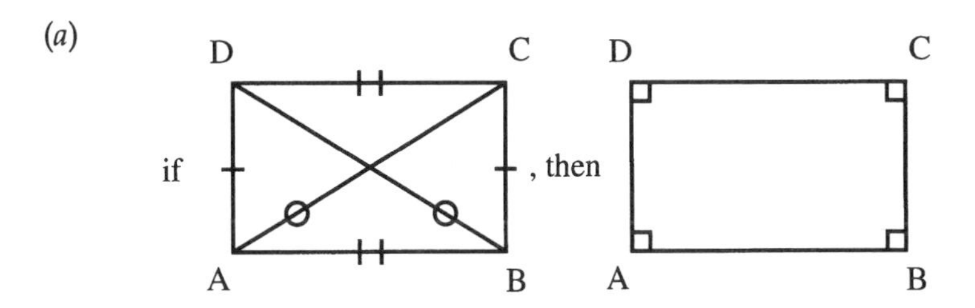

that is, if AD = BC, AB = DC and AC = BD, then A, B, C and D are right angles. In other words, an equidiagonal parallelogram is a rectangle.

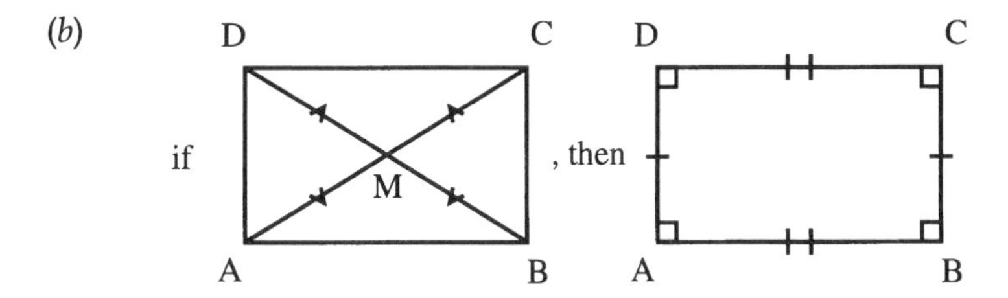

that is, if M = AC ∩ BD and AM = BM = CM = DM, then A, B, C and D are right angles, AD = BC and AB = DC. In other words, an equisemidiagonal quadrilateral is a rectangle. "After all, our peasants know something about geometry," remarks a student. Another, more doubtful: "But these axioms are theorems, aren't they?" This classroom session leads to a more profound understanding by the student of the relationships between experience, the possible choices of axioms, between axioms and theorems at the first stages of alternate axiomatic constructions. It prepares the future teachers for discussions later in their study on *which* methods of teaching geometry seem to be the most appropriate in our cultural context. It contributes to cultural-mathematical confidence.

An Alternate Construction of Regular Polygons

Artisans in the north of Mozambique weave a funnel in the following way. One starts by making a square mat ABCD, but does not finish it: with the strands in one direction (horizontal in our figure), the artisan advances only until the middle.

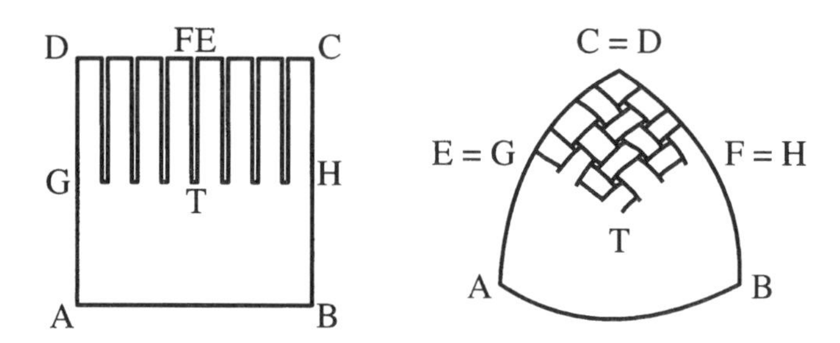

Then, instead of introducing more horizontal strips, he interweaves the vertical strands on the right (between C and E) with those on the left (between F and D). In this way, the mat does not remain flat, but is transformed into a "basket." The center T goes downwards and becomes the vertex of the funnel. In order to guarantee a stable rim, its edges AB, BC, and AC are rectified with little branches. As a final result, the funnel has the form of a triangular pyramid. So far about this traditional production technique.[27]

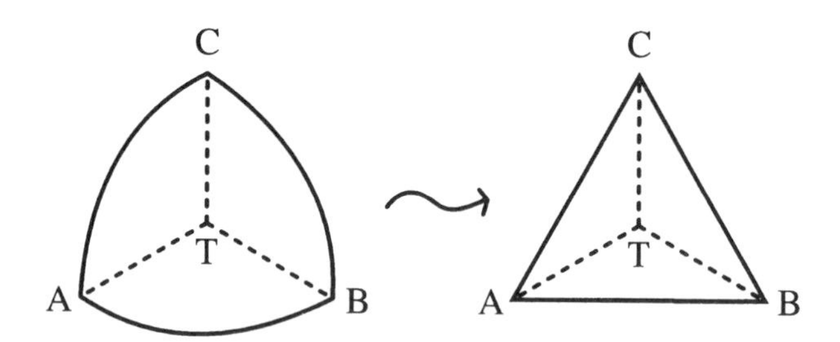

We posed our students the following question: "What can we learn from this production technique?" "The square ABCD has been transformed into a triangular pyramid ABC.T, whose base ABC is an equilateral triangle.

Maybe a method to construct an equilateral triangle?". . . . Some reacted sceptically: "A very clumsy method to do so." Counterquestions: "Avoid overhasty conclusions! What was the objective of the artisan? What is our objective?" "Can we simplify the artisans' method if we only want to construct an equilateral triangle?' "How to construct such a triangle out of a square of cardbord paper?" An answer to these questions is given in the following diagrams:

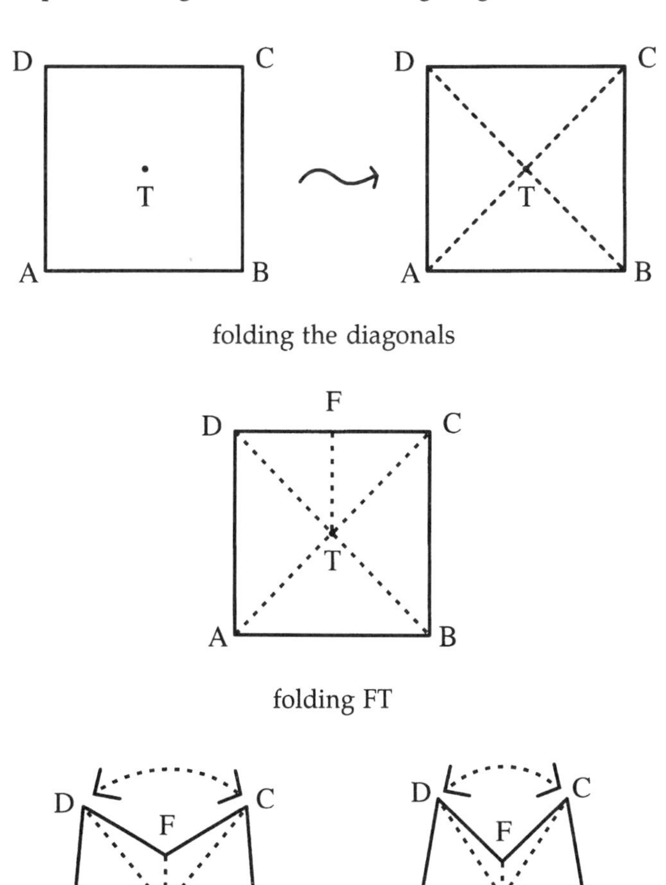

join the triangles DFT and CFT until C and D coincide,
F goes up, T goes down

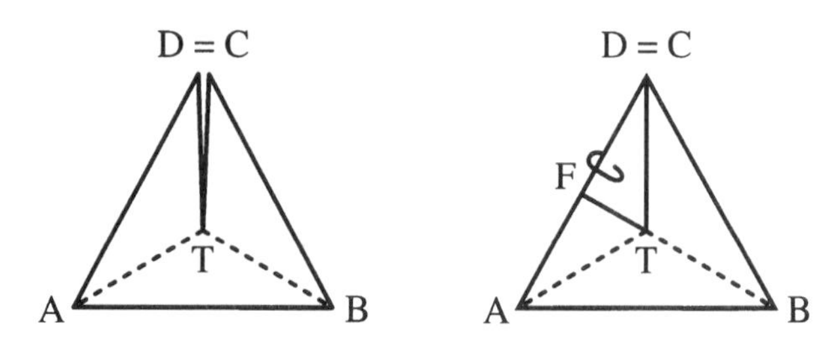

fix the "double triangle" DFT to the face ATC, e.g. with a paperclip

"Can this method be generalized?" "Starting with a regular octagon, how to transform it into a regular heptagonal pyramid?" "How to fold a regular octagon?"

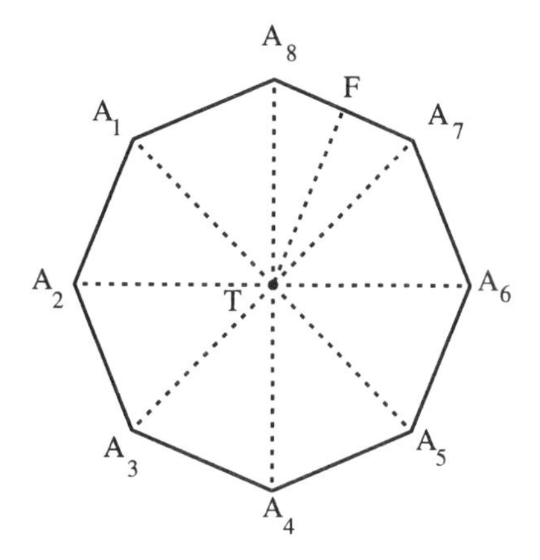

folding the diagonals and FT

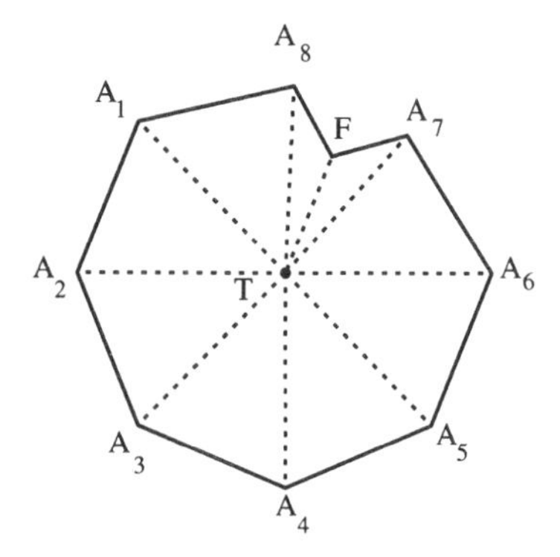

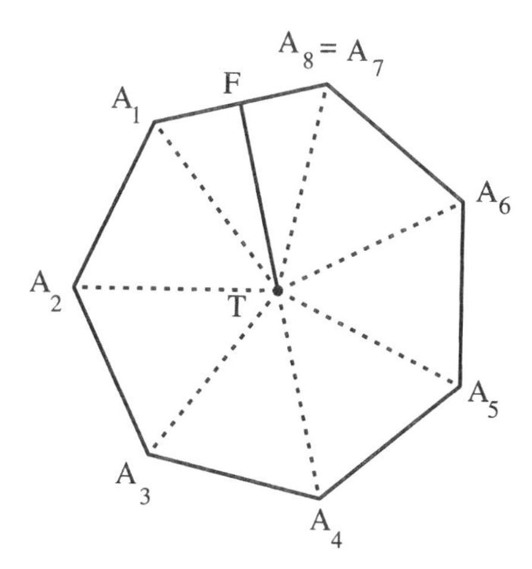

F goes up, T goes down and A_7 and A_8 approximate until they coincide

"How to transform the regular heptagonal pyramid into a regular hexagonal pyramid?" As 2^n-gons are easy to fold (by doubling the central diagonals when one starts with a square) and each time that the simplified "funnel-method" is applied, the number of sides of a regular polygon (or of the regular polygonal base of a pyramid) decreases by 1, it can be concluded that all regular polygons can be constructed in this way.[28] Once arrived at this point, it is possible to look back and ask: "Did we learn something from the artisans who weave funnels?" "Is it possible to construct a regular heptagon using only a ruler and a compass?" "Why not?" "And with our method?"

"What are the advantages of our general method in relation to the standard Euclidean ruler and compass constructions?" "What are its disadvantages?" "Which method has to be preferred for our primary schools?" "Why?"

From Woven Buttons to the "Theorem of Pythagoras"[29]

By pulling a little lassoo around a square-woven button, it is possible to fasten the top of a basket, as is commonly done in southern parts of Mozambique. The square button, woven out of two strips, hides some remarkable geometrical and physical considerations. By making them explicit, the interest in this old technique is already revived. But much more can be made out of it, as will now be shown.

When one considers the square-woven button from above, one

observes the pattern (a) or the pattern obtained (b) after rectifying the slightly curved lines and by making the hidden lines visible:

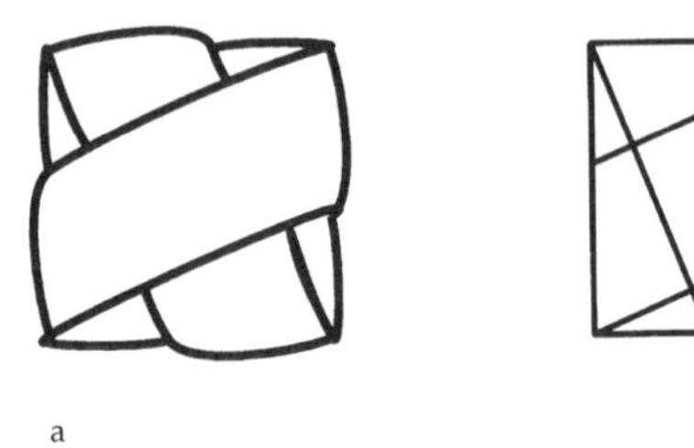

a b

In its middle there appears a second square. Which other squares can be observed, when one joins some of these square-woven buttons together? Do there appear other figures with the same area as (the top of) a square-woven button?

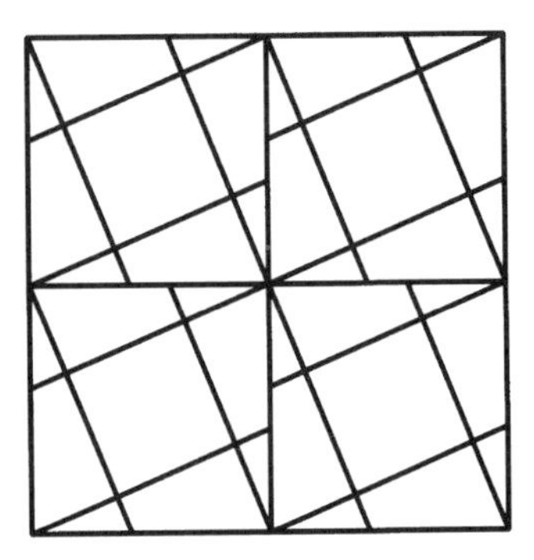

Yes, if you like, you may extend some of the line segments or rub out some others.

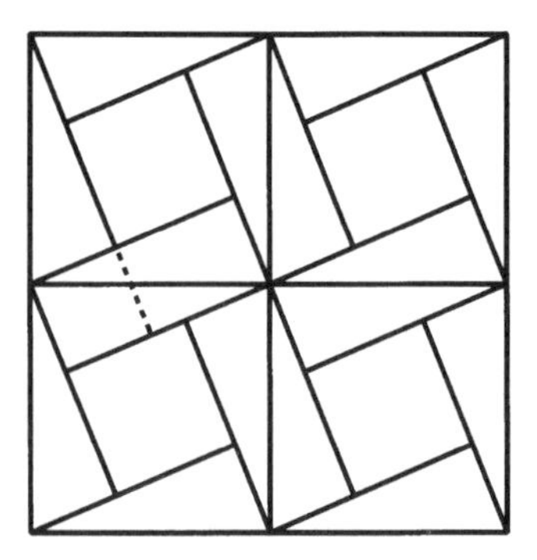

What do you observe? Equality in areas?

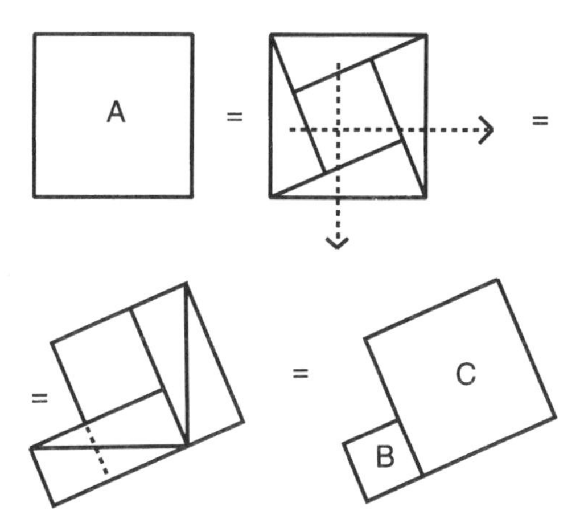

Hence A = B+C:

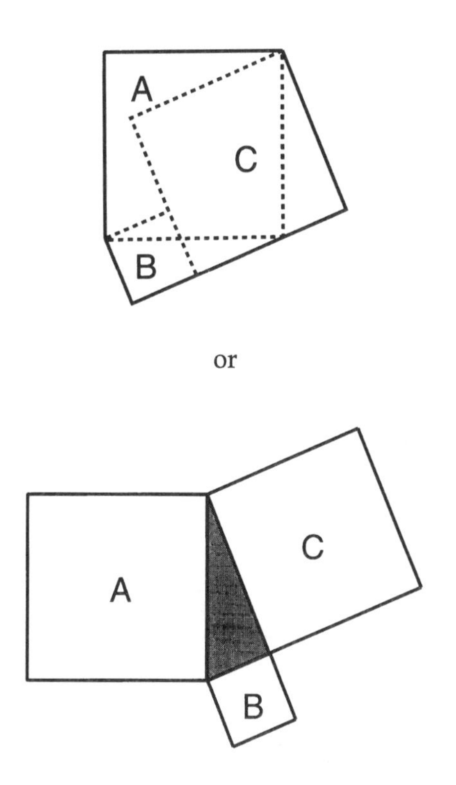

that is, one arrives at the so-called "Theorem of Pythagoras."

The teacher-students themselves rediscover this important theorem and succeed in proving it. One of them remarks: "Had Pythagoras—or somebody else before him—not discovered this theorem, we would have discovered it". . . . Exactly! By not only making explicit the geometrical thinking "culturally frozen" in the square-woven buttons, but by exploiting it, by revealing its full potential, one stimulates the development of the above mentioned necessary cultural-mathematical (self)confidence. "Had Pythagoras not . . . *we* would have discovered it."The debate starts. "Could our ancestors have discovered the 'Theorem of Pythagoras'?" "Did they?" . . . "Why don't we know it?". . . . "Slavery, colonialism . . . ." By "defrosting frozen mathematical thinking" one stimulates a reflection on the impact of colonialism, on the historical and political dimensions of mathematics (education).

From Traditional Fishtraps to Alternate Circular Functions, Football, and the Generation of (Semi)regular Polyhedra

Mozambican peasants weave their light transportation baskets "Litenga" and fishermen their traps "Lema" with a pattern of regular hexagonal holes. One way to discover this pattern is the following. How can one fasten a border to the walls of a basket, when both border and wall are made out of the same material? How to wrap a wallstrip around the borderstrip?

What happens when one presses (horizontally) the wallstrip? What is the best initial angle between the border- and wallstrip?

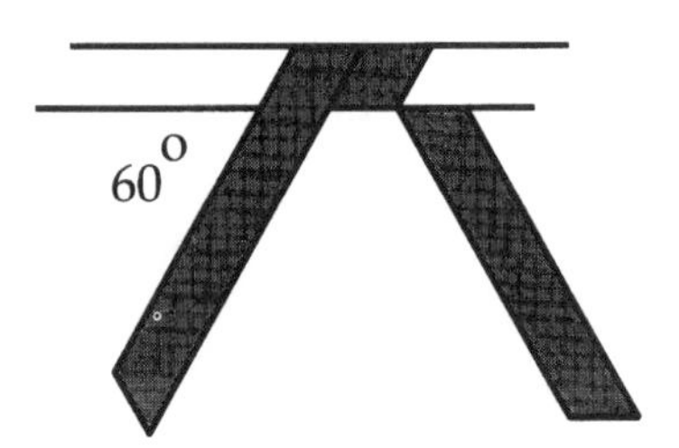

In the case that both strips have the same width, one finds that the optimal initial angle measures 60°. By joining more wallstrips in the same way and then introducing more horizontal strips, one gets the "Litenga" pattern of regular hexagonal holes.

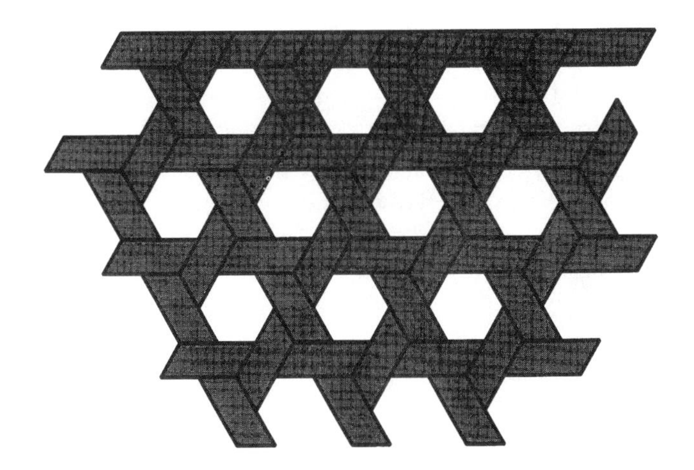

By this process of rediscovering the mathematical thinking hidden in these baskets and fishtraps—and in other traditional production techniques—the future teachers feel themselves stimulated to reconsider the value of their cultural heritage: in fact, geometrical thinking was not and is not alien to their culture. But more than that. This *"unfreezing of culturally frozen mathematics"* can serve, in many ways, as a starting point and source of inspiration for doing and elaborating other interesting mathematics. In the concrete case of this hexagonal-weaving-pattern, for example, the following sets of geometrical ideas can be developed.

A. TILING PATTERNS AND THE FORMULATION OF CONJECTURES. Regular hexagonal and other related tiling patterns can be discovered by the students.

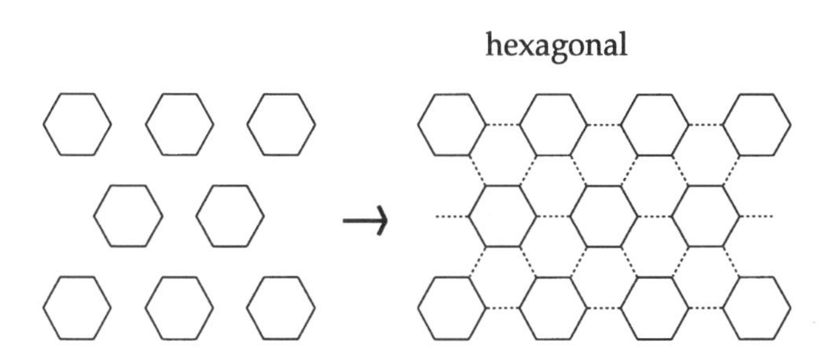

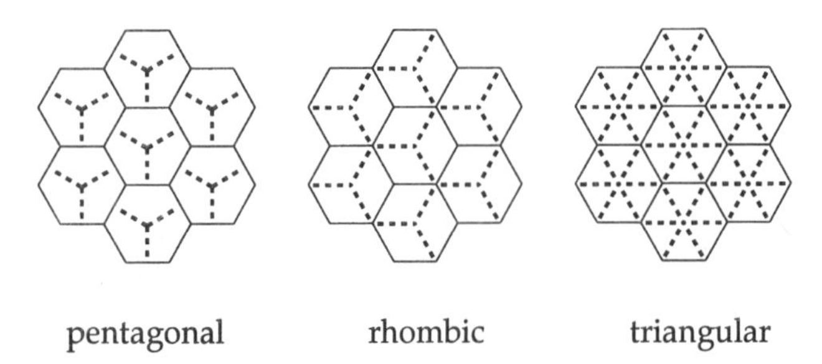

With the so-found equilateral triangle, many other polygons can be built. By considering these figures, general conjectures can be formulated, for example,

* the sum of the measures of the internal angles of a n-gon is equal to $3(n - 2)\ 60°$;

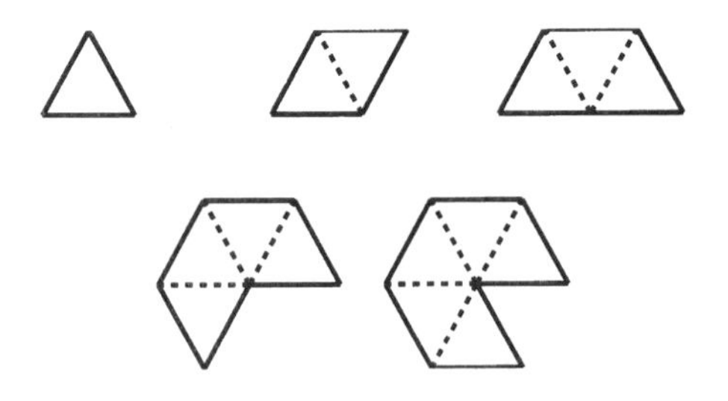

* areas of similar figures are proportional to the squares of their sides;

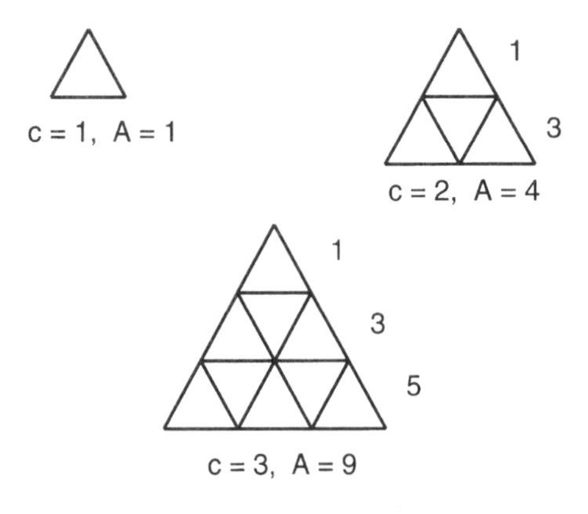

side = 1; area = 1 = 1^2 side = 2; area = 4 = 2^2 side = 3; area = 9 = 3^2

* the sum of the first n odd numbers is n^2.

Once these general theorems are conjectured, there arises the question of justifying, how to prove them.

B. AN ALTERNATE CIRCULAR FUNCTION. Let us return to the weaving of these "Litenga" baskets. What happens when the "horizontal" and "standing" strips are of different width, for example, 1 (unity of measurement) and *a*?

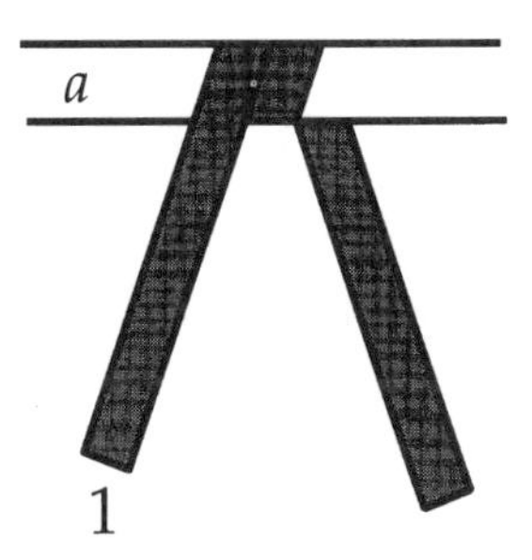

One finds a semi-regular hexagonal pattern. How does the optimal angle
α depend on *a*?

$$\alpha = \text{hex}(a)$$

How does α vary? Both α and *a* can be measured. One finds:

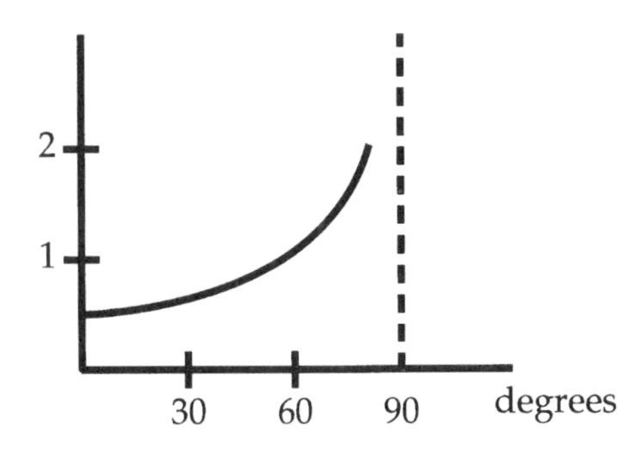

We have here a culturally integrated way to introduce a circular function. After the study of the "normal" trigonometric functions, their relationships can be easily established, for example,

$$a = \text{hex}^{-1}(\alpha) = \frac{1}{2\cos\alpha}.$$

C. *FOOTBALLS AND POLYHEDRA* The faces and edges of the "Lema" fishtrap display the regular-hexagonal-hole-pattern. At its vertices the situation is different. The artisans discovered that in order to be able to construct the trap, "curving" the faces at its vertices, it is necessary, for example, at the vertices A, B, and C to reduce the number of strips. At these points, the six strips that "circumscribe" a hexagon, have to be reduced to five. That is why one encounters at these vertices little pentagonal holes.

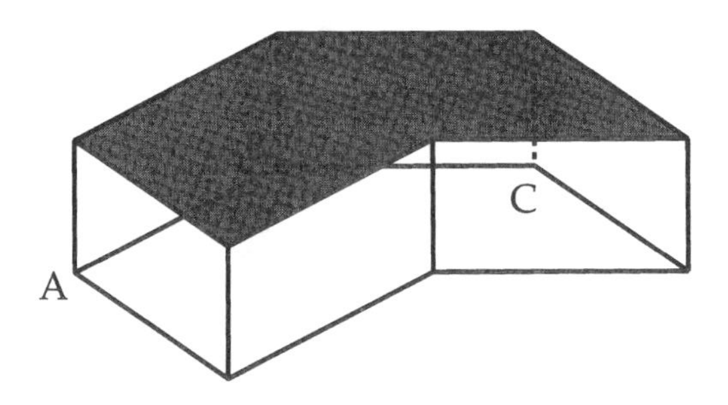

What can be learnt from this implicit knowledge? What types of baskets can be woven, that display at all their vertices pentagonal holes?

It comes out that the smallest possible "basket," made out of six strips, is similar to the well-known modern football made out of pentagonal and hexagonal pieces of leather.

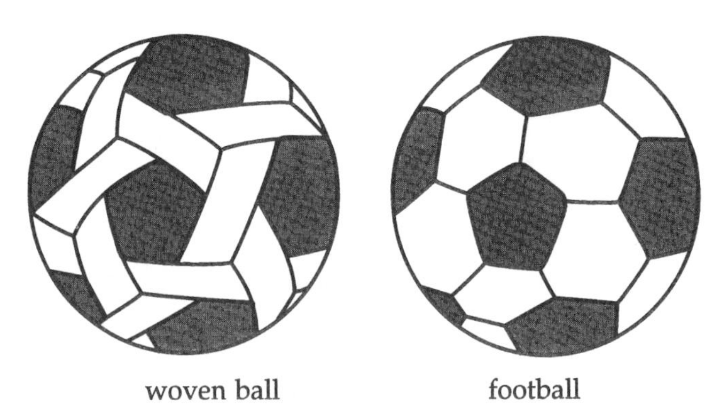

When one "planes" this ball, one gets a truncated icosahedron, bounded by 20 regular hexagons and 12 regular pentagons. By extending these 20 hexagons, one generates the regular *icosahedron*. On the other hand, when one extends the 12 pentagons, the regular *dodecahedron* is produced.

What type of "baskets" can be woven, if one augments their "curvature?" Instead of pentagonally woven "vertices," there arise square-hole-vertices. By planing the smallest possible "ball," one gets a truncated octahedron, bounded by 6 squares and 8 regular hexagons. Once again, by extension of its faces, new regular polyhedra are discovered, this time, the *cube* and the regular *octahedron*. When one augments still more the curvature of the "ball," there appear triangular-hole-vertices and by "planing" the "ball," one gets a truncated tetrahedron, bounded by 4 regular hexagons and 4 equilateral triangles. By extension of its hexagonal or triangular faces one obtains a regular *tetrahedron*.

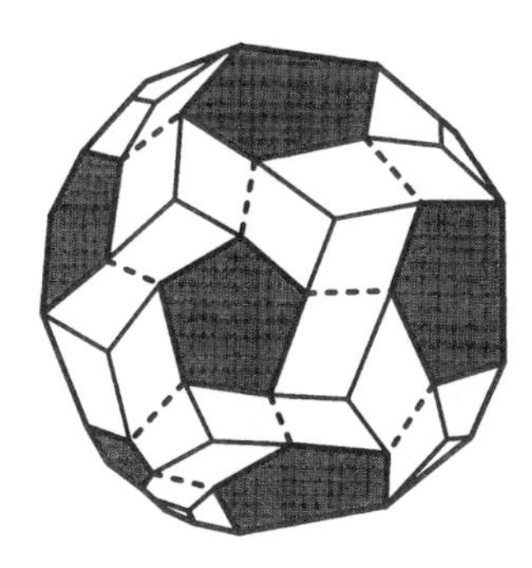

woven truncated icosahedron

truncated icosahedron

Many interesting questions can be posed to future teachers, for example,

* is it possible to "weave" other semi-regular polyhedra? Semi-regular, in what sense?
* did we generate all regular polyhedra? Why?
* what happens if one, instead of reducing the material at a vertex of the basket, augments it?

Concluding Remarks

Of the struggle against "mathematical underdevelopment" and the combat of racial and neo-colonial prejudice, a cultural-mathematical reaffirmation makes a part. A "cultural conscientialization" of future mathematics teachers, for example, in the way we described, seems indispensable.

Some other conditions and strategies for mathematics education to become *emancipatory* in former colonized and (therefore) underdeveloped countries have been suggested elsewhere.[30]

Acknowledgments

The author is grateful to Dr. A. J. Bishop (Cambridge) for his invitation to write this article and to Dr. W. Humbane (Maputo) for proofreading this paper.

Notes

1. Cf. e.g. Eshiwani (1979), Nebres (1983) and El Tom (1984).
2. Broomes and Kuperes (1983, p. 709).
3. El Tom, 1984, p. 3.
4. D'Ambrosio (1983, p. 363).
5. Gay and Cole (1967, p. 6).
7. Gay and Cole (1967, p. 66).
8. Luna (1983, p. 4).
8. Carraher a.o. (1982).
9. D'Ambrosio (1984, p. 6). Cf. D'Ambrosio (1985b).
10. D'Ambrosio (1984, p. 6), italics P. G.
11. D'Ambrosio (1984, p. 8), italics P. G.
12. D'Ambrosio (1984, p. 7).
13. Pinxten (1983, p. 173).

14. These terms are used by different authors, among them are D'Ambrosio: spontaneous; Carraher, D., Carraher, T.; and Schliemann, A,: natural; Posner, Ascher: informal; Gay and Cole: indigenous; Mellin-Olsen: folk;

Ascher and Ascher: implicit; Carraher, D., Carraher, T., and Schlieman: non-standard; Gerdes: hidden, "frozen;" D'Ambrosio: ethno-.

15. Gay and Cole (1967, p. 94).

16. Mmari (1978, p. 313).

17. Quoted by Nebres (1984, p. 4).

18. D'Ambrosio (1984. p. 10).

19. D'Ambrosio (1985a, p. 47).

20. Machel (1978, p. 401).

21. Machel (1978, p. 402).

22. Cf. Gerdes (1982, 1985a).

23. D'Ambrosio (1985, p. 45).

24. Ascher and Ascher (1981, p. 159), italics P. G.; cf. Gerdes (1985b, §2).

25. The first results are summarized in Gerdes (1985b). Cf. Gerdes (1986a, f). By bringing to the surface geometrical thinking that was hidden in very old production techniques, like that of basketry, we succeeded in formulating new hypotheses on how the ancient Egyptians and Mesopotamians could have discovered their formulas for the area of a circle [cf. Gerdes (1985b,c, 1986d)] and for the volume of a truncated pyramid [cf. Gerdes (1985b)]. It proved possible to formulate new hypotheses on how the so-called Theorem of Pythagoras could have been discovered [cf. Gerdes (1985b, 1986c, e)].

26. Experimental course developed for secondary schools in the USSR (1981) by a team directed by the academician A. Alexandrov.

27. The implicit geometrical knowledge that it reveals, is analyzed in Gerdes (1985b).

28. For more details, see Gerdes (1986b).

29. Another "culturally integrated" introduction to the "Theorem of Pythagoras" is presented in Gerdes (1986c, g).

30. Cf. e.g. Gerdes (1985a, 1986a), D'Ambrosio (1985b) and Mellin-Olsen (1986).

References

D'Ambrosio, U. (1983). Successes and failures of mathematics curricula in the past two decades: A developing society viewpoint in a holistic framework. In: *Proceedings of the Fourth International Congress of Mathematical Education* (pp. 362–364). Boston: Birkhäuser.

———. (1984). *The intercultural transmission of mathematical knowledge: Effects on mathematical education.* Campinas: UNICAMP.

———. (1985a). Ethnomathematics and its place in the history and pedagogy of mathematics. In *For the Learning of Mathematics.* Vol., no. 1, p. 5: 44–48.

———. (1985b). *Socio-cultural Bases for Mathematics Education.* Campinas: UNICAMP.

Ascher, M., and R. Ascher. (1981). *Code of the Quipu: A Study in Media, Mathematics, and Culture. Ann Arbor: University of Michigan Press, Ann Arbor.*

Broomes, D. and P. Kuperes. (1983). Problems of defining the mathematics curriculum in rural communities. In *Proceedings of the Fourth International Congress of Mathematical Education.* Boston: 708–711.

Carraher, T., D. Carraher and A. Schliemann. (1982). Na vida, dez, na escola, zero: os contextos culturais da aprendizagem de matematica. In *Cadernos de pesquisa. Vol. 42: 79–86.*

El Tom, M. (1984). *The role of Third World University mathematics institutions in promoting mathematics.* Adelaide, Australia.

Eshiwani, G. (1979). "The goals of mathematics teaching in Africa: A need for re-examination." In *Prospects,* Vol. 9, no. 3: pp. 346–352.

Gay, J. and M. Cole. (1967) *The new mathematics and an old culture: A study of learning among the kpelle of Liberia. New York: Holt, Rinehart and Winston.*

Gerdes, P. (1982). *Mathematics for the benefit of the people.* Paramaribo: CARIMATHS.

———. (1985a). "Conditions and strategies for emancipatory mathematics education in underdeveloped countries." In *For the Learning of Mathematics.* Vol. 5, no. 1: 15–20.

———. (1985b). *Zum erwachenden geometrischen Denken.* Maputo: Eduardo Mondlane University.

———. (1985c). "Three alternate methods of obtaining the ancient Egyptian formula for the area of a circle. In *Historia Mathematica,* Vol. 12, 261–268.

———. (1986a). On culture, mathematics, and curriculum development in Mozambique. In place: Mellin-Olsen & Johnsen Hoines, 15–42.

———. (1986b). *Um método geral para construir polígonos regulares, inspirado numa técnica mocambicana de entrelaçamento.* Maputo: TLANU-booklet, no. 4.

———. (1986c). A widespread decorative motif and the Pythagorean theorem. In *For the Learning of Mathematics*. Vol.8, no.1 (1988): 35–39.

———. (1986d). *Hypothesen zur Entdeckung des alt-mesopotamischen Näherungswertes* $\pi = 3\frac{1}{8}$ Maputo: TLANU-preprint, no. 1986–4.

———. (1986e). *Did ancient Egyptian artisans know how to find a square equal in area to two given squares?* Maputo: TLANU-preprint, no. 1986–5

———. (1986f). How to recognize hidden geometrical thinking? A contribution to the development of anthropological mathematics. In *For the Learning of Mathematics*. Vol. 6, no. 2: 10–12, 17.

Gerdes, P. and H. Meyer (1986g). Pythagoras, einmal anders. In *Alpha*, Vol.24, no.6 (1990): 128–129

Luna, E. (1983). *Análisis curricular y contexto sociocultural*. Santiago.

Machel, S. (1970). "Educate man to win the war, create a new society and develop our country." In *Mozambique, Sowing the Seeds of Revolution*, Harare: Zimbabwe Publishing House, 1981, 33–41.

Machel, S. (1978). "Knowledge and science should be for the total liberation of man." In *Race and Class*. Vol. 19, no. 4: 399–404.

Mellin-Olsen, S. and M. J. Hoines (1986). *Mathematics and Culture. A Seminar Report*. Radal: Caspar Forlag.

Mmari, G. (1978). The United Republic of Tanzania: Mathematics for social transformation. In *Socialist Mathematics in Education*. F. Swetz (Ed.). Southampton: Burgundy Press.

Nebres, B. (1983). *Problems of mathematical education in and for changing societies: problems in Southeast Asian countries*. Tokyo.

Nebres, B. (1984). *The problem of universal mathematics education in developing countries*. Adelaide, Australia.

Pinxten, P., I. van Dooren and F. Harvey. (1983). *The anthropology of space: explorations into the natural philosophy and semantics of the Navajo*. Philadelphia: University of Pennsylvania Press.

Section V

Ethnomathematical Praxis in the Curriculum

Arthur B. Powell and Marilyn Frankenstein

The curricular praxis of ethnomathematics can be developed by investigating the ethnomathematics of a culture to construct curricula with people from that culture, and by exploring the ethnomathematics of other cultures to create curricula so that people's knowledge of mathematics will be enriched.

Instructors acquire an important starting point from which to build and extend mathematical structures understood by their students when students reflect on their ethnomathematics, their ways of conceiving mathematics and methods of solving problems, and express and exchange their reflections through, for example, class discussions, writing, and student-teacher interviews. Fasheh (1982/reprinted here as chapter 13) stresses the importance of the experiences and culture of mathematics teachers and learners: "Teaching math without a cultural context, by claiming that it is absolute, abstract, and universal, is the main reason, I believe for the alienation and failure of the vast majority of students in the subject" (p. 8). Gilmer, president of the International Study Group on Ethnomathematics, argues that students value mathematics when they "sense [their] power and ownership in the products of that discipline" (1990, p. 4). Further, while students reflect on and reconceive their mathematical knowledge, they can discover that they know more mathematics than traditional evaluations reveal and acquire confidence that they can learn even more. Just as ethnomathematics can increase the respect and confidence that students have for themselves, it can also help them solve problems more effectively. Borba (1990/reprinted here as chapter 12) defines a real problem, as opposed to a pseudo, textbook-type problem, as a "situation which involves an impasse in the flow of life and which is important to that person's existence" (p. 40). He argues that the ethnomathematics of a group of people is more effective for solving real problems because it is rooted in their culture. Moreover, D'Ambrosio (1987) contends that

> [i]t is in the realm of one's own ethnomathematics that one's creativity will manifest itself. . . . The source of authentic mathematical and scientific creativity is not in formalized mathematics and science, but in mathematics and science in the making, fed by the creative process itself (p. 4).

In addition, Winter (1991) argues that ignoring mathematics "in the making" of children and forcing formalizations too early is responsible for math "anxiety." He feels that the difficulty so many people experience learning mathematics is not due to the inherent difficulty of its abstractions, but rather "is a problem of mathematics as a cultural form" (p.82), such as formal mathematics textbook presentation. He presents examples from his work with children that challenge the taken-for-granted curricular logic that "the relationship between experience and analytical thought may be conceived as a progression from fact to concept, from concrete to abstract, and from simple to complex" (p.85). His examples show that five-year olds can understand "abstract" math concepts such as the probability implications of various ratios, with no *formal* teaching on the subject, and before they have completely mastered supposedly lower-level skills such as counting. So when the curriculum starts from our students knowledge and respects their ways of understanding we can be much more creative in our choice of mathematics to be learned.

In learning about the ethnomathematics of students, teachers gain respect for and understanding of the kinds of mathematical ideas that they possess. However, we are not suggesting that the curriculum should be composed solely of those ideas. As Freire asserts, dialogical education does not mean teachers are merely "passive, accidental presences," attempting to avoid domination. Freire (1981) points out that

> [t]he opposite of manipulation is not an illusory neutrality, neither is it an illusory spontaneity. The opposite of being directive is not being non-directive—that is likewise an illusion. The opposite both of manipulation and spontaneity is critical and democratic participation by the learners in the act of knowing, of which they are the subjects (p. 28).

Teachers listen to students to discover themes and present related problems to challenge and extend previous perceptions of students. Also, teachers can suggest new themes, ones they judge important, and be strong influences without being superiors, constraining and controlling the learning environment.

In learning about the mathematics of students, it is also important for teachers to explore how the semantic and syntactic structure of particular languages may facilitate mathematics learning. For example, Miura (1987) discusses how features of Chinese, Japanese, and Korean are congruent with certain concepts in arithmetic. On the other hand, students can experience difficulties learning mathematics when respect is denied either by uncritical adoption or by insensitive imposition of distinct and distant cultural and linguistic conventions from one milieu to another. For example, in the People's Republic of China, even graduates from senior-middle school experience difficulties reading many-digit numerals, such as 7,612,439, without first pointing and naming from right to left the place value of each digit before knowing how to read the "7" in the millions place and the rest of the numeral. Through an investigation of this problem, Powell (1986) discovered that, during instruction, little attention is given to the linguistic regularity of naming numerals in Chinese. Furthermore, he found that this state of affairs is compounded by the adoption or imposition of a foreign convention of delimiting digits in a many-digit numeral which is at variance with the linguistic structure of Chinese numeration. Unlike in certain Romance and Germanic languages of the West where commas, spaces, or points are used as delimiters between groups of three digits, the linguistic structure of naming numerals in Chinese is based on groups of four digits. Powell suggests that, at its base, the problem that Chinese students have in reading numerals longer than four digits "points to a consequence of the cultural and political domination of so-called developed societies' scientific norms over those of developing nations" (p. 20).

This particular difficulty that Chinese students experience in learning mathematics is part of a general problematic arising from the interaction of language and mathematics. Research exists on the nature and causes of mathematics learning difficulties manifested when curricula are adopted for use in a cultural and linguistic milieu distinct and distant from the one for which the curricula were developed (see, for example, Philp, 1973; Berry, 1985; Orr, 1987). Berry (1985), in an analysis of problems in second language mathematics learning in Botswana, suggests a general theory of types of language associated learning problems, consisting of two categories. Of interest here is

Berry's second category of problems which "result from the 'distance' between the cognitive structure natural to the student and implicit in his mother tongue and culture, and those assumed by the teacher (or designer of curriculum or teaching strategies)" (p. 20). Adding to the notions of semantic and cultural differences with which Berry defines the term "distance," Powell's example suggests that we include the notion of *syntactic* difference.

An example of how semantic, syntactic, and cultural differences can be used to disempower a cultural, or linguistic, group is found in *Twice as Less: Black English and the Performance of Black Students in Mathematics and Science* (Orr, 1987). Orr taught at a white, middle-class private high school in Washington, DC to which a group of urban, African-American students were given places for a number of experimental years. When these students performed poorly in mathematics and science, she questioned why and focused on linguistic features of work done in class and at home. She and her colleagues found "explicit evidence" that African-American "students were using one kind of function word, prepositions, in a manner different from other students; their misuses were different even from the misuses with which [they] were familiar" (p. 21). That is, the semantic and syntactic use of words similar to Standard American English (SAE) by students speaking Black English Vernacular (BEV) were different from those used by students who belonged to the culture with power. Orr concluded that this linguistic difference *is* the reason why her students did poorly: "For students whose first language is BEV, then, language can be a barrier to success in mathematics and science" (p.9).[1]

This view distorts the connections between conceptual understanding and semantic and syntactic differences. As linguistics, like Labov (1972), have demonstrated BEV and SAE are both capable of generating labels for concepts attended to by the culture of the speakers. The effect, if not the object, of Orr's approach is to confer privilege on the culture and language (SAE) of the dominant power and, thereby, to deny legitimacy to other culturally generated linguistic (BEV) and cognitive experiences. In addition to her incorrect linguistic analysis, she does not adequately probe the effects of the sociopolitical structure on her students. By judging students' linguistic-related mathematical ability out of context of the power relations which must have played themselves out in that situation, she misses the opportunity to teach her students mathematics.

Instead, an ethnomathematical approach shifts the focus away from her students' misunderstandings. It starts by assuming her students have mathematical knowledge and focuses on how to learn about and build upon that knowledge.[2] Instead of Orr's approach, teachers can engage students in critically analyzing both theirs and the dominant culture's language from mathematical, sociological, and political perspectives. Then, as opposed to uncritically adopting an alternate code, the language of students is validated. In their analysis, students come to understand the value and use of their and the dominant code, explore different formulations of mathematics problems in each, and realize the power relationship which necessitates that they also acquire linguistic facility in the dominant culture.[3] However, as Delpit (1988) states "[i]f you are not already a participant in the culture of power, being told explicitly the rules of that culture makes acquiring power easier" (p. 282). To do otherwise, as Orr does, is to victimize those who are victims of the culture of power. She judges her students' linguistic-related mathematical ability against the culture of power.

An example of how ethnographic research leads to an empowering curriculum, can be found in the work of Gerdes and members of the Ethnomathematics Project of the Universidade Pedagógica in Mozambique. They are developing curriculum from ethnographic research into various aspects of traditional, material Mozambican culture that reveal "hidden moments of geometrical thinking" (1988, p. 140/reprinted here as chapter 11). They illustrate how recognizing and studying the geometry behind certain Mozambican baskets and fish traps stimulates students to reconsider the value of their cultural heritage—"geometrical thinking was not and is not alien to our culture"—and serves as a "starting point and source of inspiration for doing and elaborating other interesting mathematics" such as tessellations, trigonometry, and polyhedra (pp. 153–160). When his Mozambican students, for another example, analyze the construction of a woven button, commonly used in southern parts of Mozambique to fasten the top of a basket, they uncover the Pythagorean theorem. And they realize that

> Had Pythagoras—or somebody else before him—not discovered this theorem, *we* would have discovered it!' . . . By not only making explicit the geometrical thinking "culturally frozen" in the square-woven buttons, but . . . by revealing its full potential, one stimulates the development of . . . cultural-mathematical (self) confidence. . . . The debate starts. "Could our ancestors have discovered the 'Theorem of Pythagoras'?" "Did they?" . . . "Why don't we know it?" . . . "Slavery, colonialism . . ." By "defrosting frozen mathematical think-

> ing" one stimulates a reflection on the impact of colonialism, on the historical and political dimensions of mathematics [education] (pp. 151–152).[4]

Another category of praxis has been the development of curricula to inform students about the ethnomathematics of other cultures. Most notably, these curricular experiments have taken shape in Western societies, often in an attempt to counter myths and Eurocentric views about nonexistent contributions of non-Western peoples. For a number of equally important reasons, we think this is vital. First, the additional examples obtained by considering the mathematics of non-Western peoples provides a rich source for illustrating and applying mathematical concepts and theorems. Second, it gives a more accurate account of the history of mathematics and the contributions of peoples all over the world to that knowledge. Third, through learning more mathematics, students and teachers acquire rich and powerful mathematical insights. Finally, students of racially and culturally diverse backgrounds are culturally affirmed, gaining an appreciation for the contributions of their communities and of other people's to the history of mathematics (Frankenstein and Powell, 1989, p. 110).

We agree with Gerdes (1985) that cultural affirmation is a key factor in struggles against the mathematical underdevelopment caused by racism, sexism, and imperialism. "For mathematics to become emancipatory, it is necessary to stimulate confidence in the creative powers of every person and of every people, confidence in their capacities to understand, develop and use mathematics" (p. 17), and confidence is developed greatly through cultural affirmation. However, we also stress that we are not advocating the curricular use of other people's ethnomathematical knowledge in a simplistic way, as a kind of "folkloristic" five-minute introduction to the "real" mathematics lesson. To do so "implies that such a culture is not credited the role as a decisive and organic frame for the lives of the individuals, but rather the role as a static and illustrative frame which mainly is of archaeological importance" (Mellin-Olsen and Holnes, 1985, p. 106).[5] Mathematics should be studied in a way that uncovers its connections to the development of human societies, "showing the necessity of any given piece of calculation, measure or pattern for the particular society of which it was a part" (Singh, 1991, p. 21). Finally, we also stress that the broader historical context in which this ethnomathematical knowledge was developed, after imperial devastation, suppressed or stolen, needs to be included in the curriculum.

In the United States, the publication of Zaslavsky's *Africa Counts* in 1973 inspired a number of liberal arts mathematics courses. Zaslavsky has created numerous curricular applications from her research into the mathematics of various cultures (1985, 1991a, 1991b/reprinted here as chapter 15, 1993, 1996). Ascher (1984), implementing collegiate curriculum in ethnomathematics counters the Eurocentric bias that mathematical ideas from other cultures are "curiosities" or "a historically earlier part of our own mathematics." Ascher's work demonstrates that sophisticated mathematical ideas do not only exist as expressed in academic mathematics texts but can be embedded in everyday cultural activities. Anderson's (1990/reprinted here as chapter 14) curricula counters another myth of Eurocentrism—that all the mathematics called "academic mathematics" was created solely by Europeans. He connects the development of mathematics with "humanity's ongoing struggle to understand Nature, and capitalism's attempts to control and dominate Nature" (p. 354), showing how all people have contributed to mathematical knowledge. Frankenstein (1991) incorporates research about other culture's ethnomathematics in her pedagogy by having students reflect on ethnomathematical research to appreciate the mathematical contributions and the logic systems of other peoples. They can then use this appreciation and knowledge to develop confidence and insights that help them with their own mathematics learning.

Notes

1. Further, she makes the unsubstantiated claim that, unlike the grammar of BEV, "the grammar of standard English [SAE] has been shaped by what is true mathematically" (p. 158). She uses this ethnocentric, elitist claim to vault the supposed intrinsic superiority of the language of the culture of power.

2. Such an ethnomathematical approach, connecting language and mathematics, is precisely at the heart of the Algebra Project curricula, a U.S.-based project, designed by Robert Moses to insure that all middle-school children have access to studying algebra. Moses analyzed both the content and the difficulties that students were experiencing in elementary schools. For instance, he found that students understood the arithmetic meaning of numbers in terms of quantity ("how many?") and understood direction ("which way?"); but, they did not associate their knowledge of direction with their knowledge of numbers. To teach this and other pre-algebra concepts, he devised a five-step process: an activity, a physical event the children would experience; a picture or diagram they would draw of this event; an intuitive linguistic de-

scription of this event; a more regimented linguistic description of this event; a symbolic representation of this event, first through students' invented symbols, then finally in standardized mathematical symbols. Moses recognizes that there will be a variety of idiomatic linguistic descriptions that need to be expressed before moving to the more structured language (a language which is neither BEV nor SAE) that facilitates the move toward algebraic symbols.

3. For an example of this approach in the teaching of the dominant variety of American English to Native Alaskans, see Delpit (1988, p. 293).

Aside from this particular issue, focused on the violation of a group's ethnomathematics expressed in its linguistic particularities, there are other points of criticism of Orr's work. For more thoroughgoing critiques see O'Neil (1990) and Baugh (1988).

4. Another interesting example of the development of mathematics curriculum from ethnographic research into the mathematics of particular communities is the work of the Direction Nationale de l'Alphabetisation (DNALFA) in Bamako, Mali. They investigated why, in spite of daily preoccupation with figures and calculations, farmers and stockbreeders who bought and sold in a fluid market encountered difficulties learning such operations as multiplication and division in a conventional curriculum. They discovered that the arithmetical process taught in the formal adult classes differed significantly from the processes which the students had already acquired informally from their daily market experiences. In short, the attempt to impose classic, foreign written calculations upon "students experienced in mental arithmetic confused even the most apt among them, and it so distracted their weaker classmates that it posed a severe obstacle to their acquisition of writing and calculation skills, and the young adults found their competence reduced, rather than increased, by their educational experience" (Clarke, 1989, p. 3).

In Sierra Leone, Bockaire (1988) developed a project to "investigate the mathematics that existed in Mende culture, the strengths and limitations of such mathematics, the mathematics that probably existed long before formal schooling was introduced in the Mende land" (p. 1). Bockaire is aware that, as a consequence of engaging in and reflecting on traditional activities, "at the village levels, on the farms, at the rivers and on the foot paths or roads to their work place," people develop mathematical understandings, albeit not in the same codified form of "Western" mathematics. Cognizant of severe limitations in the colonialist-inspired school system, the aim of the ethnomathematical investigation of Bockaire and his team was to create a culturally appropriate curriculum to educate those in the illiterate and innumerate sectors since "mathematics courses in [Sierra Leone's] formal school system are structured without recognition and exploitation of the wealthy mathematical activities within the culture" (p. 1). In addition to contributing to the education of illiterate farmers, investigators used the Mende's approach to adding two sets of objects to teach, at Njala University College, the concept of homomorphism to a sophomore-level course in abstract algebra and noticed marked improvement in understanding and interest (p. 63).

5. In a recent conversation with one of the authors, Bob Lange, a physics professor at Brandeis, shared an experience which underlines how important it is for us to be aware of the particular lens through which we view other cultures. Even when that lens is respectful, it can be static and false. In order to give students and educators in Zanzibar an experience with computers based on their mathematical knowledge, Lange designed a computer program that plays their version of the game *oware* or *bao ki swahile*. He felt nervous about introducing the program for he did not want to violate the aesthetics of the beautiful board or the tactile pleasure of actually removing the smooth, rounded seeds. He spent a lot of time planning ways to have the computer serve as a mere adjunct to the actual presence of the board and seeds. Still, he was concerned that the players would find the computer inappropriate or offensive. Instead, from the first move they made, the students completely ignored the board and seeds. They educated Lange that for them the ancient African game of *oware* is not primarily about aesthetics, but about abstract strategies. In the context of high technology, they found the computer was an engaging, appropriate representation of their contests of reasoning. Lange, and Michael Savage of the African Forum for Children's Literacy in Science and Technology in Nairobi, are currently writing an article in which they will be discussing this and other observations that have arisen from this unexpectedly appropriate use of technology.

References

Anderson, S. E. (1990). Worldmath curriculum: Fighting eurocentrism in mathematics. *Journal of Negro Education*. 59(3): 348–359.

Ascher, M. (1984). Mathematical ideas in non-western cultures. *Historia Mathematica* 11: 76–80.

Baugh, J. (1988). Review of Twice as less: Black English and the performance of Black students in mathematics and science. *Harvard Educational Review* 58(3): 395–404.

Berry, J. W. (1985). Learning mathematics in a second language: Some cross-cultural issues. *For the Learning of Mathematics* 5(2): 18–23.

Bockaire, A. (1988). *Mathematics used and needed by Mende farmers of Moyamba and Kailahun districts in Sierra Leone*. Dakar, Sénégal: International Development Research Centre.

Borba, M. C. (1990). Ethnomathematics and education. *For the Learning of Mathematics* 10(1): 39–43.

Clarke, H. (1989). *Project completion report: Teaching arithmetic to illiterates*, No. 3–P–80–0034. Dakar, Sénégal: International Development Research Centre.

D'Ambrosio, U. (1987). Reflections on ethnomathematics. *International Study Group on Ethnomathematics Newsletter* 3(1): 3–5.

Delpit, L. D. (1988). The silenced dialogue: Power and pedagogy in educating other people's children. *Harvard Educational Review* 58(3): 280–298. Fasheh, M. (1982). Mathematics, culture, and authority. *For the Learning of Mathematics* 3(2): 2–8.

Frankenstein, M. (1991). Breaking down the dichotomy between learning and teaching mathematics. In unpublished manuscript.

Frankenstein, M., and Powell, A. B. (1989). Empowering non-traditional college students: On social ideology and mathematics education. *Science and Nature* 9/10: 100–112.

Freire, P. (1981). The people speak their word: Learning to read and write in São Tomé and Principe. *Harvard Educational Review* 51: 27–30.

Gerdes, P. (1985). Conditions and strategies for emancipatory mathematics education in underdeveloped countries. *For the Learning of Mathematics* 5(1): 15–20.

———. (1988). On some possible uses of traditional Angolan sand drawing in the mathematics classroom. *Journal of the Mathematical Association of Nigeria* 18(1): 107–125.

Gilmer, G. (1990). An ethnomathematical approach to curriculum development. *International Study Group on Ethnomathematics Newsletter* 5(1): 4–6.

Labov, W. (1972). The logic of nonstandard English. In L. Kampf and P. Lauter (eds.). *The politics of literature: Dissenting essays in the teaching of English* (pp. 194–244). New York: Pantheon.

Mellin-Olsen, S., and Holnes, M. J. (Ed.). (1985). *Mathematics and culture: A seminar report*. Raidal, Norway: Caspar Forlag.

Miura, I. T. (1987). Mathematics achievement as a function of language. *Journal of Educational Psychology* 79(1): 79–82.

O'Neil, W. (1990). Dealing with bad ideas: Twice is less. *English Journal* 79(4): 80–88.

Orr, E. W. (1987). *Twice as less: Black English and the performance of Black students in mathematics and science*. New York: W. W. Norton.

Philip, H. (1973). Mathematics education in developing countries: Some problems of teaching and learning. In A. G. Howson (ed.). *Developments in mathematics education* (pp. 154–180). New York: Cambridge.

Powell, A. (1986). Economizing learning: The teaching of numeration in Chinese. *For the Learning of Mathematics* 6(3): 20–23.

Singh, E. (1991). Classroom practice: Anti-racist mathematics. Unpublished manuscript.

Winter, R. (1991). "Mathophobia," Pythagoras and roller-skating. *Science and Culture* 10: 82–102.

Zaslavsky, C. (1973). *Africa counts: Number and patterns in African culture.* Boston: Prindle, Weber & Schmidt.

———. (1985). Bringing the world into math class. *Curriculum Review* 24(3): 62–65.

———. (1991a). Multicultural mathematics education for the middle grades. *Arithmetic Teacher* 38(6): 8–13.

———. (1991b). World cultures in the mathematics class. *For the Learning of Mathematics* 11(2): 32–36.

———. (1993). *Multicultural mathematics: Interdisciplinary cooperative-learning activities*. Portland, ME: Walch.

———. (1996). *The Multicultural math classroom: Bringing in the World.* Portsmouth, NH: Heinemann.

Chapter 12

Ethnomathematics and Education

Marcelo C. Borba

Editors's comment: Marcelo C. Borba, a mathematics educator, discusses relationships between ethnomathematics and academic mathematics. He claims that, in context, ethnomathematical methods are probably more efficient than academic mathematics. Further, he argues for the efficacy of ethnomathematics in providing access for students learning academic mathematics. This chapter first appeared in *For the Learning of Mathematics* 10(1): 39–43, in 1990.

Introduction

In this paper, I will discuss the notion of ethnomathematics, which can be seen as an epistemological approach to mathematics, and will relate ethnomathematics to education. This discussion will lead to a view of how mathematics should be incorporated into school curricula and to suggestions regarding how mathematics should be pedagogically practised.

1 Philosophical Background to Ethnomathematics

In this section, I will summarize a view of human beings on which the idea of ethnomathematics is based, a view of how people relate to other human beings and to the world. I will then focus on two particularly important ideas: "dialogue" and "problem." Finally, I will establish connections between the idea of ethnomathematics and this view of human beings.

1.1 Human Beings and Their Dialogical Relations

This view of humans is based on a phenomenological approach in which a person is seen as a "being-in-the-world-with-others."[1] She/he is a "be-ing" since her/his essence is manifested in her/his daily ways of existing in the world. She/he is "in-the-world," not in the sense of water "in" a glass, but in the sense of being in a relationship with the world which expands to fill a space without dimensions. This relationship expands further into the world as s/he comprehends new meanings about this relationship with the world. She/he is "with others" because she/he always works with something and/or talks with someone (even if only to her/himself).

In this phenomenological view of human beings, a human is only seen in connection with the world. She/he cannot be seen without the world; neither can the world be seen without her/him. Moreover, the concepts "human" and "world" themselves are intrinsically linked since both terms reflect meanings which have been constructed by humans. Each human relates to other humans based on comprehensions: understanding existing meanings and making new meanings. Each person is also always in a *place* in the world and living in a historical *moment*.

In her/his existence, a person experiences events in which she/he is also involved. These experiences can be seen as a "chain of consciousness" which is in continuous and indivisible flux, like a river, where thinking is both changeable and constantly flowing. However, experience is lived in its own time, different from "official," chronological time. A reflection on an experience is no longer in the original flux of experience, but is in a new part of the ongoing flux, looking back at an earlier time. Thus, consciousness is an endless, recurrent process which embodies, in a broad sense, reflecting, knowing, and thinking.

In the terminology used by Paulo Freire (1981), consciousness can be "intransitive" or "transitive." A person with intransitive consciousness doesn't link her/his experiences together; she/he always lives in the present moment and therefore cannot make important connections. She/he is likely to change only superficially, for example, in response to fashions. A person with transitive consciousness develops a more reflective perspective which allows her/him to make connections between her/his different experiences and therefore to make significant changes in response to these experiences. Freire argues that reaching a "critical transitivity" is necessarily an active and dialogical (that is, in dialogue with other people) process. Therefore transitive consciousness and dialogue are both fundamental to the processes of personal and educational growth.

1.2 Dialogue

Dialogue can be seen as a horizontal relationship between two or more human beings, in which the "being" of each person opens her/himself to the other(s) in an authentic way. Dialogue is an intersubjective relationship in which human beings try to know each other and reveal their true selves to each other (Bicudo, 1979). The subjects involved in the dialogue communicate using not only intentional signs (for example, words), but also using unconscious signs such as pauses, ways of walking or breathing, gestures, and so on. In this context, the meanings of words cannot be limited to those stored in the dictionary. However, just giving something a name shows the importance this thing has in a given culture. According to Alfred Schutz [in Wagner, 1979], words are bounded by past and future elements of someone's speech; words also have emotional and irrational values which are not explicit. Meanings of signs also change from one cultural[2] group to another, since each group "shapes" the meaning of words to their context. Finally, it is important to remark that dialogue cannot take place if the realms of concern of the human beings involved in the dialogue have no intersection. In other words, if the problems which involve them are completely different, the dialogue cannot occur.

The word "problem" has been used just above in a very different way than in most mathematics education literature. The next section will focus on the key idea of "problem."

1.3 Problem

What is a problem?[3] If I ask an adult who is standing in front of me, "What is the color of the pants I'm wearing?" is that a problem? In education it is important to distinguish a problem from a simple question to which the answer is known without any need for reflection. Another common misuse of the term problem is when it is associated with simply "not knowing." If I ask someone how many universities there are in the U.S.A., is it a problem for that person if she/he does not know the answer? This is probably not a problem for her/him because she/he isn't likely to care about the question. Whether or not the answer is already known, whether or not the answer can be easily obtained, if she/he doesn't care about the answer, it is not a problem for her/him. In this approach, what is of interest to someone is important to the idea of "problem." If an obstacle occurs in the course of someone's own existence *and* if she/he does not know how to overcome the obstacle, then she/he has a problem.

A problem can be authentic or it can be imposed. An imposed obstacle or puzzle would be a pseudo-problem, a situation which occurs frequently in mathematics teaching. Students are usually asked to solve problems which are not problems for them personally; they only attempt to solve these pseudo-problems in order to get a good grade.

Although the discussion so far may have implied that the definition of problem I have been developing is too subjective, Demerval Saviani (1985) is very clear when he argues that:

> A problem, as any other aspect of human experience, has a subjective side and an objective one, closely connected by a dialectical unity. . . . The concept of problem implies a consciousness of a situation of necessity (subjective aspect) and a situation that puzzled his consciousness (objective aspect). (Saviani, 1985, p. 21, author's translation.)

The objective and subjective aspects of the definition of problem are both culturally bounded since what is interesting for someone, the aspect of subjectivity, depends partially on the cultural traditions of that person. Obstacles (the objective aspect) are also culturally bounded, because what is an obstacle in a given culture might not be one in another culture.

A problem then can be seen as a situation which involves an impasse in the flow of life and which is important to that person's existence. When a problem results in a mathematical treatment, it can lead to the generation of mathematics by the person(s) who was (were) puzzled by this situation.

2 *Ethnomathematics*[4]

In the last section, it was seen that a person is a cognizant being who functions within the language and interpretative codes of their sociocultural group. A language is a code understandable only to people who have participated in common past experiences. Each language expresses a way of knowing developed by a group of human beings.

One way of knowing is mathematics. Mathematical knowledge expressed in the language code of a given sociocultural group is called "ethnomathematics." In this context, "ethno" and "mathematics" should be taken in a broad sense. "Ethno" should be understood as referring to cultural groups, and not as the anachronistic concept of race; "mathematics" should be seen as a set of activities such as ciphering, measuring, classifying, ordering, inferring, and modeling. As defined by Ubiratan D'Ambrosio [1985]: "Ethnomathematics is the mathematics practised among identifiable cultural groups, such as national-tribal societies, labor groups, children of a certain age bracket, professional classes, and so on" [D'Ambrosio, 1985, p 45]. Even the mathematics produced by professional mathematicians can be seen as a form of ethnomathematics because it was produced by an identifiable cultural group and because it is not the only mathematics that has been produced.

This view of professional mathematics is consistent with George Joseph's statement that, because of the Eurocentric bias of most academicians, there is a "misrepresentation of the history and cultures of societies outside the European tradition." However, Joseph's statement that "mathematics can be looked at as an international language, with a particular kind of logical structure" is not consistent with an ethnomathematical view of professional mathematics. [Joseph, 1987, p 14] While Joseph recognized that each mathematics has a particular kind of logical structure, he says that mathematics is international. In doing so, Joseph is assuming mathematics is independent of culture rather than being an historical construction which is socially and culturally bounded since the way it is organized and the way it is expressed represent the codes and understandings of professional mathematicians who are also culturally bounded. Therefore "academic mathematics" is not universal (in the sense of being independent of culture) any more than "Quipu mathematics" is, or "carpenter mathematics," or "Shantytown mathematics,"[5] and so on, nor is it international in the way Esperanto was intended to become a language common to all people. Although academic mathematics may be international in that it is currently in use in many parts of the world, it is not international in that only a small percentage of the population of the world is likely to use academic mathematics.

However, mathematics can be considered universal in the way that Alan Bishop uses the term. Based on his analysis of different cultures, Bishop argues that activities such as counting, locating, measuring, designing, playing, and explaining ". . . are both universal, in that they appear to be carried out by every cultural group ever studied, and also necessary and sufficient for the development of mathematical knowledge" [Bishop, 1988, p 182]. Bishop also believes that ". . . mathematics has a cultural history, but also that from different cultural histories have come what can only be described as different mathematics" [1988, p 180].

Even though Bishop does not use the terminology ethnomathematics, his view comes towards the approach developed in this paper in arguing that every culture does mathematics, although the mathematics is expressed in ways unique to that culture. Thus, "ethnomathematics can be seen as a field of knowledge intrinsically linked to a cultural group and to its interest, being in this way tightly linked to its reality[6] and being expressed by a language, usually different from the ones used by mathematics seen as science. This language is umbilically connected to its culture, to its ethnos" [Borba, 1987, p 38].

2.1 Efficiency of Ethnomathematics

The ethnomathematics developed by different groups are likely to be more efficient at solving problems related to their cultures than academic mathematics is (unless, perhaps, the problem is in a school context) because the ethnomathematics developed by a given cultural group is linked to the obstacles which have emerged in this group. An obstacle and the need to overcome it draws people's attention to a situation which can be described as a problem as discussed in this paper. When the solution of this problem involves a mathematical treatment, the solution contributes to the development of ethnomathematics in this culture. Over time, this ethnomathematics is probably going to be more efficient than the models stored in textbooks and written in codes not always accessible to a given cultural group, because it is connected to the culture where the problem was generated. Hence, ethnomathematics should not be misunderstood as "vulgar" or "second class" mathematics, but as *different* cultural expressions of mathematical ideas.

3 Ethnomathematics, Education, and Ideology

The notion of ethnomathematics has clear implications for education. If different people produce different kinds of mathematics, then it is not possible to think about education as being a uniform process to be developed in the same way for different groups. Instead mathematics education should be thought of as a process in which the starting point would be the ethnomathematics of a given group and the goal would be for the student to develop a multicultural approach to mathematics.

For educators to develop an educational approach based on ethnomathematics, it is important to consider the concept of the problem discussed above. Problems could be found and developed which were based in ethnomathematics, thus avoiding the use of pseudo-problems. Students should actively participate in the design of their pedagogical program, as proposed by Freire: "The content of an education for critical consciousness must be developed by searching with the students for experiences which give meaning to their lives." [Freire, 1970, p 28] Therefore problems to be solved would be chosen by both students and teachers in a dialogical relationship which fosters a critical consciousness (as discussed in section 1.1). Knowledge can be seen as a product of this dialogical relationship. Each partner is going to be learning from the other in a dialectical way.

Mechanical views of dialogical educational processes should be avoided; one should not expect eleven-year-old boys or girls to develop a sophisticated comprehension of the contradictions of the political-economic system. Children do develop a consciousness of relationships in their world out of their reflections on the ways they play: on the rules of a game, on the friendships among the partners of this game, and even on the mathematical relationships of this game.

A pedagogy with students as partners with the teachers doesn't mean that the educational process is value-free. The incorporation of sociocultural aspects in mathematics education and the dialogical way of doing it each have a role to play. A dialogue where the teacher speaks through her/his ethnomathematics (usually developed in college) and students speak with theirs, is not neutral. Such a dialogue can allow students to strengthen their sociocultural roots, since their (ethno) knowledge is legitimized (recognized as valuable) in the educational process. This pedagogy can also emphasize that mathematics is not a single, unique expression and cannot be seen as a "straight line." A forest might be a better image of the whole set of ethnomathematics, in which each tree would be considered as a different expression of ethnomathematics, socioculturally produced.

Dialogue, which should be seen as a horizontal rather than vertical or hierarchical relationship, doesn't mean that the role played by the teacher is the same as the one played by the students. An equal relationship doesn't mean a uniform one. The teacher is different from the student because, among other reasons, she/he has an explicit intention of educating. She/he has worked and studied towards various goals as an educator, one of which may be developing a democratic educational relationship between teacher and students which can facilitate the students' development of a critical consciousness. In order to foster this development, such a teacher believes she/he has to share power with the students in the educational process.

3.1 Ethnomathematics and Education: Are They Really Compatible?

The accepted mathematics in this educational proposal ranges from ones developed by students to the one accepted/developed/intended by the teacher. In the classroom dialogue, the teacher can learn from the ethnomathematics "spoken" by the students, just as the students are learning from the academic ethnomathematics of the teacher. This dialogical process has no dichotomy between education and research, between teacher and researcher. The one who educates is also the one who researches the ethnomathematics developed by students. Therefore research influences educational praxis, and vice-versa.

The ethnomathematics of a cultural group is part of the group's life; the mathematics is generated by the culture in an "umbilical" way. Ethnomathematics is developed by the cultural group's interest in their problematic situations, which then further develop the group's interest in their ethnomathematics. This interest in ethnomathematics is natural because it was generated by the members of the cultural group in response to their own situations. However, this awakened interest in ethnomathematics does not automatically transfer to an interest in learning/developing any other ethnomathematics, such as academic mathematics. Students may not have much interest in investigating deeply the concepts which underlie their ethnomathematics. If the teacher forces students to work on problems, even problems based on the underlying ideas of the students' own ethnomathematics, they will be pseudo-problems, just as so often happens in regular schools with academic mathematics.

This argument could lead us to a belief that there is no way out of the dilemma of the use of pseudo-problems in the classroom. However, the previously discussed idea of dialogue offers a potential solution, since dialogue in its authentic form implies a mutual speaking and hearing. Hopefully the people involved in a dialogue can find convergent points and intersections in their realms of meaning. The teacher/researcher has a particular ability and responsibility to help the students find the intersections between their realms of meaning and the teacher's.

4 Final Comments: Ethnomathematics in Current School Situations

Using this same framework, educators such as Borba, Frankenstein, Gerdes, and Skovmose[7] have been developing pedagogical proposals along the lines supported by this paper. However, most of these pedagogies have been applied, with encouraging results, in "non-formal" schools and in adult education. Thus, the question still remains whether this kind of proposal makes sense in current formal school situations. Although there is still a long way to go in developing such a pedagogy for formal classrooms, it can be argued that such a framework can be tried in school situations and initial answers can be developed.

The ideas developed in this paper indicate that curricula cannot easily be changed by simply substituting some content for others. It is necessary to consider more fundamental kinds of change. In traditional curricula the use of pseudo-problems is unavoidable since students do not participate in choosing the themes which are going to be developed by the teacher during the school year.

"Thematization" and "project organization," to use Skovsmose's terminology [1985], are ways which many authors have found of both breaking the atomization of traditional curricula and building a new view of mathematics. In this approach, the themes and/or the projects to be developed are decided by both students and teachers. The themes are not necessarily "mathematical," or "biological," or "artistic;" themes developed jointly with students will probably not closely match the academic disciplines. They are just researches to be undertaken by the group, where the role played by teachers is to help the students develop a critical view of the world, a "transitive consciousness" in Freire's words [1981].

In this educational proposal, the ethnomathematics, the ethnobiology, ethnochemistry, and so forth, practiced by different groups of students would be the starting point of the pedagogical process. This "ethnoknowledge" developed by groups of students should be compared with the (ethno) knowledge developed by the academic disciplines in a way that this academic knowledge can be also seen as culturally bounded. The students and the teachers should discuss the efficiency and relevance of different kinds of knowledge in different contexts. With this approach, mystification about science might be avoided and mathematics might no longer be seen as an oppressive and all-powerful realm of knowledge.

Acknowledgement

Although they are not responsible for the content of this paper, I would like to thank Marcia Ascher, Maria Bicudo, Ubiratan D'Ambrosio, David Henderson, Anne Kepple, Margaret McCasland, Jan Rizzuti, and John Volmink for comments made on this paper.

Notes

1. This view of man is based on Heidegger [1981], Schutz [in Wagner, 1979] and Borba [1987 a or b].

2. Culture should be understood in this paper as what was added to the world by humans (as the result of human work, of human's creative and recreative struggles) and as meanings which are shared by a cultural group.

3. This view of problem is based on Saviani [1985] and Borba [1987 a or b].

4. A discussion of the reason for using the term "ethnomathematics" instead of "oral," "informal," "non-standard," "cultural," "natural," or "everyday," etc, mathematics can be found in Borba [1988a] and Borba [1988b].

5. I'm referring here to the papers written by Ascher [1986], Schliemann [1986], and Borba [1987b], respectively about the Quipu, about carpenters, and about a Shantytown.

6. Reality should be understood in this paper as the human dimension of the world; where the natural components of the environment and those elaborated by humans are present, then reality doesn't end in empirical data.

7. For instance, see Borba [1987a, 1987b], Frankenstein [1983], Gerdes [1985] and Skovsmose [1985].

References

Apple, M. (1979). *Ideology and curriculum.* Boston: Routledge and Kegan Paul.

Ascher, M. and R. Ascher. (1986). Ethnomathematics. *History of Science* 24.

Bicudo, M. A. V. (1979). Intersubjetividade e educaçao, *Revista Didatica* 15.

———. (org) (1987). *Educaçao matematica.* São Paulo: Editora Moraes.

Bishop, A. J. (1988). Mathematics education in its cultural context. *Educational Studies in Mathematics* 19.

Borba, M. C. (1987a). Um estudo de etnomatematematica: sua incorporaçao na elaboraçao de uma proposta pedagogica para o "Nucleo-Escola" da vila nogueira-sao quirino. Master's thesis. Rio Claro, São Paulo, Brasil: UNESP.

———. (1987b). Etnomatematica: a matematica da favela em uma proposta pedagogica. In P. Freire, A. Nogueira, and D. Mazza. *Na Escola que fazemos: uma reflexao interdisciplinar em educaçao popular.* Brasil: R. J. Petropolis, Editora Vozes.

———. (1988a). Etnomatematica: Uma discussao teorica sobre esta noçao. *Proceedings of the II National Brazilian Meeting of Mathematics Education.* Maringa, Parana, Brasil: Sociedade Brasileira de Educaçao Matematica.

———. (1988b). Etnomatematica: o homem tambem conhece o mundo de um ponto de vista matematico. BOLEMA—*Boletim de Educacao Matematica* 3, No. 5: pp19–34. Rio Claro, S. P., Brasil: Universidade Estadual Paulista (UNESP).

D'Ambrosio, U. (1984). Culture, cognition, and science learning. *Proceedings of Science Education in the Americas.* Panama: NSTA-OAS.

———. (1985). Ethnomathematics and its place in the history of pedagogy of mathematics. *For the Learning of Mathematics: an International Journal of Mathematics Education.*

———. (1986). *Da realidade a açao: Reflexoes sobre educaçao e matematica.* Campinas, S.P, Brasil: Summus and Campinas, Editora da UNICAMP.

Davis, P. J., and Hersh (1981). *The mathematical experience.* Boston: Birkhaauser.

Frankenstein, M. (1983). Critical mathematics education: An application of Paulo Freire's epistemology. *Journal of Education* 165, no. 4.

Freire, P. (1970). *Pedagogy of the oppressed.* New York: Seabury.

———. (1981). *Educação como pratica da liberdade.* Rio de Janeiro: Editora Paz e Terra.

Gerdes, P. (1985). Conditions and strategies for emancipatory mathematics education in underdeveloped countries. *For the Learning of Mathematics: an International Journal of Mathematics Education.*

———. (1981). "Sobre o despertar do pensamento geometrico." Maputo, Mozambique: Universidade Eduardo Mondlane.

Heidegger, M. (1981). *Todos nos . . . ninguem—um enfoque fenomenologico do social.* São Paulo, Brasil: Editora Moraes.

Joseph, G. G. (1987). Foundations of eurocentrism in mathematics. *Race and Class 28.*

Martins, J., and M. A. V. Bicudo. (1989). *A pesquisa qualitativa em psicologia: fundamentos e recursos basicos.* São Paulo, Brasil: Editora Moraes and EDUC.

Pavao, A. M. B. (1981). *O pricipio de autodeterminação no serviço social: Visão fenomenologica.* 2nd ediçao. São Paulo, Brasil: Cortez Editora.

Saviani, D. (1985). *Do senso comum a consciencia filosofica.* São Paulo, Brasil: Cortez Editora.

Schliemann, A. D. (1986). Escolarização formal versus experiência pratica na resolução de problemas: um estudo com marceneiros e aprendizes de marcenaria. *Psicologia: Teoria e Pesquisa* 3 (3): set/dez.

Skovsmose, O. (1985). Mathematical education versus critical education. *Educational Studies in Mathematics* 16.

Wagner, H. R. (1979). *Fenomenologia e relações sociais: Textos escolhidos de alfred schutz.* Rio de Janeiro, Brasil: Zahar Editores.

Wilder, R. (1981). *Mathematics as a cultural system.* New York: Pergamon Press.

Chapter 13

Mathematics, Culture, and Authority

*Munir Fasheh**

Editors's comment: Munir Fasheh, a mathematics educator, draws from his experiences in Palestine to emphasize the importance of using culture, considered broadly to include societal as well as personal experiences, to make mathematics learning more effective and meaningful. He also discusses resistance to this approach from educational authorities. This chapter first appeared in *For the Learning of Mathematics* 3(2):2–8, in 1982.

This article deals with the interaction between mathematics instruction on the one hand and established cultural patterns of belief, thinking and behavior on the other hand, especially in Third World countries. The article points to the importance of culture in influencing the way people see things and understand concepts, and to the importance of using cultural and societal sources and personal experiences in making the teaching of mathematics more effective and more meaningful, as well as, to the ways in which mathematics can be used to deal with some drawbacks in one's own culture and society. In addition, the article points out the conflict that usually arises between existing authorities and the teaching of mathematics when the latter is taught in such a way as to enhance critical thinking, self-expression, and cultural and social awareness. The region under consideration is the West Bank of the River Jordan (eastern Palestine) where I spent

* This is a revised version of a short communication that was presented at the Fourth International Congress on Mathematical Education in August 1980, at Berkeley, California.

my school years and over fifteen years as a mathematics teacher and educator.

Some Questions

Is it true that mathematics is a neutral subject independent of culture with its existing patterns of belief and behavior, and its intellectual structures?

Is the teaching of mathematics different from the teaching of history?

Should mathematics be taught in an abstract and detached manner, or in a way which is more subjective, personal, and full of meaning?

Is it possible to teach mathematics effectively, that is, to enhance a critical attitude of one's self, society, and culture; to be an instrument in changing attitudes, convictions, and perspectives; to improve the ability of students to interpret the events of their immediate community, and to serve its needs better—without being attacked by existing authorities whether they are educational, scientific, political, religious, or any other form?

Why is mathematics never, or at least rarely, taught to be useful in Third World countries?

Why are most students who major in mathematics in these countries usually "conservative" in their social outlook and their behavior and "timid" in their thinking and their analyses?

What should be the objectives of teaching mathematics in Third World countries and what can be done to achieve them?

In this paper, I will discuss answers to some of these questions in the context of what I experienced in one small region.

Background

This communication is an outcome of a personal experience in teaching mathematics for over fifteen years at different levels, but mainly out of the experience I had as the Head Supervisor of Mathematics Instruction in the West Bank for five years (between 1973 and 1978). My job was mainly to introduce and implement a new curriculum in mathematics in that region with its very unique circumstances. The region included over 800 schools and more than 1,600 mathematics teachers. The population consisted of about three-fourths of a million Palestinian Arabs. The region was under the British rule prior to 1948, under the Jordanian rule till 1967, and since 1967 under Israeli occupation. Politics and political problems are a way of life, and the social patterns of thinking and behaving, on the whole, are traditional and conservative. This made my work very difficult, but very interesting and not without problems and troubles. As an example, an educator usually has to deal with one authority; in the West Bank I had to deal with four "authorities": the Israeli military occupation; the Jordan government (through the syllabus, general exams, and the fact that the population carry Jordanian passports); the traditional and conservative institutions and segments of the population; and the national aspirations and needs of the people in the region.

The first part of the new syllabus to be implemented was a "blind" copy of the New Math materials in Western countries. It was written (through the initiative of UNESCO) for the tenth, eleventh, and twelfth grades in the Arab states, by people many of whom are foreign to the area and know nothing of its culture. I was in charge of implementing the new syllabus in the schools of the West Bank. Six regional mathematics supervisors and many interested and enthusiastic teachers actively assisted in implementing the new syllabus. The first course of training teachers (which included over 200 teachers) was initiated and run by Birzeit University in the summer of 1972. It was followed later by similar courses for teachers in lower grades through the Technical Educational Office in Ramallah.

Objectives of Teaching Mathematics in Third World Countries

Mathematics in Third World countries (at least in mine) is usually taught as a set of rules and formulas that students have to memorize, and a set of problems—usually nonsensical to students—that they must solve.[1] The only reason for studying mathematics for most students is to pass the examinations. Though the cited objectives of teaching mathematics usually assert knowing certain mathematical facts and being able "to think correctly, logically, and scientifically," among other objectives, I came to believe that the main objective of teaching mathematics (or any other subject) in developing countries, is to doubt, to inquire, to discover, to see alternatives, and, most important of all, to construct new perspectives and convictions. One of the main objectives of teaching mathematics should be to realize that there are different viewpoints and to respect the right of every individual to choose his/her own viewpoint. In other words, mathematics should be used to teach tolerance in an age which is full of intolerance. The objective of teaching mathematics should be to discover new "facts" about one's self, society, and culture, to be able to make better judgments and decisions; and to build the links again between mathematical concepts and concrete situations and personal experiences. All these, in my opinion, are necessary for a balanced development of any country or society.

Mathematics Teaching in Action

In this section, I will mention some examples from my experience that bear out the nature of the objectives cited in the previous section.

1. A first grade teacher put a chart with days, and squares next to each day, at the corner of the classroom. Every day he would cross a number of squares, next to that day, equal to the number of students absent on that day. After about a month, the six to seven-year-old students noticed that the biggest number of absentees occur on Saturdays. The teacher asked about the "reason." An interesting and heated discussion followed. One student said, "because it comes after Friday," (Friday is the official weekly holiday). Another student said: "The kids like to spend Saturday at home because their fathers are at home." (Some men from that village work in Israel and so Saturday is their day off.) A third student gave as a reason, "the poor transportation on Saturdays due to the fact that some workers don't go to work on that day."

Now, why is such an example worth more than a whole book of routine and dull exercises?

First, such an example deals with a problem that is familiar and interesting to the students simply because they are living it. Second, the problem is new to all of them, including the teacher. Such an experience makes the students feel that they are "equal to the teacher"; they are both dealing with the unknown, so to speak. Third, it breaks a certain cultural belief which exists very strongly in our society, the belief is that there is only one "correct" answer to every question or problem, and that answer is given by "authority" (in this case the teacher). All that the students have to do is to memorize these answers at least until the day of the examination. According to this belief or pattern, there is no dialogue and there are no different alternatives or viewpoints. Through an example like the 'biggest number of absentees" problem, the students actually share in the education process: they give and not merely receive ideas and opinions.

Fourth, the students, through such an example, realize at this early age the importance and usefulness of collecting data and putting it in an orderly form. This is especially important in a community that doesn't believe in experimentation and collecting information as a means to knowledge. They learn to be patient if they want to get results and arrive at conclusions. Such an experience convinces them that mathematics can be used to discover "facts" about the community and to make judgments.

Fifth, the children realize early in their lives that there is a big difference between a "fact" and its interpretations. "The biggest number of absentees occurs on Saturdays" is a "fact." But that fact has many interpretations and explanations, as the answers of the children themselves showed.

Sixth, such an experience helps change the attitudes of students towards knowledge and learning in general and towards mathematics in particular. It helps create a healthy relationship between students and teachers which is not authoritarian or parochial but rather dynamic, interactive, in addition to instilling confidence and self-respect in the students.

2. Two examples follow from mathematics for illiterate adults.[2]

(*a*) A class, which consisted of women only, was asked (early in the course) to keep a daily record of the time each woman spent in cooking, cleaning, washing, taking care of the children, and so forth, for at least a month. What would the women learn from such an experience?

First, they learn to put the information in a tabular form. They use addition and multiplication to get to certain conclusions. In short, they learn and use some topics in mathematics in a practical setting.

Second, the women have a much clearer picture of themselves, their lives, their roles, and the meaning of being a housewife. If a husband shouts at his wife at the end of the day: "What have you been doing all day?" she can show him details of the many jobs that a woman usually has at home every day, seven days a week, all year around, all her life, and in almost all cultures.

(*b*) In the summer of 1979, I taught mathematics to a class of illiterate workers in Birzeit University. They had had some classes of Arabic but none in mathematics. In the first period of teaching that class, I started by asking some general questions. One question was, "Suppose a friend wants to come and visit you here but he doesn't know the place. All he knows is where the post office is. Draw a map for him showing the way from the post office to this building." What took place in that period still thrills me. As a response to the question mentioned above, each worker drew a map from the post office to the classroom. The maps were all different from each other (four of the maps are shown in figure 13–1).

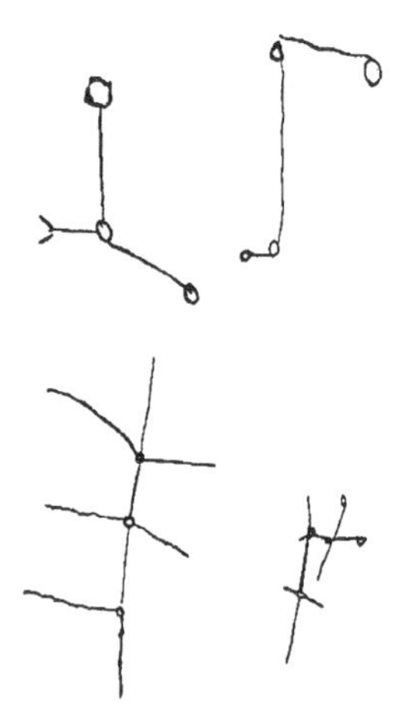

Figure 13–1.

One of the workers commented, "How come they are all different although we have all drawn the same road!" Another worker replied, "But each of us understands his own map." A third worker said to the one who drew the lower right map, "Why did you draw the road to be straight although it is not straight?" To which the other replied, "When I walk from the post office to where we are, I walk straight here!" The discussion that took place in that first period pointed to things that I never thought of or realized when I first asked that question. We talked about the meaning and importance of conventions in understanding and communicating with each other. We talked about the different meanings and uses of the word "straight" in Arabic (which is also true about the different uses of this word in English, and I suppose in many other languages too). We talked about the importance of being able to use a map, and so on. Later that day I felt that that class was probably the richest class I gave in my life; and supposedly to illiterates! In the process of trying to teach, I was busy learning and reacting. I was completely involved. I was convinced more than ever that students (whether they are children or adults) are not empty shells to be filled with our wisdom and knowledge; rather, they are full of experiences and ideas, and they have their own personal ways of looking at things. As teachers, we should start with these experiences and personal viewpoints in our teaching. The first reaction I had when I saw the drawings was to say, "They are wrong." I am glad I didn't.

3. Teaching sets to a seventh grade class for the first time.

I brought cards that had five holes punched in a row along the top. Each hole corresponded to one of five questions that I wrote on the board. The answer to each question was either "yes" or "no." If the answer was "yes," each girl was instructed to keep the hole on her card corresponding to that question unchanged. If the answer was "no," she had to make a notch above the corresponding hole on her card. As an example, one of the questions was, "Are you a member of the public library?"

The thirty-two girls in that class, and on that first day, learned a lot about each other. They learned for example, that only six girls were members of the public library in town. An interesting discussion about the reason for the small number of members followed. In addition to the information they got about their small "community," they also learned about the uses of such "computer" cards in companies, firms, and so forth. (Remember this took place in the 1970's.) Also, the girls were introduced to some mathematical concepts about sets. For example, through the question, "How many girls who reside outside town are members of the library?" intersection of sets was introduced. And through the question, "How many of the new girls in the class are members of the library?" the empty set was discussed. And so on.

4. Deductive thinking,[3] historically, has helped to support the belief in "absolute truths" and the existence of one "correct" answer to each problem. The axioms of geometry, for example, were considered for a long time to be true innately, naturally, and a priori. They were considered to be familiar to every thinking creature and true in all possible worlds, thus making the possibility of seeing alternatives very difficult, if not impossible. Although mathematics and most mathematicians have moved away from such an "arrogant and naive" viewpoint, still mathematics is being taught as if its statements

are absolute and eternal. Unfortunately, this same attitude exists also in domains other than math, such as the social, religious, and political domains. Thus, the teaching of deductive thinking traditionally supported the "dogmatic" dimension in education.

However, through my experience with schools and in Birzeit University, I came to believe that deductive thinking can be used very effectively to create new attitudes and awarenesses towards knowledge in general and towards math in particular—awarenesses and attitudes that are very much needed in our society and culture and I imagine in other societies in Third World countries too.[4] First, students learn that math is man made. They learn that axioms are not God-given or Nature-given, but rather they are statements that evolve with time and through a long and hard process. They learn not only that basic statements evolve, but also how the meanings of words and concepts (such as axiom) evolve. Second, students learn to see similarities among things that do not seem similar at first sight. Realizing that the two diagrams in figure 13–2, for example, could be considered as models of one and the same abstract system (by interchanging the meanings of "line" and "point" in the axioms of the system) was always shocking and interesting to the students and a source of a serious, deep, and involved discussion that usually lasted several periods. In my experience, this kind of interaction broke many prejudices and rigid ways of thinking. Third, students learn an intellectual model or structure—the axiomatic model—which is missing in our culture. Fourth, it helps students see alternatives and the meaning of "relatively true." They learn that axioms can be completely or partially changed to produce new systems and models. Fifth, it helps students relate a certain event or phenomenon from the real world to many possible abstract models; and vice versa, one abstract system can have twenty "concrete" models or applications in the real world. Sixth (which I believe to be the most important point), it increases the awareness, on the part of the teachers and students, that most, if not all people are logical. The difference among different people lies either in their basic assumptions or in the "logic" they use or in both. Accusing a student of being illogical leads to feelings of worthlessness in the student and makes the teacher miss the opportunity of understanding that student and of expanding his/her (i.e. the teacher's)

Figure 13–2.

"treasure" of mental images and structures. With that in mind, the question becomes, not whether a certain person is logical or not, but rather what are the assumptions and type of logic that that person is using. The question also becomes, not only what type that person is using, but also how that person came to accept or adopt that logic (e.g. by force, by custom, by critical reflection, and so on). That does not mean that all logics and all assumptions are equally effective in understanding and dealing with a specific problem in a certain situation. All it means is that different assumptions and different logics are needed at different times and in different situations; there is no perfect logic that is good for all times and for all situations. Thus, in our teaching, we should try to arm the students with different types of logic and with the confidence to choose which they feel to be appropriate.

Culture, the Individual, and the Teaching of Mathematics

A common misconception in the teaching of math has been, and still is, the belief that math can be taught effectively and meaningfully without relating it to culture or to the individual student. This, and not the difficulty of the subject, in my opinion, was, and still is, the main reason why math is considered meaningless, incomprehensible, and not a popular subject by the vast majority of students. In this section, I will discuss further the interaction between culture, the learner, and the teaching of math.

It has been a general belief that the teaching of math is different from the teaching of history, or sociology, or political science. Such a belief asserts that in the latter subjects there are different viewpoints, while in math "facts" are true irrespective of culture, of the individual, or of time. I came to believe that this is a misleading belief that negatively affects our teaching of math. "The First World War took place in the period between 1914 and 1918" is a historical fact, but its description and interpretation differ from one person to another and from one nation to another. Similarly, I believe, "one equals one" is a mathematical fact, but its description and interpretation and application differ from one situation to another and from one culture to another. A fresh and delicious apple is not equal to a rotten apple. A certain chair is not equal to another chair in all its details no matter how identical they seem to be. (I asked this question about two similar chairs to my son who was ten years old, "In what way are they equal?" After some discussion he said, "They are equal in name," that is, in being called "chairs.") No person is equal to himself the next day. One dollar in 1970 is not equal to one dollar in 1980. And so on. Strictly speaking, then, "one equals one" does not have true instances or applications in the real world.

The truth of the matter is that in schools and in all our teaching we keep the world of reality separate from the world of abstraction (with the exception of some trivial and irrelevant examples that are scattered in textbooks under the misleading title of "applications"). In the world of abstraction, we usually agree about "facts"; but in the real world, we face many interpretations and meanings and ways of looking at these facts; so we argue and we fight. People, for example, agree that one equals one is true in abstraction, but antagonistic feelings and different opinions emerge when we say for example that "women are equal to men" or when we say that "one vote for Jordan (with two million people) in the UN is equal to one vote for the U.S. (with 200 million people) in the UN." Teaching with meaning and by relating the abstract world to the real world makes math more relevant and more useful. In addition, it helps students appreciate remarks such as Einstein's often cited remark, "As far as the laws of math refer to reality, they are not certain; and as far as they are certain, they do not refer to reality."

Culture influences the way people see things and understand concepts. In Arabic, for example, there are more than one hundred names for "camel" (each name describes the camel in a different position or mood). In English there is but one word for camel. On the other hand, there are hundreds of words in English for flowers (each word describes a certain kind of flower) while there are only two or three words in Arabic for flowers. Similarly, there are many words for ice in the Eskimo languages (each word describes ice in a different form, use or setting), one or two words for it in English or Arabic, and no word for it in some tropical languages. The Arab's concept of camel is much richer than the others; the Briton's concept of flower is much richer than the others, and the Eskimo's concept of ice is much richer than the others.

I read once about a place where people do not differentiate between yellow and green. They have only one word to describe "both" colors. I was amazed, "Can't they see the difference?" Then one day I was describing the color of a car as green to a Frenchman. He said, "But it is turquoise!" For him there were two colors with two different names. For me there was one color with one name.

Another example. Once, a number of us at Birzeit University were trying to find a word equivalent to the word "privacy." Five people agreed on one term, the other three held the view that, strictly speaking, "privacy" has no equivalent. We debated the issue for a long time. Later we noticed that the three who had difficulty in finding an equivalent for "privacy" in Arabic had all lived for some time in the U.S. and thus had an "experiential" meaning of "privacy" as it is understood in the American context, while the other five who found the term satisfactory got it from a dictionary or learned it in school (and have never lived in the West). The two words evoked the same meanings and images and experiences in the minds of those who never lived in the West, and different meanings in the minds of the three people who lived in both cultures. "Privacy" in the way it is practiced in America, is never experienced in an Arab society, which is essentially a "communal" type of living.

One more example. There is one word (*raquam*) in Arabic for the two words in English: "numeral" and "digit." This created a confusion between the two concepts in the minds of most teachers and students whom I worked with in the West Bank, reflected in the statements that they expressed in relation to these two concepts.

The Arabic language, on the other hand, can be used very effectively to help the students to think critically and within context. To borrow a remark made by an Arab thinker, "Usually in other languages you read in order to understand; in Arabic you have to understand what you are reading in order to read it correctly." Many words in Arabic can be read or pronounced in as many as eight or ten different ways depending on the context. Thus, one has to understand the meaning of the word in order to read it correctly. It is unfortunate, however, that Arabic is taught in schools as a set of ready-made statements that are repeated for hundreds or thousands of times. Under such conditions, the ears of the students—and at best their tongues—work, but their minds are kept "safe" from thinking.

If culture determines the way we see a camel, and the number of colors that exist, and how accurate our perception of a certain concept is, may it not also determine the way we think, the way we prove things, the meaning of contradiction, and the logic we use?

Just because the same word or symbol is used for camel by different individuals or by different nations, we cannot conclude that they have exactly the same concept. And just because we use the same symbol for the number "one" in the same classroom or in different classrooms, that does not mean that the same images pop up in the minds of the children. We unify the symbol but mistakenly conclude that the meaning and the images are unified. This fact is often ignored by the teachers and educators of math. When a word such as "area" or "proof" or "axiom" is used in a math class, teachers do not pose questions to know what are the meanings and the images that are created in the minds of their students. Math teachers are usually satisfied if students use such words "correctly" in a purely mechanical way.

The world is heading to a peak of cultural changes and cultural awarenesses. Math can be used to stress one's own culture with its special and beautiful characteristics. At the same time, math can be used to make one aware of the drawbacks in one's own culture and try to overcome them. In other words, math could and should be used to point out the strengths and weaknesses of one's own culture. (I have always read, for example, that the Arabs and Moslems contributed a lot to math and the sciences, one of their contributions being the solution of the general cubic equation. However, the curriculum never showed me how they did it and Western historians in general have denied this contribution by the Moslems.[5]) Teaching math in a way detached from cultural aspects, and in a purely abstract, symbolic and meaningless way, is not only useless, but also very harmful to the student, to society, to math itself, and to future generations.

It should not be understood from the above that math should or could be taught within one culture separate from other cultures. Advances in thought in one culture should be understood and welcomed by other cultures. But these advances should be "translated" to fit the "borrowing" culture.[6] In other words, to import ideas is acceptable and should be encouraged, but the meanings and implications of these ideas should be "locally made." That is what we do, for example, with a refrigerator when we import one from France to our country: we fill it with Arabic food rather than with French food.

Not only local and cultural meanings should be encouraged, but also personal feelings and interpretations, which are just as important, especially with little children. We should encourage in the children "subjective" ways of looking at mathematical expressions and concepts as much as objective ways of understanding them. We should not stress one and forget the other. One of the most beautiful and revealing definitions of a point I ever heard came from a six-year old girl. When asked how she saw a point, she said that it is a circle without a hole. That definition involves the concept of limit in math. A teacher who lacks the imagination of that child may be unable to understand what she was talking about.

We have to encourage children to "see what they mean" and not only "what we mean." Numbers and symbols and words we use with children are never meaningless to them, and these symbols don't mean only what we mean by them when we mention them or use them. Children have their own personal likes and dislikes about symbols. These likes and dislikes become, in some cases, strong emotions and convictions. These things are usually ignored by math teachers and educators. We have to ask questions, to little children in arithmetic classes, such as, "Which do you like better, five or two, and why?" And not only questions like, "Which is greater, five or two, and why?"

In Third World countries we should be careful not to follow the Western way of interpreting objective knowledge as being purely abstract, absolute and detached. In teaching a mathematical concept or "fact," we should ask for examples where that concept or fact is applicable or true and where it is not; we should ask about some of the uses, misuses, and abuses of that concept or fact. We should ask for personal and cultural meanings of that concept or fact rather than just ask the students to memorize it or to solve routine problems concerning it. Math can be used to help students relate, organize, see alternatives, and make better decisions. We as teachers and educators of math should find ways to accomplish that.

Authority and the Teaching of Mathematics: Conflict and Trouble

One very important aspect of any culture is what constitutes authority in that culture and how that authority reacts to and deals with people when they think in a critical way or in a way that deviates from the "correct" path. My own experience, and the experience of many others that I knew or read about, made me increasingly believe in the following conviction: in spite of the claim by educational institutions and by authorities that control these institutions that they encourage free and critical thinking, such educational institutions, in general, discourage critical, original, and free thinking and expression, especially when that touches upon "important" issues in the society. Students who ask relevant questions about important events in the immediate community and see new alternatives and seek new interpretations of what exists are usually considered to be very "dangerous." Teaching people to question, to doubt, to argue, to experiment, and to be critical, and teaching that increases the awareness of students constitute, in my opinion, a real threat to existing and established institutions, beliefs, and authorities everywhere and of every kind.

People who engage in teaching in this way are the subject of all kinds of familiar accusations, from disturbing law and order, to teaching corruptive and immoral ideas, to being a threat to national security; and eventually they are forced to stop teaching, to say the least. Socrates was accused of this in "democratic" Athens in ancient times, and Oppenheimer and Eldridge Cleaver were similarly accused in "democratic" U.S. in modern times. This is also true for socialist and Third World countries. The rule underlying such a response seems to be that if a questioning attitude is fostered in teaching, this may lead to a questioning of other things in society, including authority and the underlying assumptions and structures of that society.

Anyone who has never experienced this and who doubts it should have a second look at his/her own teaching and try to find out how much it is related to important issues. I mean by "important" those which are related to matters that control the economy, technology, politics, religion, ethics, the flow of information, and the suppression of certain groups—whatever is considered to be of prime importance in that society. I have encountered this type of experience many times. One such situation arose in forming math and science clubs in high schools in the West Bank. Students were free to choose any experiment to perform or any topic to gather information about and to discuss. The clubs lasted beautifully and successfully in many schools for almost two years. They were much more successful, however, in girls schools than in boys schools. Female students were, in general, more receptive to new ideas and more inquisitive, sincere, independent, persistent, interested, and original than male students. Most administrators and teachers bet that the clubs would die or cease to exist in few months. They were right as far as the boys schools were concerned: within six months all the clubs in boys schools ceased to exist; but in some girls schools clubs continued to be active for about two years. They only ceased to exist because of the constant attacks, harassments, and hostile attitudes that began to mount from two directions. Both the Israeli authorities and fanatic individuals among the local Arab population fought the existence of these clubs—each for its own reasons and in its own ways.[7]

In short, I came to believe that the teaching of math, like the teaching of any other subject in schools, is a "political" activity. It either helps to create attitudes and intellectual models that will in their turn help students grow, develop, be critical, more aware, and more involved, and thus more confident and able to go beyond the existing structures; or it produces students who are passive, rigid, timid, and alienated. There seems to be no neutral point in between.

Schools in Third World countries (at least in mine) in their present structure and form help produce students of the second type mentioned above. The classroom is highly organized; the syllabus is rigid; and the textbooks are fixed. Math is considered as a science that does not make mistakes; and its truth is considered timeless and absolute. There is one correct answer to every question and one meaning for every word and that meaning is fixed for all people and for all times. "Wrong" answers are not tolerated; students are usually punished severely (one way or another) if they make "mistakes." Teachers, in their turn, are also expected to perform according to a certain set of rigid expectations and they are punished if they don't. One second grade math teacher, for example, in order to evade punishment, explained the inability of some of his students to answer the inspector's questions by labeling them in front of the whole class as mentally retarded. When the inspector objected, the seven-year-old children volunteered to accuse each other of being retarded when any one of them made a mistake.

Organizing a number of ideas and statements in a coherent way so as to be able to see new order and new relations among these ideas and statements is strongly discouraged in schools. Causal statements and statements that express relations are usually to be memorized, but never to be discussed or questioned. Some examples are, "If we don't get enough rain, it is because some girls wear sleeveless dresses."; "Losing the war (in 1967) is a lesson for us from God." (That was in fact what Nasser of Egypt said after the war.) Students give correct definitions of the square, rectangle, and rhombus, but the relationship among them is hardly discussed or asked for. When my son was eight years old, I asked him one day what he had learned in school that day. One of the things he mentioned was, "If you want to avoid getting sick, then you have to wash your hands before you eat." After a short time, he started eating without washing his hands. When I reminded him of what he had just said, he told me he had merely to memorize that fact for the test.

Under all these conditions, it is not hard to see why most teachers and students who are attracted to math and the sciences, at least in countries that I'm familiar with, are "conservative" in their outlook, traditional in their behavior, and timid in their thinking. The same conditions also explain why math is taught, almost always, in a detached and irrelevant way.

Summary and Suggestions

1. The math teacher, the learner, their experiences, and their culture are extremely important factors in the teaching of mathematics and in making it more meaningful and more relevant. Teaching math without a cultural context, by claiming that it is absolute, abstract, and universal is the main reason, I believe, for the alienation and failure of the vast majority of students in the subject.

   In addition, teaching math through cultural relevance and personal experiences helps the learners know more about reality, culture, society, and themselves. That will, in turn, help them become more aware, more critical, more appreciative, and more self-confident. It will help them build new perspectives and syntheses, and seek new alternatives, and, hopefully, will help them transform some existing structures and relations.

2. Teacher training courses and programs form a good starting point to move in the direction mentioned above. Learning new topics in math or new methods of teaching them is not enough to acquire insight and relevance—the two most precious qualities in the teaching profession. Teacher training courses and programs, I believe, should include also courses on culture, society, the relationship between language and thought, and the history of evolution of mathematical concepts among other things. No change in math curriculum is effective unless the teachers understand the change in all its dimensions.

   We have to distinguish between superficial or formal success of a new plan or program or curriculum and a real success; between superficial change and real change. The new syllabus in math that was adopted by the Arab countries in the 1970s was "successful." You can see the change: new books, new topics, new symbols, new terms, and lots of training courses. But that is superficial change in my opinion. The real change which means change in attitudes, values, assumptions, relations and structures were completely missing.

3. Changing attitudes, values and basic assumptions and relations, however, are very costly to the teacher. There is a price for teaching math in a way that relates it to other aspects in society and culture which may result in raising the "critical consciousness" of the learner. And the price that the teacher usually pays varies directly

with the power of authority (regardless of whether that authority asserts its power in a subtle or direct way) and with the effectiveness of the teacher. The fear of paying the price is one main factor, in my opinion, that diverted education from its "natural" course and forced it to take detached and meaningless forms.

4. To achieve some of the objectives cited above, I would like to suggest the following as one possibility. The suggestion is directed to any world organization with an interest in developing new programs in math such as UNESCO or the International Congresses on Mathematical Education. Twenty or thirty educators from different cultures who are convinced of the importance of relating the teaching of mathematics with cultural aspects can start working on developing a syllabus based on this relationship. The strength, weaknesses, misuses, abuses (and not only the uses) of mathematics in several cultures should be included in the syllabus. Different interpretations, perspectives, and examples of certain concepts in different cultures should also be discussed. Students who go through such a syllabus will, I believe, be able to understand themselves, their beliefs, and their culture better. They will also be able to understand other people and other cultures better. In addition, such a syllabus, I believe, will help "humanize" math by helping to bridge the gap between science and technology and other social and cultural aspects in society. Most important, it will help, I hope, in fighting three of the biggest evils in our time: absolutism, intolerance, and ignorance.

Notes

1. This does not mean that mathematics in technologically advanced countries is necessarily taught in a much better way. Many math books that I have seen used in elementary classes in the U.S., for example, require the children to "fill in the blanks" without any trace of real learning taking place.

2. Since 1976, 1 have also been involved in the teaching of mathematics to illiterate adults.

3. For one reason or another, the chapter (in the new syllabus) on axiomatic deductive thinking was omitted in many Arab countries.

4. As a result of these and other experiences, I devised a course that deals with some of the deficiencies in our schools and in our culture. The

course attempts to relate axiomatic, deductive, and inductive modes of thinking to experiences that the students encounter in their daily life or in their intellectual questioning. The course was first taught to freshmen science students in Birzeit University in 1978 A textbook for the course (in Arabic) was published by Birzeit University in 1981.

5. I would like here to thank David Henderson of Cornell University who was a visiting professor at Birzeit University during the second semester of 1980/81 who first pointed out to me that it was Omar Khayyam and not Italian mathematicians who first found a general solution for the Cubic Equation. In fact, Dr. Henderson wrote a detailed exposition of Khayyam's geometric solution of the General Cubic Equation which was included in the course mentioned in note 4.

6. There are, however, some cases in history (specially in relation to some conventions) where blind imitation proved to be better. Arabic numerals formed one such case. Originated in the East, they were written from right to left. (In expressing a natural number, we start with the units position at the extreme right and move leftwards.) And that is the way it is in almost all cultures that exist now. That makes it easier for different people to communicate "numerically." Can you imagine what would have happened if the Europeans decided to express natural numbers in a way consistent with their own cultures by starting with the units position on the extreme left? On the other hand, the number line, being originated in the West, "grows" from left to right, and it is that way in all societies, which makes it much easier for people from different cultures to communicate "graphically."

7. I would like to mention one reason that I believe to be important in making the clubs succeed in girls schools but not in boys schools. The girls, in general, are outside the mainstream of society (which is true for most societies); so they found more meaning and relationship in "unorthodox" ways in education like the clubs' activities than did the boys.

Chapter 14

Worldmath Curriculum: Fighting Eurocentrism in Mathematics

S. E. Anderson

Editors's comment: S. E. Anderson, a mathematics educator, shows how including historical information about the mathematical contributions of people from Africa, Asia, and Latin America leads students of color and working-class students in the United States to develop self-confidence and replace previous feelings of alienation toward mathematics with an attitude that mathematics is intellectually stimulating. This chapter first appeared in *The Journal for Negro Education* 59(3): 348–359, in 1990.

A Few Grim Statistics

Before we go into the nuts and bolts (or chips and rams) of how mathematics can be taught without the racist and sexist bias of Eurocentrism, we must be clear about the sociopolitical demographic terrain in which this curricula struggle takes place. According to the Commission on Professionals in Science and Technology's report, Scientific Manpower, 1987 and Beyond: Today's Budget—Tomorrow's Work Force (pp. 10–17):

- by 2010—just twenty years from now—at least one-third of the eighteen-year-olds in the U.S. will be Black or Latino;
- by the year 2000—in ten years—85 percent of the new members of the U.S. work force will be Blacks, Latinos, and White

women—only 15 percent of the new workers will be White males;

- by 2000, 75 percent of all goods manufactured in the U.S. will be produced by automation; the fewer jobs (in absolute numbers) that will be available will either be no-skills, McDonald's-type jobs or hi-tech skills jobs;
- 619 U.S. citizens received the Ph.D. in mathematics in 1978; however, by 1988 that number was down to 341—half those doctorates went to foreign citizens while only three went to Latinos, two to Native Americans, and one to an African American;
- in the past eleven years, only 78 African Americans received doctorates in physics and/or astronomy, along with 97 Latinos and 11 Native Americans;
- the median mathematics level of Black, Latino, and Native American high school graduates is less than sixth grade; that is, the majority of these graduates have no knowledge or only minimal knowledge of elementary algebra.

The Eurocentric Basis of Mathematics

We can go on and on with grim statistics, but those given above are sufficient to convey the overall crisis of capitalist education. That crisis is currently on the verge of imploding precisely because of its elitist, racist, and sexist nature. As the American Association for the Advancement of Science's (AAAS) (1989) report on literacy goals for science, mathematics, and technology states:

> When demographic realities, national needs, and democratic values are taken into account, it becomes clear that the nation can no longer ignore the science education of any students. Race, language, sex, or economic circumstances must no longer be permitted to be factors in determining who does and who does not receive a good education in science, mathematics, and technology. To neglect the science education of any (as has happened too often to [white] girls and minority students) is to deprive them of a basic education, handicap them for life, and deprive the nation of talented workers and informed citizens—a loss the nation can ill afford. (pp. 156–157)

However, as Alkalimat (1990) makes clear:

> . . . strategic interests in modern society [have] structured the university as a tool of societal power to train skilled mental workers for the

> ever changing technical division of labor in advanced capitalism. Further, the university is used for the purpose of indoctrinating an elite with the metaphysical myth of eternal Eurocentric domination of the world. The first point is for class rule, the latter for race rule. (p. 2)

The task of progressive educators is not to try to fix or prop up the capitalist system of education. Our duty is to lay the seeds for a more egalitarian educational system based on the assumption that any person can learn anything and that we study Nature to love and appreciate its complexities, beauty, and offerings, not to control and defile it. The true beauty of mathematics comes from our ability over time to weave subjective and objective factors into a variety of logical systems that help explain the complexities of Nature. It is from this reference point that we must examine an alternative mathematics curriculum and the specific case that will be presented herein.

According to Joseph (1987, p. 22–26), the present structure of mathematics education and research—on all levels—is based on four historiographic pillars which have been placed in the sand foundation of Eurocentrism:

1. the general disinclination to locate mathematics in a materialistic base and thus to link its development with economic, political, and cultural changes;
2. the confinement of mathematical pursuits to an elite few who are believed to possess the requisite qualities or gifts denied the vast majority of humanity;
3. the widespread acceptance of the view that mathematical discovery can only follow from a rigorous application of a form of deductive axiomatic logic believed to be a unique product of Greek mathematics; hence, intuitive or empirical methods are dismissed as having little mathematical relevance;
4. the belief that the presentation of mathematical results must conform to the formal and didactic style devised by the Greeks over 2,000 years ago and that, as a corollary, the validation of new additions to mathematical knowledge can only be undertaken by a small, self-selecting côterie whose control over the acquisition and dissemination of such knowledge has a highly Eurocentric character.

Institutionalized Eurocentric curricula constantly reinforce the racial and sexual inferiority complexes among people of color and women. The dominant curriculum in use today throughout the United States is explicit in asserting that mathematics originated among men in Greece and was further developed by European men and their North American descendants. According to Ball (1922), the history of mathematics "cannot with certainty be traced back to any school or period before that of the Ionian Greeks" (p. 1). Klein, writing in 1953, claimed the following:

> Mathematics finally secured a firm grip on life in the highly cogenial soil of Greece and waxed strongly for a short period . . . with the decline of Greek civilization, the plant remained dormant for a thousand years . . . when the plant was transported to Europe proper and once more imbedded in fertile soil (p. 10).

From generation to generation for centuries this type of Eurocentric "scholarship" has been reproduced in the objective and subjective pursuit of justifying racism and imperial rule. When there is mention in mathematics textbooks of African, Indian, Chinese, or Mayan contributions, that mention is relegated to passing sentences, paragraphs, or, on rare occasions, a non sequitur chapter.

Systematically, the vast majority of mathematics curricula in colleges and high schools are "turning off" generations of young people from pursuing careers in mathematics and science in general and, in the process, reinforcing racist assumptions about people of color (Steen, 1987, p. 251). Those of us who want to see true mathematical and scientific knowledge flourish among our youth and workers must break from the Eurocentric perspective and begin the reconstruction of the rich and complex fabric of world mathematical and scientific knowledge. This requires breaking with most of the pedagogical and curricular traditions that exist today. There are risks, however. If one is white and breaks with the traditions, one risks being viewed by one's more conservative racist peers as a "crackpot" liberal on a suicide mission. If one is not White, one risks being seen as some sort of atavistic militant bent on painting the ivory tower black and destroying the icons of "rigor" and "proof" and replacing them with "simplicity" and "fun."

Six Pedagogical Disasters in Mathematics Education

Before looking at my own practical experiences in teaching from a radical pedagogy, I will outline six major pedagogical disasters in mathematics teaching. Popular belief and practice dictates that before one can learn mathematics one has to:

1. Separate arithmetic from algebra;
2. Teach mathematics without any historical references;
3. Use textbooks that are elitist and cryptic;
4. Do work and be tested as an individual as opposed to working and being tested as study groups;
5. Accept the myth that mathematics is pure abstraction and, therefore, antithetical to one's cultural and working environment; and
6. Memorize, memorize, memorize.

Separating Arithmetic from Algebra

Because we teachers pursue pedagogy undialectically, we tend to compartmentalize knowledge rather than show the interrelatedness of it. As a result, we spend years pounding arithmetic skills into very young folk, chastising those who attempt—before we are ready to "teach" it to them—algebraic processes such as taking a bigger number from a smaller one. Additionally, most elementary and secondary school teachers just do not understand mathematics. They are akin to the one-eyed man (or woman) leading the blind as opposed to keen-sighted teachers intent on developing mathematical vision and perspective in young people. When algebra is imposed upon our youth during that dramatic pubescent period when so many other new and challenging life-altering developments are also occurring, it is retrofitted onto worn-out arithmetical computational drills. Thus, algebra becomes a traumatic tearing apart—a world construct grafted onto six or seven previous years of proselytizing a limited arithmetic world disinfected from everyday experiences and the historical development of people.

By junior high school the overwhelming majority of our youth are convinced that mathematics teachers are their enemies and, even worse, that mathematics is some sort of poison or mind controling drug that teachers try to force upon them. Those of us who teach mathematics at the junior high level reinforce ourselves by saying: "Oh well, mathematics is for the chosen few—the gifted—anyway." By the time most typical college freshmen enter a college mathematics class, they are filled with fear, apprehension, and misinformation about mathematics. In an accelerated scramble, they run away from any mathematics or mathematics-related subjects for the rest of their college experience. Usually contributing to this scramble is the "erudite" male professor who stands arrogantly before his classes, teaching mathematics as if the students were merely reviewing the material—or as if there were no students before him at all!

Teaching Mathematics as an Ahistorical Development

Mathematics and the natural sciences are the only areas of study presented with little or no historical, cultural, or political references. This ahistorical approach is essential for what Alkalimat (1990) identifies as a process of "indoctrinating an elite with the metaphysical myth of eternal Eurocentric domination of the world" (p. 2). This pedagogical approach reinforces the institutionalization of Eurocentrism, class elitism, and sexism. European names such as Pythagorus, Euclid, Cauchy-Rieman, Fourier, and Newton are tossed about sans flesh, bones, and personalities; then they are attached to various levels of abstractions as if they always existed.

Elitist And Cryptic Textbooks

We have all had at least one experience with a mathematics text so botched, convoluted, and elitist in its explanation of concepts covered in class that we dreaded opening it. Cryptic mathematics texts are the norm at all levels of American education. Capitalist/racist education not only mass-produces these intimidating, incomprehensible mathematics texts, but it also demands usage of them. Such texts ensure the class rule that Alkalimat describes. In other words, every aspect of the educational process and every facet of industrial and technological activity must convince the majority of people—especially people of color—that they are biologically, intellectually, and/or psychologically incapable of understanding mathematics.

The Individual Uber Alles

The typical secondary school or college mathematics class is structured such that individuals must compete ruthlessly with each other for "knowledge." At the beginning of most courses, the teacher, who assumes that most students will fail this most rigorous and demanding of subjects, usually lays out the failure parameters (although students usually already know those parameters from their own or other students' previous disastrous encounters with the subject and/or teacher). In such settings, cooperation among a few students evolves not as a result of its organic development in the course, the materials, or the teacher but rather out of desperation. Individualism is reinforced within the mathematics class, further strengthening and legitimizing class, race, and sexual divisions within it and within society at large. Although the student-as-loner syndrome contradicts the way most people historically and presently have solved problems, it reaffirms the foundations of capitalism's totalitarian rule.

"It's Not Real Math Unless it's Abstract"

Systematically, from kindergarten to college, from private schools to public schools, and from coast to coast, mathematics is presented as an abstraction with little or no relationship to the real world; as divorced from subjectivity; as pure, blindingly white, and, therefore, beautiful in its sinosoidal-like monotony. Only a chosen few are given the "keys" to enter this secret society of purity and beauty. Those few are selected on the basis of their class/race position and by sheer probability (because all those privileged to access mathematics will not pursue it).

Erudition, abstraction, and compartmentalization—the cornerstones of Eurocentric mathematics and science—are the products of capitalism's need to further subdivide the world, its people, and their activities (i.e., to alienate and distance people from their creative source and their creativity). This process allows capital to extract more surplus value from human labor and gain more control over the minds and sociopolitical activities of the people. (It is also one of several reasons why mathematicians, engineers, and scientists—as a group—have tended to be more politically conservative than their social science or humanities counterparts. Thus, our students suffer through years of math-as-"mythamatics," never internalizing it or seeing it as a natural human endeavor.

Rote Memorization is a Dead End

We are all witnesses to the disastrous reality of rote learning. How many of us have seen students who, by rote memorization alone, have taken—and passed—a series of mathematics courses, yet they lack even fundamental mathematics skills? How many times have we asked students questions just slightly removed from the subject they have memorized only to get silence and a glazed look as a response. Why?

We teachers and professors have reduced mathematics problem-solving and testing to memorization of definitions, formulae, theorems, and the like. Thus, the student who can memorize the most and who has access to the most reference material and/or problem-solving technology will most likely do well. This kind of learning is solely for immediate recall; little is internalized or retained for use in learning the next level of mathematics.

An Alternative Curriculum and Pedagogy

Reforms in mathematics education have not solved the inherent problems precisely because of their reformist nature. Reforms merely rearrange the same content into a "new and improved" format. However, there are a few of us in the mathematics education field who have broken with tradition and brought every aspect of teaching mathematics into question: our pedagogy, our assumptions about how and why people learn, our texts, and ourselves. Witness Bob Moses and his Algebra Project for elementary schools, in which youngsters learn algebra from their rides on the subway; Arthur Powell (1986) of Rutgers University-Newark and his Writing in Math project; and Marilyn Frankenstein of the University of Massachusetts and her concept of "radical math" (Frankenstein and Powell, 1989).[1] These and other progressive educators have suggested and use alternatives to the lock-step, nuts-and-bolts approach to the arithmetic-algebra-precalculus-calculus "learning" sequence that is so pervasive yet so devastating.

There are certain "radical" actions that can be taken to improve, quantitatively and qualitatively, students' knowledge and appreciation of mathematics as real, do-able, and theirs. What Bob Moses does at the elementary school level should be expanded upon. For example, students could be assigned to work collectively on a problem involving the study of turbulence in a tub or the chaotic state of a dripping faucet; their investigations could extend over a whole school semester or term, whereupon they would research and develop mathematical rules, definitions, and techniques to solve the problem. Short of a major overhaul of the first two years of college mathematics, college-level courses should be developed that show the interconnectedness of mathematics and real-world problems. These courses should also show how people throughout history have created mathematical techniques to solve problems.

How and What I Teach

The following is a discussion of what I have been doing and developing over the past twenty years with a full spectrum of college students, but principally working-class Black and Latino adults (young and old) precisely because they are the most alienated from and fearful of mathematics. Rarely are these students intellectually challenged or stimulated, but once they are, and once teachers understand that they can understand, I have found their pursuit of knowledge to be unstoppable.

Most of my pedagogical work is aimed at psychological upliftment through emphasizing that ordinary people create mathematical ideas and "do" mathematics. In my basic mathematics courses (calculus and its prerequisites), I assume the role of confidence builder; I tell my students that they all start out with an "A," yet they must struggle to maintain it. I let them know that I assume they all have the intellectual capabilities to understand the material, and if they do not I attribute it to either my own or the textbook's failure to communicate clearly. In my view, the quality of mathematics knowledge is central and the quantitative aspect secondary. Therefore, although I may set out to cover six chapters in a semester, I will not kill my students to achieve it. If they complete only three or four chapters and learn those well, then I am confident that they can pick up the rest.

A Typical Algebra Class

I use the same approach and structure in my elementary, intermediate, and precalculus algebra classes that I use in my finite mathematics classes or my "Math as a Human Endeavor" classes. The first two sessions are lecture-discussions on the historical, cultural, and sociopolitical implications of mathematics. In these sessions, I outline the interrelationships between the development of mathematical ideas and techniques, humanity's ongoing struggle to understand Nature, and capitalism's attempts to control and dominate Nature.

From the outset, I emphasize that some of the very first mathematical/scientific thinkers and innovators were African women. Humanity originated in Africa, and women, for the most part, were the first farmers; both logic and scientific evidence support the assertion that the rudimentary development of mathematical and scientific inquiry and experimentation began with them. As women and, later, men furthered the domestication of foodstuffs, they created more leisure and thinking time for themselves; thus, a few men and women began to study Nature in its qualitative and quantitative modes. Over time, however, the domestication of foodstuffs also played a direct role in dividing the sexes, bringing forth social classes and its concomitant differences in mathematics knowledge.

My lectures include discussion of the ancients' rudimentary methods of scientific observation, inquiry, and experimentation and their complementary mathematical achievements. I show how early mathematics and science led to the building of the pyramids, the Great Wall of China, and the road to Kathmandu as well as the development of astronomy and astrology, the techniques of iron smelting and metal-plating, and surgery—thousands of years before European civilization. I also show how humankind's mathematical ideas and techniques grew out of people's need to understand Nature, to build and domesticate, and to communicate and record results from various types of social interactions.

I point out to my students that the Egyptians, Chinese, and Indians used different styles of mathematical generalizations in algebraic problem solving. Pre-Hellenic algebraic proofs such as those found in two of the most important Egyptian mathematical documents, the Ahmes Papyrus (c. 1650 B.C.) and the Moscow Papyrus (c. 1850 B.C.), while not deductively axiomatic, were and still are valid proofs. As Joseph (1987) notes: "Egyptian proofs are rigorous without being symbolic, so that typical values of a variable are used and generalizations to any other value is immediate" (p. 24).

During these sessions, I stress that the Egyptians created a tremendous civilization that lasted for thousands of years. I also stress that Egypt is in Africa and that the people who inhabit the land were and are Africans. I point out that Egyptian civilization produced complex technological innovations and forms of communication and engaged in an extensive interchange of goods and ideas with other people thousands of years before they helped bring forth Greek civilization. I note that at least half of the Greek language is African in origin and that the Greek cosmological and mythological constructs were founded upon Egyptian constructs, as was Greek shipbuilding, architecture, and mathematics. I explain that Euclid—the so-called "father" of plane geometry—spent twenty-one years studying and translating mathematical tracts in Egypt. Pythagoras also spent years studying philosophy and science in Egypt and possibly journeyed East to India and/or Persia where he "discovered" the so-called Pythagorean Theorem in the Indian Sulbasutras, a collection of mathematical documents (c. 800–500 B.C.). As I ask my students, how could a theorem whose proof was recorded in Babylonian documents dating 1,000 years before he was born be attributed to Pythagoras?

The intent of these first sessions is to shatter the myth that mathematics was or is a "White man's thing" and to show my students that all civilizations, though they differ and develop at different paces, have always been bound inextricably to each other. It is important that students know that Europe is not now nor was it ever the "civilizing center" of the world surrounded by wildness and chaos. To further undo this Eurocentric assumption, I then discuss the importance of the constant flow of ideas and techniques into Europe from the early Greeks through the Medieval and Renaissance periods to the rise of capitalism (see Midonick, 1965). I show how certain aspects of European mathematics could not have developed had not the Europeans traded with more advanced societies. One of the most glaring examples of this is the case of the 150-year political struggle around the incorporation of the Hindu-Arabic numeral system into common usage in Europe. For a century-and-a-half during the medieval period (specifically around A.D. 1200–1350), the dominant Roman Catholic Church's fear of a rising rival class, coupled with European racism and xenophobia impeded the spread of mathematical knowledge throughout Europe. The Vatican denounced Hindu-Arabic numerals as "the work of the devil" because it viewed the widespread use of an easy way to calculate as a means by which European merchants and craftsmen would become even more independent of the Church.

This example leads into a discussion of the myth of the "Dark Ages," which asserts that because a general retrogression occurred among the European feudal elite during the Medieval period, nothing was happening intellectually anywhere else in the world. On the contrary, ideas in mathematics, science, and philosophy flourished both inside and outside of Europe during that time. Great African and

Middle Eastern Arab scholars lived and studied in places like Toledo and Cordoba in Spain as well as in Sicily. Timbuktu, Cairo, Baghdad, and Jundishapur were also key centers of scholarly learning and research. The rich and complex Arab culture that dominated southern Europe, most of Africa, the Middle East, and parts of India and China during this period brought forth such intellectual centers as Caliph al-Mansur's House of Wisdom (Bait al-Hikma) in Baghdad, where documents, scholars, and researchers integrated the great astronomical studies of Indian, Chinese, Greek, and Babylonian scholars.

Bait al-Hikma was also a research university out of which a key mathematician evolved: Mohammed ibn-Musa al-Khwarizmi (c. A.D. 825–??). My classroom presentations include discussion of al-Khwarizmi because he authored two foundational mathematics texts: The first, Hisab al-djabr wa-al Muqabala (The Science of Equations or The Science of Reduction and Cancellation) not only gave Europe its first systematic approach to algebra, but also was the source of the name for the subject matter (al-djabr or "algebra"). His second book (now found only in the Latin original), *Algorithmi de Numero Indorum,* explains the Indian numeral system. Although al-Khwarizmi emphasized the Indian origins of the numeral system, subsequent European translations of this book attributed the system to him; hence, schemes using these numerals came to be known as "algorithms" (a corruption of the name "al-Khwarizmi"), and the numerals became known as "Arabic numerals" (Joseph, 1987, p. 21).

I also mention other important Arab mathematicians such as the algebraist Omar Khayyam (c. A.D. 1050–1122), known in the West solely as a hedonist poet; and Nasir Eddin al-Tusi (A.D. 1201–1274). It took 500 years for al-Tusi's ideas of geometry on a non-flat surface to catch on in Europe; in the eighteenth century, the Italian geometrist Saccheri used al-Tusi's work to create the basis of what we now call "non-Euclidian geometry."

I then proceed to give my students an idea of the intimacy of mathematical abstraction or concepts to sociopolitical developments. I discuss the period from the eighth through the fifteenth century, when Europe, Africa, Asia, and the Americas engaged in a rich interchange of ideas and culture. I show how this interchange helped to nurture the intellectual and scientific demands of the rising European merchant class. For example, in the 1600s European merchants began making great demands on shipbuilders for more and bigger ships to carry more African slaves, raw materials, and finished products at a faster pace. Ironically, in their research to improve techniques for building very large ships, European shipbuilders turned to the wisdom of Gambian (West African), Chinese, and Indian mariners. Thus, my students come to understand the contradictions inherent in the way developments in hydrodynamics (and its attendant mathematics) contributed to the European bourgeoisie's relentless pursuit of profit and to the horror of the original Holocaust—the slave trade.

I further point out that calculus was created to facilitate the study of ballistics and the wars of land consolidation waged by England and Germany. I also show how the needs of the military in contemporary capitalist societies for more efficient weaponry and defense mechanisms have inspired and continue to inspire many mathematicians and scientists to pursue the War Machine—bringing forth new mathematical and scientific discoveries and techniques primarily for the sake of capital.

In sum, these lecture/discussions help students to understand the following:

(*a*) People of color were the original founders and innovators of mathematics and science.

(*b*) Europe was never isolated from Third World (actually First World) mathematical and scientific achievements.

(*c*) European capitalism developed because of Europe's incorporation of the mathematical and scientific ideas and techniques of the First World into their capitalist superstructure.

(*d*) Europe dominated, enslaved, and colonized Africa, Asia, and the Americas and thereby stopped and/or reversed most, but not all, forms of First World intellectual, mathematical, scientific, and technological activity.

Through this brief historical survey, I attempt to put mathematics within a human context. Although I limit term papers to my calculus and "math as a human endeavor" classes, many of my students become very interested in the historical development of mathematics. Often they wind up doing more reading and writing term papers in the field of mathematics and science history for other classes.

A Few Words on Structuring Classes

About two weeks into a class, I facilitate the creation of study groups of no more than three or four people of the students' choice. These groups allow collective study (a concept that often must be explained to most students) and in-class group work. The study groups are responsible for one or two "progress reports" (others may call them "tests" or "exams," but I prefer my term because I want to convey the message that I am optimistic about my students' performance and concerned primarily with building mathematical confidence).

I want my students to see, talk, and read about real people doing real mathematics and getting real results. Therefore, I also incorporate a weekly fifteen to twenty minute class discussion of the "Science Times" section of the New York Times to stress the direct and indirect relationship of mathematics to the social and natural sciences, the development of technology, and job market skills. These readings/discussions help "demystify" science and mathematics while they also help build students' technical and formal vocabularies.

Students are requested to note all important mathematical definitions, theorems, and formulae on note cards. They are to bring these note cards to class with them at all times, including progress report time, for use as references. However, from the very beginning I let my students know that my classes (and mathematics itself) are not about seeing who is best at memorizing; rather, they are about getting everyone to understand and use mathematics. Just as we have references available at our job or in our community work, we should have them when we are pursuing the solution to mathematical and scientific problems.

I encourage the use of scientific calculators in my classes. If time and computers are available, I encourage students to use some of the effective interactive software available. Nonetheless, I see the computer as a supplement, never as an alternative, to my direct interactions with students.

Conclusion

Primarily, the results of my two decades of evolving a non-Eurocentric approach to teaching mathematics are reflected in my students having a more positive, self-assured attitude about themselves successfully doing mathematics. For example, of the hundreds of students that took my algebra classes at the State University of New York-Old Westbury (1971–1977) and Rutgers University in Newark (1986–1989), 85 percent passed the course. Out of these about 60 percent pursued at least one more mathematics course that they initially had planned to avoid. Over that ten-year period many more African American students began taking and passing the precalculus courses at both these sites—an increase of 25 to 28 percent at Old Westbury and a 19 to 21 percent increase at Rutgers-Newark. Conversely, among students enrolled in the traditional algebra lecture and testing formats at Old Westbury, dropout and failure rates increased from 1978 to 1988. In basic algebra classes at Rutgers-Newark, the failure and dropout rates have hovered at around 50 percent for the past seventeen years!

In terms of generating mathematics majors, my successes have been small ones: to my knowledge, only seven or eight of my students have decided to pursue or have successfully majored in mathematics. Thus, it can be said that this particular non-Eurocentric approach produces the basic self-confidence students of color need to get through mathematics classes and satisfy the requirements of non-mathematics majors. To generate more mathematics majors, one needs control of the entire mathematics curriculum and faculty, not just a few classes.

In struggling against a Eurocentric curriculum in any discipline, we teachers cannot rectify the problem by trying to use the same methods that have suppressed or distorted knowledge. Nor can we merely graft discussions of the particular attributes and achievements of people of color, willy-nilly, onto the existing order; to do so would only compound the madness of class-, race-, and/or gender-based prejudice. As the Egyptian political economist Amin (1989) claims, such narrow-minded nationalism can and does generate individualism:

> In every case . . . nationalist culturalist retreat proceeds from the same method, the method of Eurocentrism: the affirmation of irreducible "unique traits" that determine the course of history, or more exactly, the course of individual, incommensurable histories (p. 77).

Those of us who are genuinely concerned with educating students for liberation rather than training them for the job market must attack, critique, and dismantle the Eurocentric educational construct while simultaneously planting the seeds for more holistic, in-tune-with-nature, popular, and egalitarian forms of learning. We cannot wait any longer. The grim statistics I mentioned at the beginning of this article indicate that a subtle but effective form of educational genocide is taking place in addition to the general moral and political decline of one of the greatest working classes in the world. The vast majority of us—African Americans and Latinos—presently in the halls of academe are here because of popular movements to open the universities to all. As a consequence, we are honor-bound to struggle against the tendency toward tunnel vision. In all fields of human endeavor, and particularly in the crucial fields of mathematics and science, to offer an alternative that is genuinely egalitarian and truthful we must open our eyes to the centrality of the contributions made by the vast majority of the world's people.

References

Alkalimat, A. (1990). *Eurocentricism: A review of Samir Amin's Eurocentrism.* Unpublished review. Department of Africana Studies. SUNY-Stony Brook.

American Association for the Advancement of Science. (1989). *Science for all Americans: A project 2061 report on literacy goals in science, mathematics, and technology.* Washington, D.C.: American Association for the Advancement of Science.

Amin, S. (1989). *Eurocentrism.* New York: Monthly Review Press.

Ball, W. W. R. (1922). *History of Mathematics.* London: Macmillan and Co.

Commission on Professionals in Science and Technology. *Scientific manpower, 1987 and beyond: Today's budget—Tomorrow's work force.*

Frankenstein, M. and Powell, A. (1989). "Mathematics Education and Society: Empowering Non-Traditional Students" In C. Keitel, P. Damerow, A. Bishop, & P. Gerdes (Ed.), *Mathematics, education and society* (pp. 157–159). Paris: UNESCO.

Joseph, G. G. (1987). Foundations of Eurocentrism in mathematics. *Race and Class* 27(3): 13–28.

Klein, M. (1953). *Mathematics in Western Culture.* New York: Macmillan.

Matthews, C. M. (1990). Science, engineering, and mathematics precollege and college education. Congressional Research Service IB 88068: 5–6.

Midonick, H. O. (1965). *The Treasury of Mathematics.* New York: Philosophical Library.

National Research Council. (1989). *Everybody counts: A report to the nation on the future of mathematics education.* Washington, D.C.: National Academy Press.

Powell, A. B. (1985). Working with "underprepared" mathematics students. In M. Driscoll and J. Confrey (eds.). *Teaching mathematics: Strategies that work* (pp. 181–92). Portsmouth, NH: Heinemann.

Steen, L. A. (1987, July 17). Mathematics education: A predictor of scientific competitiveness. *Science* 251.

Chapter 15

World Cultures in the Mathematics Class

Claudia Zaslavsky

Editors's comment: Claudia Zaslavsky, a mathematics educator, is well-known for her groundbreaking work, *Africa Counts: Number and Patterns in African Culture* (Boston: Prindle, Weber and Schmidt, 1973; reprinted Brooklyn, NY: Lawrence Hill, 1979), that stimuated interest in and the development of curricular materials in multicultural mathematics, including her own *Multicultural Mathematics: Interdisciplinary Cooperative-Learning Activities* (Portland, ME: Walch, 1993) and *The Multicultural Math Classroom* (Portsmouth, NH: Heinemann, 1996). In this chapter, she presents various mathematical practices of African peoples and indigenous peoples of the Americas and discusses why and how these practices can be incorporated into mathematics curricula for children. This chapter first appeared in *For the Learning of Mathematics* 11(2): 32–36, in 1991.

In the classrooms of the United States, we have children representing most of the cultures of the world. Some arrived in our country in the past few years, while others are descendants of immigrants who came in previous decades and centuries, and still others trace their ancestry to the original inhabitants of the Americas. At one time the United States was called a "melting pot" of many nationalities and cultures. Yet there were groups that never had the opportunity to become part of the mix—most Native Americans, people of African descent, and Latinos from the colony of Puerto Rico and from the area that was formerly Mexico, now the southwestern part of the U.S., to name but a few. They were the throwaways, like the garbage left over after the ingredients had been chosen for the melting pot. They constituted—and they still do—a disproportionately large segment of the population living in poverty.

Now we realize that the "melting pot" analogy is false; on the contrary, people have become more interested in recent times in seeking out their roots. The ethnic groups that have lived longest in the Americas—and have been most oppressed—are the Native peoples and the Africans who were brought to the New World in chains, to serve as slaves to European plantation owners. Now their descendants are determined to reassert their cultural heritage. Although their ancestors often included English, Spanish, and other Europeans, they frequently choose, or are compelled by societal pressures, to identify with the oppressed peoples. Incidentally, Native Americans and Africans also intermarried, and many African Americans can count American Indians in their ancestors. The United States Census includes people of Spanish origin with either Whites or Blacks, but also calculates the total Hispanic (of Spanish origin) population as a separate, additional group. Subdivisions and names are the subject of disagreement even among the people who belong to these groups, and discussions on the topic are often heated. In the media, in government and commission reports, the term "minorities" includes Blacks, Latinos/as (Hispanics), Native Americans, and Asian Americans. Members of these groups may resent the implication of the word "minority" as signifying "lesser." They also object to being lumped together as though they are all the same. Even within each of the groups I have named, people differ widely in geographic origin, cultural styles, and social class.

It is estimated that by the year 2000, one-third of all students will be "minority." Children growing up in these families and communities often differ from children of the dominant culture in their learning styles. They may have less access to educational opportunities, both inside and outside the classroom. Many of these "minority" students attend schools that are poorly serviced, score lowest in the all-important standardized "achievement" tests, and drop out of school at a high rate. A disproportionate number of these children are placed in the lowest track (or stream) from the earliest grades, where they are presented with a limited, outdated curriculum, taught by rote memorization methods, and tested by standardized paper-and-pencil, multiple choice tests.

The mathematics community in the United States is embarking upon a program to reach *all* students. As stated in the *Curriculum and evaluation standards for school mathematics* (NCTM, 1989):

> It is crucial that conscious efforts be made to encourage all students, especially young women and minorities, to pursue mathematics (page 68).

Recognition is given to the varied backgrounds and interests of the students:

> Students should have numerous and varied experiences related to the cultural, historical and scientific evolution of mathematics (page 5).

It is not only children of minority groups who benefit from the inclusion of topics relating to their heritage. Students in our "global village" must learn to respect and appreciate the contributions of peoples in all parts of the world.

Educators are beginning to recognize the value of infusing mathematics with the achievements of world cultures, to "multiculturalize the curriculum" (Bishop, 1988; D'Ambrosio, 1989; Gerdes, 1988).

Introducing a Cultural Perspective

Leading educators in the United States deplore the extent to which standardized tests and textbooks drive the mathematics curriculum. Tests take priority. "If it's not on the test, don't teach it," is the prevailing viewpoint of many school administrators and teachers.

Here and there some teachers are motivated to implement the mandated curriculum by introducing a cultural perspective. Teachers may even present language, social studies, or art lessons that have mathematical content without being aware of the mathematics, as the following incidents will show.

To celebrate Children's Book Week, the reading supervisor of a local school invited me to talk to eight-year-old students about my book *Count on your fingers African style* (1980, 1996). The majority of the children in this school are low-income, dark-skinned, Spanish-speaking immigrants or children of immigrants from the Dominican Republic. The teachers had already discussed the book with the children, and the students were prepared with questions about counting, about writing books, about the process of publication. Many could count in Spanish, and several, from Haiti, knew French. We compared the counting words in the three languages—English, Spanish, and French—noting the similarities and the differences.

I asked the children to pretend that they were visiting a market in an African country, where no one spoke their language. How would they ask for eight oranges? Of course, they suggested using their fingers. Then I proposed that each child imagine how he or she would indicate eight and, when I gave the signal, to raise their hands showing "eight" on their fingers. What a variety of ways! Some children used the methods described in the book, while others invented unique styles. Many ways to solve one problem, all equally valid, and a good mathematics lesson in the guise of a talk about books!

We discussed numbers and how useful they are. One boy contributed a remark about "playing the numbers," the illegal gambling game that is popular in low-income communities, in conflict with the legal state lottery. The teacher quickly interrupted: "But watch out for the cops (police)." The boy seemed bewildered by her comment, but she did not explain.

As I left the school I noticed a beautiful patchwork quilt hanging in the lobby, the work of a class of nine-year-olds. It was composed of thirty squares in a 5 x 6 arrangement. The squares were identical in construction, each consisting of small squares and triangles of print or solid color cloth sewn together. By mixing and matching the colors and patterns of the fabrics as their fancy led them, the children were able to achieve a varied and pleasing effect.

I arranged to interview the teacher to learn more about this mathematical production. "I was an art major," she said, "and I had them make the quilt so that they would get a feeling for life in colonial times (eighteenth century). I like to combine social studies with art. Now we are doing Native American bead patterns." When I added that the children were also doing very good mathematics, she seemed surprised. After all, this activity was not in the mathematics curriculum, nor were these applications included on the standardized achievement test. Later I read about an exhibit of African-American patchwork quilts. The author commented on the similarity of many standard quilt patterns to traditional African textiles and the possibility that patchwork quilting was introduced into England and America by African slave women (Barry, 1989).

I asked the teacher whether the boys had objected to sewing. "Oh, no," she replied. "I told them that tailors sew. No problem whatsoever." The photographs of the quilt and of the students working with beads appeared in the *Arithmetic Teacher* (Zaslavsky, 1990).

In this article, I shall describe some of the mathematical practices of African peoples and of the indigenous peoples of the Americas.

Numbers and Numeration

Work with numbers has dominated the mathematics curriculum since the beginning of public schooling, satisfying the needs of shopkeepers, clerks, farmers, and factory workers. All peoples have developed numeration systems to the extent of their needs. The English system of numeration and most European systems are based on grouping by tens and powers of ten. Why is ten commonly used as a base? Is it because we have ten fingers (digits)? The peoples of West Africa and Middle America, as well as the Inuit and other Eskimo peoples of the far north, group by twenties. In some languages, such as Mende of Sierra Leone, the word for twenty means "a whole person"—all the fingers and toes.

Children can learn about numeration systems by examining the construction of larger numbers. In the Yoruba (Nigeria) language, for example, the name for forty-five means "take five and ten from three twenties," using the operations of multiplication and subtraction, rather than multiplication and addition, as in most European languages. Different solutions to the same problem, one just as good as the other (Zaslavsky, 1979a: page 207).

Finger gestures to express numbers are commonly used by people who do not speak each other's languages. These gestures may be related to the number words; or, again, they may be quite different. Even before the indigenous peoples of North America were pushed westward by European settlers, tribes speaking different languages met on the Great Plains. Of necessity, they developed systems of finger signs, including signs for numbers (Zaslavsky, 1979b).

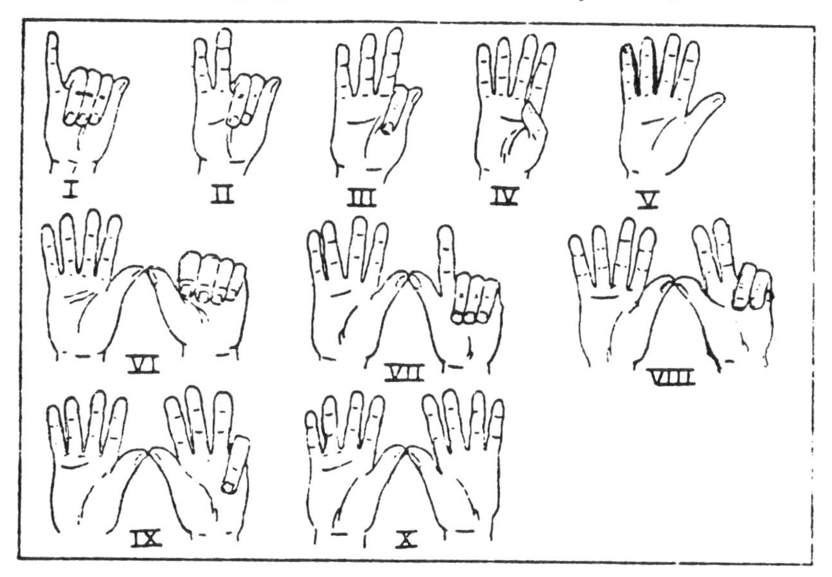

Figure 15–1. Illustration from William Tomkins: *Indian Sign Language.*

Source: (Dover, New York).

Ancient Egypt provides a rich source of material about numbers, numeration systems, written numerals, computational methods, and applications (Gillings, 1979, Joseph, 1991). Less well known is the African origin of this great civilization in the upper Nile valley of the land that is now Sudan and in the once fertile Sahara region. Prior to the rise of modern imperialism, historians had acknowledged that the ancient Egyptians were dark-skinned people. However, Europeans could not admit that the peoples they had conquered and enslaved, the black people of Africa, were kin to the inventors of ancient Egypt's high civilization. Africans had to be portrayed as less than human and denied their history, in order to justify their enslavement, while Egypt was detached from the African continent. In his book *Black Athena*, Martin Bernal (1987; see also Davidson, 1987) discusses this issue in great detail and with voluminous documentation. Furthermore, many of the Greeks whom we revere as the "fathers of mathematics" either studied in Egypt or were Egyptians themselves—Pythagoras, Eudoxus, Euclid were but a few (Joseph, 1987). Beatrice Lumpkin (1983a) discusses the three great periods of African participation in the development of mathematics: the ancient period of the pyramids and temples, the classical Hellenistic period, and the Islamic period.

The peoples of Middle America developed their own systems of written numerals, in the case of the Maya dating back at least two thousand years. The systems were based on twenty and powers of twenty, and included the use of zero, positional notation, addition, and the repetition of symbols. When applied to Mayan chronology, the groupings proceeded in this fashion: 20, 18×20, $18 \times 20 \times 20$, and so forth, to represent the twenty-day months, eighteen-month year, and larger groupings. The secular year consisted of eighteen months of twenty days each, plus five additional days to make 365, with cycles of fifty-two years. The sacred year consisted of thirteen twenty-day months. Both counts were used simultaneously. For an excellent reference work on the mathematics of the Americas, see Closs (1986).

The Inca *quipu* represents a unique system of record-keeping. A positional numeration system, based on ten and powers of ten, is embodied in a collection of colored strings resembling a tangled mop. Yet the *quipu* can encompass a whole census. For an entrancing discussion of *quipu* construction and usage in the context of the Inca Empire, which included all of Peru and a large section of the Andean region five hundred years ago, see Ascher (1981).

Another aspect of number is the ability to do mental arithmetic. The year 1990 marked the two hundredth anniversary of the death of the slave Tom Fuller, known as the African Calculator. Shipped to North America in 1724 at the age of fourteen, he developed remarkable powers of calculation, although he was forbidden access to any kind of schooling, as were all slaves, and he could neither read nor write. Late in his life he was used by antislavery advocates to demonstrate the mental capacity of Black people (Fauvel & Gerdes, 1990).

Design and Pattern

Most cultures have developed characteristic designs, which they incorporate into their cloth, basketry, wooden objects, and buildings. For example, the Navajo of the southwestern United States are known for their beautiful rugs, the artistic creations of Navajo women. The intricate patterns and fine workmanship of these rugs, sometimes

Figure 15–2. The seven one-color one-dimensional patterns (Native American).

Source: Le Roy H. Appleton, *American Indian design and decoration* (Dover, New York, 1971).

called the "first American tapestries," earn for them a well-deserved place in museums. When the Spanish conquistadores introduced sheep in the sixteenth century, the southwest tribes, traditionally weavers of cotton, turned to the use of wool. The Navajo learned weaving from the Pueblo tribes, probably in the late seventeenth century. From weaving plain cloth and blankets with simple designs for their own use, they turned to creating richly patterned blankets and rugs for the commercial market. By the mid-1800s these weavings commanded a good price. In recent years, individual women have become famous for their fine weaving and innovative designs. Navajo rugs, with their geometric motifs and repeated patterns, provide an excellent medium for the analysis of symmetry (Zaslavsky, 1990).

The Bakuba people of Zaire are justly famous for their embroidered raffia cloth, known as "Bakuba velvour," and for their marvelous wood carvings. Characterized by the repetition of traditional geometric motifs, these art works, too, have found their way into many museums in Europe and the United States. Some of the pieces in the British Museum date back more than two centuries. A visit to a museum to study these works of art might inspire students to create their own patterns (Zaslavsky, 1979a; Washburn and Crowe, 1988).

Figure 15–3. The seven one-color one-dimensional patterns (Bakuba, Zaire).

Source: Adapted from Geoffrey Williams, *African Designs from Traditional Sources* (Dover, New York, 1971).

Architecture

Most of us are so accustomed to living in a rectilinear environment that it is difficult for us to imagine different surroundings. Our furniture and accessories are designed to fit along straight walls and into right-angled corners. Yet, people of other cultures would be just as uncomfortable if they had to give up their circular houses for our rectangles. The Native peoples of the Great Plains (northcentral U.S.) traditionally lived in conical *tipis,* portable tents made of animal skins and decorated with meaningful symbols—a dwelling appropriate to the life style of the people. Early in this century, Black Elk, the Oglala Lakota leader, lamented:

> We made these little gray houses of logs that you see, and they are square. It is a bad way to live, for there can be no power in a square. You have noticed that everything an Indian does is in a circle, and that is because the Power of the World always works in circles and everything tries to be round (Neihardt, 1961, p. 198).

Why has the round house been traditional in some societies? One must consider the available materials and technology, as well as the way the society makes its living. A settled agriculturalist builds for the future, while a pastoral nomad may abandon a shelter after a few months, or pack it up and move on. Consider this aspect, too: a family that builds its own home, using hard-to-find materials, wants to achieve the largest possible floor space for a given quantity of materials for the walls. In other words, maximize the area for a given perimeter. The circle is the answer.

Circular homes can take a variety of forms: the conical *tipi,* the hemispheric *igloo* of the northern Inuit, the beehive-shaped thatched dwelling on Mt. Kilimanjaro, the tall mud-brick, thatch-roofed cylinders of West Africa. For suggested lessons based on these ideas, see Zaslavsky (1989, 1993, 1994, 1996).

Who has not been impressed by the splendor and accurate construction of the Egyptian pyramids, now nearly five thousand years old? The early inhabitants of the land that is now Mexico also built pyramids, some over two millennia ago (Lumpkin, 1983b, 1987).

Less well known than the Egyptian pyramids is the African city-state, Great Zimbabwe ("great stone house"), with its complex stone architecture. Started perhaps eight or nine centuries ago, it served for several centuries as the seat of government for the rulers of a vast kingdom in southern Africa. Several hundred smaller stone structures, spread across the land, are a memorial to the former power of the realm (Asante, 1983).

Sufficient information is available about the pyramids and Great Zimbabwe to enable students to analyze their measurements, to compare the labor time and the quantity of materials that went into building these edifices, and to construct models.

Games of Chance and Skill

From time immemorial human beings have tried to divine the future. Some divining practices led to games of chance, and eventually to the important and growing field of mathematical probability and statistics. Tossing a coin is one of the simplest forms of gambling. The Igbo (Nigeria) game of *Igba-ita* ("pitch and toss") involved tossing cowrie shells, still used as currency into the twentieth century, and noting whether they landed with the openings up or down. With the adoption of coinage, the game became known as *Igba-ego* (*ego* means "money"). The British commentator G. T. Basden observed groups of men gambling in the marketplace. The challenger tosses twelve cowries. "Quick as lightning the players note the positions and forfeit their stakes or collect their gains. The play becomes exceedingly fast, and soon a cloud of dust encircles each group of gamblers. I have watched players at this game, and it has always been quite beyond me to note the positions of the fall, the cowries have been counted and snatched up again long before I could begin to count" (quoted in Zaslavsky 1979a: page 114). Students can play the game with macaroni shells or other asymmetric objects, and compare these outcomes with the results of tossing symmetric coins. The Native American Bowl Game involves tossing four peach or plum pits that have been marked on one side, and noting the outcomes.

It is claimed that diagrams for three-in-a-row games were chiseled into the roof slabs of the temple to the Egyptian pharoah Seti I about the year 1300 B.C.. No doubt the evidence, if it ever existed, has since been worn away by the elements and by pollution. The British play Noughts and Crosses and Nine Men's Morris, children in the United States play Tic-Tac-Toe, while "Mill" is the name of the game in several European countries. One of the most complex versions of three-in-a-row games is the Lesotho *Murabaraba*. In simple or complicated form, such games of strategy help children to acquire the necessary skills in problem-solving and decision-making (Zaslavsky, 1982, 1996).

In the British Museum is a beautifully carved wooden statue of the ninety-third king of the Bakuba (Zaire) people. Early in the seventeenth century, he brought the peaceful arts to his people, and taught them the game they called *Lela,* a variation of the universal African game of transferring, usually known by its Arabic name, *Mankala.* To celebrate his reign, the king is portrayed with a model of the gameboard in front of him.

This ancient African stone game, played in different versions in most of the continent and in parts of Asia, is considered among the world's best games of strategy. In its simplest form, the game is appropriate for children just entering school, and affords practice in counting and in the concept of one-to-one correspondence. At a more advanced level, all four operations of arithmetic come into play. Yet the game is so sophisticated as to challenge adults in national competitions (Zaslavsky, 1979a: pages 116–136). African teenagers play with such speed that it is virtually impossible for an inexperienced onlooker like me to follow, much less understand, what is going on.

The game has proved its usefulness in several ways. African captives brought it to the Americas, and social scientists have analyzed the rules of the versions popular in the United States, the Caribbean islands, and Brazil, in order to trace the ancestry of Black people living in those regions. In the month I wrote this, I received a letter from a teacher in the state of Texas requesting permission to use the sample game in *Africa counts* in her award-winning project entitled "Lasers from the Jungle: Turning Primary Students on with African Legends." I readily granted permission, but asked her to substitute "rain forest" for "jungle" in the title. For one thing, Africa does not have jungles. Second, the word "jungle" in connection with Africa has negative connotations, evoking the Tarzan image.

Children can invent new versions of a game by changing the rules, varying the shape of the gameboard, or using a different quantity of playing pieces. Games of strategy encourage young people to develop skill in logical inference, as the following incident will illustrate. In his book *Games of the North American Indians* (1907), Stewart Culin describes a three-in-a-row game called *Picaría,* played by Pueblo youngsters in the Southwest (Zaslavsky, 1982: pages 46–50). Culin's description of the rules is ambiguous. I had seen one version in several recent publications, but it seemed to me that an alternative set of rules would make a better game. I asked a group of eleven- and twelve-year-old students to play according to the first set of rules. Within a few minutes they complained: "The first person to move always wins." Then I suggested that they try the other version, without telling them that it was the one I favored. They were unanimous in declaring it the better game.

Conclusions

The introduction of multicultural, interdisciplinary perspectives into the mathematics curriculum has many points in its favor:

- Students become aware of the role of mathematics in all societies. They realize that mathematical practices arose out of people's real needs and interests.
- Students learn to appreciate the contributions of cultures different from their own, and to take pride in their own heritage.
- By linking the study of mathematics with history, language arts, fine arts, and other subjects, all the disciplines take on more meaning.
- The infusion into the curriculum of the cultural heritage of "minority" students builds their self-esteem and encourages them to become more interested in mathematics. As one eleven-year-old boy wrote in his evaluation of a classroom activity based on African culture: "As you probably don't know I feel very strongly and am in deep thrust with my black people, and the math has made me feel better." There is little that one can add to this heartfelt comment!

References

Asante, M., and K. Asante. (1983). "Great Zimbabwe: An ancient African city-state." In I. Van Sertima (ed.). *Blacks in science* (pp. 84–89). New Brunswick, NJ: Transaction Books.

Note: For lessons incorporating some of the concepts described in this article, as well as other activities, see Krause (1983), Seattle Public Schools (1984), and Zaslavsky (1993, 1994, 1996), as well as references in the text of the article.

Ascher, M., and R. Ascher. (1981). *Code of the Quipu.* Ann Arbor, MI: University of Michigan.

Barry, A. (1989). "Quilting has African roots, a new exhibition suggests." *New York Times* (November 16): p. C12.

Bernal, M. (1987). *Black Athena: The Afroasiatic roots of classical civilization.* New Brunswick, NJ: Rutgers University.

Bishop, A. J. (1988). *Mathematical enculturation.* Kluwer, Dordrecht.

Closs, M. P. (Ed.). (1986). *Native American mathematics.* Austin, TX: University of Texas.

D'Ambrosio, U. (1989). Project: A research program and a course in the history of mathematics: Ethnomathematics. *Historia Mathematica* 16: 285–286.

Davidson, B. (1987). The ancient world and Africa: whose roots? *Race & Class* 29 (2): 1–15.

Fauvel, J., and P. Gerdes. (1990). African slave and calculating prodigy: Bicentenary of the death of Thomas Fuller. *Historia Mathematica* 17: 141–151.

Gerdes, P. (1988). On culture, geometrical thinking and mathematics education. *Educational Studies in Mathematics* 19: 137–162.

Gillings, R. J. (1975). *Mathematics in the time of the Pharoahs.* Cambridge, MA: MIT Press.

Joseph, G. G. (1987). Foundations of Eurocentrism in mathematics. *Race and Class* 28 (3): 13–28.

———. (1991). *The crest of the peacock: non-European roots of mathematics.* London: I. B. Tauris.

Krause, M. (1983). *Multicultural mathematics materials.* Reston, VA: NCTM.

Lumpkin, B. (1983a). Africa in the mainstream of mathematics history. In I. Van Sertima (ed.). *Blacks in science* (pp. 100–109). New Brunswick, NJ: Transaction Books.

———. (1983b). The pyramids: Ancient showcase of African science and technology. In I. Van Sertima (ed.). *Blacks in science* (pp. 67–83). New Brunswick, NJ: Transaction Books.

———. (1987). Pyramids—American and African: A comparison. In I. Van Sertima (ed.), *African presence in early America.* (pp. 169–187). New Brunswick, NJ: Transaction Books.

NCTM. (National Council of Teachers of Mathematics) (1989). *Curriculum and evaluation standards for school mathematics.* Reston, VA: NCTM.

Neihardt, J. G. (1961). *Black Elk speaks.* Lincoln, NB: University of Nebraska.

Seattle Public Schools. (1984). *Multicultural mathematics posters and activities.* Reston, VA: NCTM.

Washburn, D., and D. W. Crowe (1988). *Symmetries in culture.* Seattle, WA: University of Washington.

Zaslavsky, C. (1979a). *Africa counts: Number and pattern in African culture.* New York, NY: Lawrence Hill. French edition: *L'Afrique compte! Nombres, formes et démarches dans la culture africaine.* Argenteuil, France: Éditions du Choix, 1995.

———. (1979b) It's OK to count on your fingers. *Teacher* 96 (February): 54–56.

———. (1980). *Count on your fingers African style.* New York: Crowell; republished (1996), New York: Black Butterfly.

———. (1982). *Tic-tac-toe and other three-in-row games.* New York: Crowell.

———. (1989). People who live in round houses. *Arithmetic Teacher* 37 (September): 18–21.

———. (1990). Symmetry in American folk art. *Arithmetic Teacher* 38 (September): 6–12.

———. (1991). Multicultural mathematics education for the middle grades. *Arithmetic Teacher* 38 (February): 8–13.

———. (1993). *Multicultural mathematics: Interdisciplinary cooperative-learning activities.* Portland, Maine: J. Weston Walch.

———. (1994). *Multicultural math: Hands-on activities from around the world.* New York, NY: Scholastic Professional.

———. (1996). *The multicultural math classroom: Bringing in the world.* Portsmouth, NH: Heinemann.

Section VI

Ethnomathematical Research

Arthur B. Powell and Marilyn Frankenstein

As the field of ethnomathematics develops, we will need to continue reinterpreting conclusions others have drawn about various people's mathematical knowledge, and continue uncovering and disseminating the distorted and hidden histories of mathematical knowledge. This research will expand and deepen the knowledge we create and recreate about our world. It will also lead us to reexamine how we generate knowledge. So, for example, we suggest coinvestigation between students and teachers into discovering each others' ethnomathematical knowledge. This will improve our teaching, as we discuss below, and will also point the way to new research methodologies. Finally, in support of Freire's theory that the purpose of knowledge is for people to resolve the fundamental contradiction of our epoch between domination and liberation, we need to explore the connections between the cultural action involved in teaching and learning ethnomathematics and the economic and political action needed to create a liberatory society.

Not only does ethnomathematical research force a reconsideration of mathematical knowledge, but also through this reconsideration significant epistemological questions emerge. Are the mathematics found in different cultural processes and artifacts different mathematics or different manifestations of one universal mathematics? When Gerdes and his colleagues in Mozambique find graph theoretic ideas in the *lusona* of Angola, are they finding a different mathematics or just a different expression of what we know as academic mathematics? On the other hand, is the value that we bestow on the *lusona* tradition solely due to connections we see between it and academic mathematics? Further, do we miss truly different mathematics because we examine different cultural traditions through the lens of academic mathematics? As Pinxten discusses in chapter 17, when, indeed, we confront a different concept of space, it is difficult for academic mathematicians to see. Moreover, as we reconsider many cultural practices and artifacts as the product of mathematical activity, some worry that we will loose what is unique about "mathematics." Aside from the elitism of this concern, what is lost by a blurring of disciplines? The process of making meaning and understanding the world may require that we see complicated interconnections, overlappings, and fuzzy boundaries among disciplines as well as between rationality and other ways of knowing. Finally, as Audre Lorde (1984) states

> as we come more into touch with our own ancient, non-european consciousness of living as a situation to be experienced and interacted with, we learn more and more to cherish our feelings, to respect those hidden sources of power from where true knowledge and, therefore, lasting action comes (as quoted in Weiler, 1994, p. 28). . .
>
> Rationality is not unnecessary. It serves the chaos of knowledge. It serves feelings. It serves to get from this place to that place. But if you don't honor those places, then the road is meaningless. Too often, that's what happens with the worship of rationality and that circular, academic analytic thinking. But ultimately, I don't see feel/think as a dichotomy. I see them as a choice of ways and combinations (as quoted in Weiler, 1994, p. 29).

Further as ethnomathematical research continues to force us to reconsider what counts as mathematical knowledge, it also forces us to reconsider all of our knowledge of the world. Henderson (1990) argues that although "formal, symbolic expressions are often excellent ways of capturing certain aspects of our experience" the erroneous view that "formal" mathematics is the ultimate, *real* mathematics "limits the understandings which we construct of our human experience . . . [and] damages the human spirit" (p. 5). Pinxten, van Dooren, and Harvey (1983)[1] argue that the mathematics education of Navajos start from a fully developed knowledge of their spatial system (a dynamic system which is in direct contrast to the static concepts of academic mathematics), not solely because this is the only way to avoid sociocultural and psychological alienation of Navajos, but also because "[a]s long as science cannot pretend to have valid answers to all basic questions . . . it is foolish to exterminate all other, so-called primitive, prescientific, or otherwise foreign approaches to world questions" (p. 174). Adams (1983) reminds us that in Eastern societies such as India and Africa, "[t]here are no distinct separations between science and religion, philosophy and psychology, history and mythology. All of these are viewed as one reality and are closely interwoven into the fabric of daily life" (p. 43).[2] He argues that Western science would deepen its knowledge of the world by reconsidering the value of emotional, intuitive, and spiritual knowledge. He quotes Einstein who claimed "there is no inductive method which could lead to the fundamental concepts of physics . . . there is no logical path to these laws; only intuition, resting on a sympathetic understanding of experience, can reach them" (p. 41). We feel much more research needs to be done to uncover how the logic of all peoples can interact with each other to help us all understand and act more effectively in the world.

One place for mathematics teachers to start this research is with their students' ethnomathematics. Frankenstein (1989) encourages her students to uncover their ethnomathematics through evaluating their own learning. She stresses that there is always some correct reasoning involved in any attempt to solve a problem. By probing what they do understand about a particular problem, they discover their everyday mathematics knowledge and are empowered and motivated to find out what they need to learn in order to understand more. Powell and López (1989), teacher-student co-researchers, use journals to help students explore their ethnomathematics.

These explorations suggest the importance of developing methodologies that effectively and ethically probe our students' mathematical knowledge. Lave (1988) and her colleagues in the Adult Math Project used an ethnography of supermarket shopping as the basis from which to design their experiments in cognitive psychology, and then searched for explanations of their findings back in the supermarket. Gerdes (1988) has devised ways of uncovering Mozambican mathematical traditions from material culture which were buried as a consequence of slavery and colonialism. He and his colleagues and students

> looked to the geometrical forms and patterns of traditional objects like baskets, mats, pots, houses, fishtraps, etc. and posed the question: why do these material products possess the form they have? In order to answer this question, we learned the usual production techniques and tried to vary the forms. It came out that the form of these objects is almost never arbitrary, but generally represents many practical advantages and is, quite a lot of times, the only possible or

> optimal solution of a production problem. The traditional form reflects accumulated experience and wisdom. It constitutes not only biological and physical knowledge about the materials that are used, but also mathematical knowledge, knowledge about the properties and relations of circles, angles, rectangles, squares, regular pentagons and hexagons, cones, pyramids, cylinders, etc. (p. 140).

Frankenstein uses an interview technique in which she encourages students not to answer with a vague "I don't know," insists they don't censor themselves in their attempts to solve problems, and asks them to evaluate their own understandings.

Knijnik (1992/reprinted here as chapter 18), discussing the theory of her work with the Landless People's Movement in Brazil, defines an ethnomathematical approach as "the research of the conceptions, traditions, and mathematical practices of a specific social group and the pedagogical work involved in making the group realize that they do have knowledge; they can codify and interpret their knowledge; they are capable of acquiring erudite knowledge; [and] they are capable of establishing comparison between these two different types of knowledge in order to choose the most suitable one when they have real problems to solve" (p.4). Powell (Powell & López, 1989; Powell & Ramnauth, 1992) uses a participatory research model combined with journal and other writing activities to prompt students to reflect on both the cognitive and affective components of learning mathematics and to engage them in analyzing critically methodological dimensions of teaching and learning. As López observes:

> [I] became interested in the study due to my poor math skills. I felt that if I took a more active role in the learning of mathematics I might be able to do better in the course. Throughout the semester I kept a journal detailing my observations of the class, course, and my learning of mathematics. . . . We met after classes and whenever our schedules allowed us to discuss what I felt that I had gained as a result of writing in a mathematics course. I was then asked to comment on the writing experience and the journals that I had kept, to see exactly how it was that I had gained a better understanding of the mathematics I was learning. I found many instances where certain ideas or concepts became clearer to me as a result of writing about them . . . (p. 172).

Extending the idea that writing is effective in learning mathematics and that teacher and student jointly can study this process, Powell, Jeffries, and Selby (1989) have inserted into the discussion the need to attend to the more general, human process of empowerment.

Their concern is for the empowerment of all actors in various settings of mathematics education. Not only did students and their instructor study student journals, but students critiqued the instructor's pedagogical approach. In analyzing the project, students defined research activities in pedagogy to be participatory and to have the potential to be empowering when they give authority to the voices of students. In many instructional settings, students generally feel, and are often considered, to be without power. To give authority to the voices of students and to incorporate their perspectives in transforming mathematics pedagogy, instructors must begin by listening to students and finding authentic ways to incorporate students' perspectives into educational research.

On the other hand, we need to avoid what Youngman (1986) calls Freire's tendency toward an "uncritical faith in 'the people' [which] makes him ambivalent about saying outright that educators can have a theoretical understanding *superior* to that of the learners and which is, in fact, the indispensable condition of the development of critical consciousness" (p. 179). While we listen to student themes, we organize them using our critical and theoretical frameworks, and we represent them as problems challenging previous student perceptions. We also suggest themes that may not occur to students, themes we judge important for shattering the commonly held myths about the structure of both society and knowledge and that interfere with the development of critical consciousness.

We need to do more research to find ways of helping our students learn about their ethnomathematical knowledge, contributing to our theoretical knowledge, without denying the unequal development of tools for producing knowledge, but as much as possible "based on cooperative and democratic principles of equal power" (Youngman, 1986, p. 179). Importantly, we need to do this in such a way that the coinvestigators include cultural workers whose products and traditions are being studied. This is not only to verify research findings, but also, and more importantly, so that the benefits of research are shared equably. Furthermore, we need to attend to praxis—the inseparability of action and reflection—to break down the dichotomies between teaching and learning, between formulating research questions and finding answers. As Lave (1988) concludes about the Adult Math Project study of supermarket mathematics, "description and analysis have been part of the project as a whole in all its phases, rather than uniquely divided between methods (or disciplines)" (p. 121).[3]

But the underlying question throughout all this work is how the cultural action involved in teaching and learning ethnomathematics can play a role in the economic and political action needed to create a liberatory society. Carby (1990) highlights this issue in her remarks on the changes in the literary canon at the universities, where African-American women have become subjects on the syllabus, but the academy continues to ignore the material conditions of most African Americans. She challenges us to think through the issues of real power: "Are the politics of difference effective in making visible women of color while rendering invisible the politics of exploitation?" (p. 85). Moreover, as Lange and Lange (1984) found, although mathematics education can be empowering in a more general way, it is not necessarily the best approach in working with people on specific empowerment issues. The piece-rate workers they were organizing in the textile industry in the southern United States were struggling with a pay system made intentionally obscure, and the Langes felt it was more empowering to create a slide-rule distributed by the union that did the pay calculations for the workers, making the mathematics problem disappear, so that the workers could "focus on the social and economic relations underlying the way they are treated and paid" (p. 14).

On the other hand, the general empowerment through ethnomathematical knowledge is, we feel, a very important part of the struggle to overcome a colonized mentality, Samora Machel (1978) argues that

> colonialism is the greatest destroyer of culture that humanity has ever known. African society and its culture were crushed, and when they survived they were co-opted so that they could be more easily emptied of their content. This was done in two distinct ways. One was the utilization of institutions in order to support colonial exploitation. . . . The other was the "folklorising" of culture, its reduction to more or less picturesque habits and customs, to impose in their place the values of colonialism (p. 400).

As we have discussed above, the connections between educational action and liberatory social change are the least developed aspects of research activities in ethnomathematics. On this point, the work of Knijnik (1992) with the landless in Brazil is a notable exception, an example of a mathematics education researcher working with the *Movemento Sem Terra* (Landless People Movement) toward a shared goal of transforming the political economy of Brazil.[4] Our practice confirms that ethnomathematical knowledge increases student's self-confidence and opens up areas of critical insight in their understanding of the nature of knowledge. But there is no confirmation that this knowledge results in action against oppression and domination. In the current historical context of an advanced capitalist society, it may be that the most critical collective change that a pedagogy of the oppressed can bring about is a subtle shift in ideological climate that will encourage action for a just socialist economic and political restructuring. This is not insignificant. Nteta (1987) argues that "revolutionary self-consciousness [is] an objective force within the process of liberation" (p. 55). He shows how the aim of Steve Biko's theories and the Black Consciousness Movement in South Africa "to demystify power relations so that blacks would come to view their status as neither natural, inevitable nor part of the eternal social order . . . created conditions that have irreversibly transfigured South Africa's political landscape" (pp. 60–61). We argue that as we more clearly understand the limits of our educational practice, we will increase the radical possibilities of our educational action for liberatory change. Thus, we feel the most important area for ethnomathematical research to pursue is the dialectics between knowledge and action for change.[5] Fasheh (1982/reprinted here as Chapter 13) points the way to the direction of investigation by hypothesizing that

> teaching math through cultural relevance and personal experiences helps the learners know more about reality, culture, society, and themselves. That will, in turn, help them become more aware, more critical, more appreciative, and more self-confident. It will help them build new perspectives and syntheses, and seek new alternatives, and, hopefully will help them transform some existing structures and relations (p. 8).

Notes

1. A revised, updated version of the last chapter of their book is reprinted in this volume as chapter 17.

2. In this essay, Adams (1983) provides an incredible example of the exaggerated distortion of the knowledge of African peoples. To explain how the Dogon of Mali acquired their extensive astronomical knowledge some Western scientists went as far as hypothesizing that alien ships landed from outer space to tell the Dogon about the stars, refusing the *vastly* more credible hypothesis that, since lenses have been shown to date back to 2000 B.C., the Dogon may have placed one lens in front of the other and created a basic telescope (pp. 36–37, 41).

3. Lave's work on the inseparability of cognition and context and an ethnomathematical perspective of the meaning of "logical thought," forces us to reevaluate Luria's (1976) conclusions about reasoning and problem solving. His political project was progressive, stemming from a desire to show that with schooling peasants were intellectually equal to people in the other classes in Russian society. But his interpretations were limited by his static view of what "abstract" reasoning is. His cognitive psychological experiments in mathematics problem solving with unschooled people in remote Russian villages led him to theorize that these people could not abstract "the conditions of the problems from extraneous practical experience . . . and [derive] the appropriate answer from a system of reasoning determined by the logic of the problem rather than graphic practical experience" (p. 120). Luria felt that "the significance of schooling lies not just in the acquisition of new knowledge, but in the creation of new motives and formal modes of discursive verbal and logical thinking divorced from immediate practical experience" (p. 133). But, if as Lave recognizes, cognition occurs always inextricably seamlessly intertwined in practice then possibly Luria's experimental findings were the result of a cultural dissonance (affective, linguistic, ideological, and so on) between the experimenters and the experimented upon. And if, as an ethnomathematical view theorizes, we cannot and do not want to dichotomize the abstract and the practical, then we can conclude that school's attempts to do this are attempts to *obscure* knowledge in the interest of the status quo. We hope some people more versed in cognitive psychology than we are, will pursue these questions. Also, we think such a reevaluation of Luria would be an important case study in how even the most progressive, critical projects occur in a historical context which limits their work. *No* knowledge is static, all knowledge must be critically interrogated, reexamined, and recreated.

4. In a different context, Powell (see Baldassarre, Broccoli, Jusinski, and Powell, 1993) worked with students who acted on their critiques of educational materials, resulting in the banning of the distribution of those offensive materials. This points to differences in the terrain of political struggles where in the United States a strong, conscious oppositional political movement is lacking.

5. To this end, John Volmink and the editors have organized a Criticalmathematics Educators Group. Contact the editors for a copy of the group's newsletter.

References

Adams III, H. H. (1983). African observers of the universe: The Sirius question. In I. V. Sertima (ed.). *Blacks in science: Ancient and modern*, (pp. 27–46). New Brunswick, NJ: Transaction.

Baldassarre, D., Broccoli, M., Jusinski, M. M., and Powell, A. B. (1993). Critical thinking and critical collective action in education: Race, gender, and the mathematical establishment. In W. Oxman and M. Weinstein (eds.). *Critical Thinking as an Educational Ideal: Proceedings of the Fifth Annual Conference of the Institute for Critical Thinking*. October 22–24, 1992 (pp. 154–158). Montclair: New Jersey, Institute for Critical Thinking, Montclair State.

Carby, H. V. (1990, Sept./Oct.). The politics of difference. *Ms. Magazine*. pp. 84–85.

Fasheh, M. (1982). Mathematics, culture, and authority. *For the Learning of Mathematics* 3(2), 2–8.

Frankenstein, M. (1989). *Relearning mathematics: A different third R-Radical maths*. London: Free Association.

Gerdes, P. (1988). On some possible uses of traditional Angolan sand drawing in the mathematics classroom. *Journal of the Mathematical Association of Nigeria* 18(1): 107–125.

Knijnik, G. (1992). An ethnomathematical approach in mathematical education: A matter of political power. *For the Learning of Mathematics* 13(3): 23–26.

Lange, B., and Lange, J. (1984). Organizing piece-rate workers in the textile industry. *Science for the People* (May/June): 12–16.

Lave, J. (1988). *Cognition in practice*. Cambridge, England: Cambridge.

Lorde, A. (1984). *Sister outsider*. Trumansburg, NY: The Crossing Press.

Luria, A. R. (1976). *Cognitive development: Its cultural and social foundations*. Cambridge, MA: Harvard.

Machel, S. (1978). Knowledge and science should be for the total liberation of man. *Race & Class* 19(4): 399–404.

Nteta, C. (1987). Revolutionary self-consciousness as an objective force within the process of liberation: Biko and Gramsci. *Radical America* 21(5): 55–61.

Pinxten, R., Van Doren, I., and Harvey, F. (1983). *The anthropology of space*. Philadelphia: University of Pennsylvania.

Powell, A. B., Jeffries, D. A., and Selby, A. E. (1989). An empowering, participatory research model for humanistic mathematics pedagogy. *Humanistic Mathematics Network Newsletter* (4): 29–38.

Powell, A. B., and López, J. A. (1989). The role of writing in learning mathematics and science: A case study. In P. Connolly and T. Vilardi (eds.), *The Role of Writing in Learning Mathematics and Science* (pp. 157–177). New York: Teachers College.

Powell, A. B., and Ramnauth, M. P. (1992). Beyond questions and answers: Prompting reflections and deepening understandings of mathematics using multiple-entry logs. *For the Learning of Mathematics* 12(2): 12–18.

Weiler, K. (1994). Freire and a feminist pedagogy of difference. In P. L. McLaren and C. Lankshear (eds.). *Politics of liberation: Paths from Freire* (pp. 12–40). London: Routledge.

Youngman, F. (1986). *Adult education and socialist pedagogy*. London: Croom Helm.

Chapter 16

Survey of Current Work on Ethnomathematics

Paulus Gerdes

Editors' comment: Paulus Gerdes, a Mozambican mathematician and mathematics educator, has been a leading researcher in uncovering mathematical ideas embedded in African cultural practices and artifacts and in presenting these findings to the mathematical community. In this chapter, he presents a comprehensive survey and analysis of research in the field of ethnomathematics. This chapter is a revised version of a paper he presented at the annual meeting of the American Association for the Advancement of Science, Boston, MA, February 11–16, 1993.

For the first time in its history the annual meeting of the American Association for the Advancement of Science has this year in its program a session on Ethnomathematics. I thank the organizers, in particular Vice-President C. Davis of the American Mathematical Society, for the recognition of ethnomathematics as a scientific endeavor and for the honor to be invited to present the first AAAS survey on current work on Ethnomathematics. This is not an easy task. All over the world ethnomathematical research is being done and already hundreds of publications have seen light. In preparing this presentation, I had to reflect on questions such as

- What is ethnomathematics all about?
- How and why did ethnomathematics emerge?
- What research trends exist?
- What to include in the survey?
- Where and when does "current work" start?
- What concrete examples should be given?

As the other speakers in this session—U. D'Ambrosio (Brazil), M. Frankenstein (U.S.A.), M. Closs (U.S.A.), and G. Joseph (UK)—will address the audience on specific themes of ethnomathematical research in the Americas, Asia, and Europe, I will concentrate in my survey on the African continent and on Mozambique in particular.[1] It seems useful to start the survey with an historical introduction.

Ethnomathematics may be defined as the cultural anthropology of mathematics and mathematical education. As such it is a relatively new field of interest, that lies at the confluence of mathematics and cultural anthropology. As the view of mathematics as "culture-free" and "universal" has been dominant and maybe still is, ethnomathematics emerged later than other ethnosciences. Among mathematicians, ethnographers, psychologists, and educationalists, Wilder, White, Fettweis, Luquet, and Raum may be registered as forerunners of ethnomathematics.

Isolated Forerunners

In his address to the International Congress of Mathematicians in 1950, entitled *"The Cultural Basis of Mathematics,"* R. Wilder states that it is not new to look at mathematics as a cultural element: "Anthropologists have done so, but as their knowledge of mathematics is generally very limited, their reactions have ordinarily consisted of scattered remarks concerning the types of arithmetic found in primitive cultures" (Wilder, 1950, p. 260). He cites the essay *"The Locus of Mathematical Reality: An Anthropological Footnote"* (1947) by the anthropologist L. White as a noteworthy exception and refers also to C. Keyser's *"Mathematics As A Culture Clue"* (1932). Keyser defended the thesis that "The type of mathematics found in any major Culture is a clue, or key, to the distinctive character of the Culture taken as a whole" (Wilder, 1950, p. 261). Wilder comments "Since the culture dominates its elements, and in particular its mathematics, it would appear that for mathematicians it would be more fruitful to study the relationship from this point of view" (Wilder, 1950, p. 261). Wilder agreed with the anthropologist R. Linton, who observed in his *"The Study of Man"* (Linton, 1936, p. 319) that "The mathematical genius can only carry on from the point which mathematical knowledge within his culture has already reached."

Wilder summarized his ideas in the following way: "In man's various cultures are found certain elements which are called *mathe-*

matical. In the earlier days of civilization, they varied greatly from one culture to another so much that what was called 'mathematics' in one culture would hardly be recognized as such in certain others. With the increase in diffusion due, first, to exploration and invention, and, secondly, to the increase in the use of suitable symbols and their subsequent standardization and dissemination in journals, the mathematical elements of the most advanced cultures gradually merged until, except for minor cultural differences like the emphasis on geometry in Italy, or on function theory in France, there has resulted essentially one element, common to all civilized cultures, known as mathematics. This is not a fixed entity, however, but is subject to constant change. Not all of the change represents accretion of new material; some of it is a shedding of material no longer, due to influential cultural variations, considered mathematics. Some so-called 'borderline' work, for example, it is difficult to place either in mathematics or outside mathematics" (Wilder, 1950, p. 269, 270).

Later Wilder elaborated his ideas in two books *"Evolution of Mathematical Concepts"* (1968) and *"Mathematics As A Cultural System"* (1981). According to Smorynski, Wilder's thesis may be summarized as "Each culture has its own mathematics, which evolves and dies with the culture" (Smorynski, 1983, p. 11).

White starts his aforementioned study *"The Locus of Mathematical Reality . . ."* by asking the question "Do mathematical truths reside in the external world, there to be discovered by man, or are they man-made inventions?" (White, 1956, p. 2349). Looking for an answer, he states that "mathematics in its entirety, its 'truths' and its 'realities,' is part of human *culture*" (White, 1956, p. 2351) and "Mathematics is a form of behavior: the responses of a particular kind of primate organism to a set of stimuli" (White, 1956, p. 2353) and concludes that mathematical truths "are discovered but they are also man-made. They are the product of the mind of the human species. But they are encountered or discovered by each individual in the mathematical culture in which he grows up" (White, 1956, p. 2357). Mathematics did not originate with Euclid and Pythagoras—nor even in ancient Egypt or Mesopotamia, but, in White's view "Mathematics is a development of thought that had its beginning with the origin of man and culture a million years or so ago" (White, 1956, p. 2361).

Wilder and White were not aware of the studies by the German mathematician, ethnologist, and pedagogue E. Fettweis (1881–1967)[2] on early mathematical thinking and culture nor of the reflections of the French psychologist G. Luquet on the cultural origin of mathematical notions (Luquet, 1929).

Apparently O. Raum's book *"Arithmetic in Africa"* (1938) was not known among mathematicians and anthropologists of his time. It contains the substance of a course given in the colonial department of the University of London's Institute of Education. T. Nunn writes in the foreword that ". . . education . . . cannot be truly effective unless it is intelligently based on indigenous culture and living interests" (Raum, 1938, p. 4). O. Raum, son of a missionary and brought up in Africa, with teaching experience in South Africa and Tanganyika (currently, Tanzania) states in his preface that: "One of the untested assumptions made by many Europeans is that Africans have no gift for arithmetic. It may be true that the Africans to whom they teach arithmetic make mistakes. But so do European children. The fault may therefore not lie in the African pupils at all: it may be that the method of teaching or the type of arithmetic taught is wrong" (Raum, 1938, p. 5). The book makes three principal suggestions: "Firstly that the ordinary behavior of Africans provides concrete examples of the generalizations concerning the logical nature of number which have been evolved by the world's mathematical philosophers. Secondly it is suggested that, for teaching the African child to handle the system of numbers and to carry out operations with it, tribal activities, both adult and juvenile, with a numerical bearing, are the most suitable media. Thirdly it is suggested that, if generalizations and abstractions are to be acquired by the pupils as lasting instruments of thought, advanced arithmetical processes must be developed from the numerical problems of their own cultural background." One of the principles of good teaching "lays down the importance of understanding the cultural background of the pupil and relating the teaching in school to it" (Raum, 1938, p. 5).

The reflections of Wilder, White, Fettweis, Luquet, and Raum did not find a lot of echo. The prevailing idea in the first half of the century was that of mathematics as an universal, basically aprioristic form of knowledge. A reductionist tendency tended to dominate mathematics education, implying on culture-free cognition models.[3]

Ubiratan D'Ambrosio, the Intellectual Father of the Ethnomathematical Program

The failures of the hasty "New Mathematics" curriculum, transplanted from the North to the South in the 1960s; the importance attributed in the new politically independent states of the Third

World to education for all, including mathematical education, in the strive for economic independence; and the public unrest in the North about the involvement of mathematicians and mathematical research in the Vietnam war. . . . are some factors that stimulated reflection about the place and implications of mathematical research and teaching.

At the end of the 1970s and beginning of the 1980s, there started to be a growing awareness among mathematicians of the societal and cultural aspects of mathematics and mathematical education.[4] Indications of this are the session on the societal objectives of mathematical education and "Why to study mathematics?" at the 1976 International Congress on Mathematical Education (ICME3, Karlsruhe, Germany), the 1978 Conference on Developing Mathematics in Third World Countries (Khartoum, Sudan) [see El Tom, 1979], the 1978 Workshop on Mathematics and the Real World (Roskilde, Denmark) [see Booss and Niss, 1979], the session on Mathematics and Society at the 1978 International Congress of Mathematicians (Helsinki, Finland), the 1981 Symposium on Mathematics-in-the-Community (Huaraz, Peru), the 1982 Caribbean Conference on Mathematics for the Benefit of the Peoples (Paramaribo, Surinam). The Brazilian U. D'Ambrosio played a dynamizing role in all these initiatives. It is in that period that he launches his *ethnomathematical program*. At the fourth International Congress of Mathematics Education in 1984 (Adelaide, Australia) he presents in the opening plenary lecture his reflections on the "*Socio-cultural bases for mathematics education*" (cf. D'Ambrosio, 1985a).

D'Ambrosio proposes his ethnomathematical program as a "methodology to track and analyze the processes of generation, transmission, diffusion and institutionalization of (mathematical) knowledge" in diverse cultural systems (D'Ambrosio, 1990, p. 78). In contrast to "academic mathematics," that is, the mathematics taught and learned in the schools, D'Ambrosio calls "*ethnomathematics* the mathematics practiced among identifiable cultural groups, such as national-tribal societies, labor groups, children of a certain age bracket, professional classes, and so on" (D'Ambrosio, 1985b, p. 45). "The mechanism of schooling replaces these practices by other equivalent practices which have acquired the status of mathematics, which have been expropriated in their original forms and returned in a codified version" (D'Ambrosio, 1985b, p. 47). Before and outside school almost all children in the world become "matherate," that is, they develop the "capacity to use numbers, quantities, the capability of qualifying and quantifying and some patterns of inference" (D'Ambrosio, 1985a, p. 43). In school "the 'learned' matheracy eliminates the so-called 'spontaneous' matheracy. An individual who manages perfectly well numbers, operations, geometric forms and notions, when facing a completely new and formal approach to the same facts and needs creates a psychological blockage which grows as a barrier between different modes of numerical and geometrical thought" (D'Ambrosio, 1985a, p. 45). As a consequence, "the early stages of Mathematics education offer a very efficient way of instilling in the children a sense of failure and dependency" (D'Ambrosio, 1985a, p. 45). In other words "the mathematical competencies, which are lost in the first years of schooling, are essential at this stage for everyday life and labor opportunities. But they have indeed been lost. The former, let us say spontaneous, abilities have been downgraded, repressed and forgotten while the learned ones have not been assimilated either as a consequence of a learning blockage, or of an early drop-out, or even as a consequence of failure or many other reasons" (D'Ambrosio, 1985a, p. 46). The question which arises then is what to do: "should we . . . give up school mathematics and remain with ethnomathematics? Clearly not . . ." (D'Ambrosio, 1985a, p. 70). In D'Ambrosio's view, one should compatibilize cultural forms, that is, ". . . the mathematics in schools shall be such that it facilitates knowledge, understanding, incorporation and compatibilization of known and current popular practices into the curriculum. In other words, recognition and incorporation of ethnomathematics into the curriculum" (D'Ambrosio, 1985a, p. 71). In order to incorporate ethnomathematics into the curriculum, it is necessary to "identify within ethnomathematics a structured body of knowledge" (D'Ambrosio, 1985b, p. 47). Researchers have to try to find "underlying structure in these ad hoc practices. In other terms, we have to pose the following questions:

1. How are ad hoc practices and solution, of problems developed into methods?
2. How are methods developed into theories?
3. How are theories developed into scientific invention? (D'Ambrosio, 1985b, p. 48).

Some characteristics of this ethnomathematics are already known:

1. It is limited in techniques since it draws on narrow resources. On the other hand, its creative component is high, since it is unbound to formal rules obeying criteria unrelated to the situation.

2. It is particularistic since it is context bound, although it is broader than ad hoc knowledge, contrary to the universalistic character of mathematics, which ideally claims and aims to be context-free.
3. It operates through metaphors and systems of symbols which are psycho-emotionally related, while mathematics operates with symbols which are condensed in a rational way" (D'Ambrosio, 1987, p. 37–38).

Let us now briefly review other concepts that have been proposed and that are related to D'Ambrosio's ethnomathematics.

Gestation of New Concepts

Colonial education presented mathematics generally as something rather "western" or "European," as an exclusive creation of "white men." With the hasty curriculum transplantation—during the 1960s—from the highly industrialized nations to Third World countries there continued, at least implicitly, the *negation* of African, Asian, American-Indian, . . . mathematics.

During the 1970s and 1980s, among teachers and didacticians of mathematics in developing countries and later also in other countries emerged a growing resistance to this negation (cf. e.g. Njock, 1985), against the racist and neo-colonial prejudices, that it reflects, against the Eurocentrism of mathematics and its history.[5] Many teachers and didacticians stressed that beyond the "imported school mathematics" there also existed and continues to exist *other mathematics*.

In this context, various concepts have been proposed to contrast with the "*academic* mathematics" / "school mathematics" (i.e., the school mathematics of the transplanted, imported curriculum):

- *indigenous* mathematics [Cf. e.g. Gay and Cole, 1967; Lancy, 1976]. Criticizing education of Kpelle children (Liberia) in "western-oriented" schools—they "are taught things that have no point or meaning within their culture" (1967, p. 7)—Gay and Cole propose a creative mathematical education that uses the indigenous mathematics as starting point;
- *sociomathematics* of Africa [Zaslavsky, 1973]: "the applications of mathematics in the lives of African people, and, conversely, the

influence that African institutions had upon the evolution of their mathematics" (p. 7);[6]

- *informal* mathematics [Posner, 1978, 1982]: mathematics that is transmitted and that one learns outside the formal system of education;
- *mathematics in the (African) sociocultural environment* [S. Touré, S. Doumbia (Côte d'Ivoire), 1984]: integration of the mathematics of African games and craftwork that belongs to the social-cultural environment of the child in the mathematics curriculum;
- *spontaneous* mathematics [D'Ambrosio, 1982]: each human being and each cultural group develops spontaneously certain mathematical methods;[7]
- *oral* mathematics [Carraher et al., 1982; Kane, 1987]: in all human societies there exists mathematical knowledge that is transmitted orally from one generation to the next;
- *oppressed* mathematics [Gerdes, 1982]: in class societies (e.g., in the countries of the Third World during the colonial occupation) there exist mathematical elements in the daily life of the populations, that are not recognized as mathematics by the dominant ideology;
- *non-standard* mathematics [Carraher et al., 1982; Gerdes, 1982, 1985a; Harris, 1987]: beyond the dominant standard forms of "academic" and "school" mathematics there develops and developed in the whole world and in each culture mathematical forms that are distinct from the established patterns;
- *hidden* or *frozen* mathematics [Gerdes, 1982, 1985a, b]: although, probably, the majority of mathematical knowledge of the formerly colonized peoples has been lost, one may try to reconstruct or "unfreeze" the mathematical thinking, that is "hidden" or "frozen" in old techniques, like, for example, that of basket making;
- *folk* mathematics [Mellin-Olsen, 1986]: the mathematics (although often not recognized as such) that develops in the working activity of each of the peoples may serve as a starting point in the teaching of mathematics;
- mathematics *codified in know-how* [Ferreira, 1991].

These concept proposals are provisional. They belong to trends that emerged in the context of the Third World and that later on found an echo in other countries.[8]

The various aspects illuminated by the aforementioned provisional concepts have been gradually united under the more general "common denominator" of D'Ambrosio's ethnomathematics. This process has been accelerated by the creation of the *International Study Group on Ethnomathematics* [ISGEm] in 1985.

Illustrative Examples

Sugar-Cane Farmers in Brazil

"The sugar-cane farmers of northeastern Brazil have their own measures and procedures of determining land area, which are not taken into account in Brazilian mathematical education. While their procedures might at first appear to be simply incorrect, our interviews show that farmers indeed understand how measures of linear magnitudes are related to areas. The imprecision accepted by farmers appears not to be due to carelessness, but rather motivated by practical considerations and the preference for working with integer values" (Abreu and Carraher, 1989, p. 70).

Cowry Games in Côte d'Ivoire

In 1980, a research seminar on *"Mathematics in the African Socio-Cultural Environment"* was introduced at the Mathematical Research Institute of Abidjan (IRMA, Côte d'Ivoire), founded in 1975 by S. Touré. The seminar is directed by S. Doumbia. One of the interesting themes analyzed by her and her colleagues is the mathematics of traditional West-African games. Their work deals with classification of the games, solution of mathematical problems of the games and exploring the possibilities of using these games in the mathematics classroom. Their conclusion—as revealing as it is—that the rules of some games, like Nigbé Alladian, show a traditional, at least empirical knowledge of probabilities, will certainly stimulate further research.

One plays Nigbé Alladian with four cowry shells. In turn, each of the two players casts the cowry shells. When all four land in the same position, that is, all "up" or all "down," or when two land in the "up" position and the other two in the "down" position, the player wins points. In the other cases, one "up" and three "down," or three "up" and one "down," A participant does not win points. The researchers of IRMA found experimentally that the chance of a cowry shell to fall in the "up" position is two-fifths. The following table shows the probabilities of four "up," four "down," and so forth:

<table>
<tr><th>position</th><th>probability</th><th></th></tr>
<tr><td>all "up"</td><td>$\frac{16}{625}$</td><td rowspan="3">$\frac{313}{625}$</td></tr>
<tr><td>all "down"</td><td>$\frac{81}{625}$</td></tr>
<tr><td>two "up," two "down"</td><td>$\frac{216}{625}$</td></tr>
<tr><td>one "up," three "down"</td><td>$\frac{96}{625}$</td><td rowspan="2">$\frac{312}{625}$</td></tr>
<tr><td>three "up," one "down"</td><td>$\frac{216}{625}$</td></tr>
</table>

One sees that the rules of the game had been chosen in such a way that the chance to win points is (almost) the same as to win no points. S. Doumbia concluded "without any knowledge of calculation of probability, the players have managed . . . to adopt a clever counting system, in order to balance their chances. The probability of scoring some points is $\frac{313}{625}$ as against $\frac{312}{625}$." (Doumbia, 1989, p. 175)[9]

Market Women in Mozambique

Lecturers and students of the master's degree program in mathematics education for primary schools at Mozambique's Pedagogical University (Beira Branch) have been analyzing arithmetic in and outside school. On interviewing illiterate women to know how they determine sums and differences, it was found that the women "solved easily nearly all the problems, using essentially methods of oral / mental computation, i.e., computation based on the spoken numerals. The methods used were very similar to those suggested by the present day mathematics syllabus for primary education, but including some interesting alternatives" (J. Draisma, 1992, p. 110). For instance, 59 percent of the interviewed women calculated mentally $62–5 = ..?$ by first subtracting 2 and then 3. Schematically:

$$\begin{array}{l} 62 - 5 = ..? \\ \hline 62 - 2 = 60 \\ 60 - 3 = 57, \end{array}$$

That is, they used the same method as is emphasized in the schoolbook. Another 29 percent of the women subtracted first 5 from 60 and then added 2:

$$\begin{array}{l} 62 - 5 = ..? \\ \hline 60 - 5 = 55 \\ 55 + 2 = 57, \end{array}$$

and 12 percent subtracted first 10 from 62, and added the difference between 10 and 5, that is, 5. Schematically:

$$\begin{array}{l} 62 - 5 = ..? \\ \hline 62 - 10 = 52 \\ 10 - 5 = 5 \\ 52 + 5 = 57. \end{array}$$

Did these women *(re)invent* their method ? Did they *learn* them? From whom and how?

When multiplying, most of the interviewed women solve the problems by doubling. An example illustrates the process $6 \times 13 = ..?$ Schematically the solution is the following:

$$\begin{array}{l} 6 \times 13 = ..? \\ \hline 2 \times 13 = 26 \\ 4 \times 13 = 2 \times 26 \\ 2 \times 26 = 52 \\ 6 \times 13 = 26 + 52 \\ 26 + 52 = 78 \end{array}$$

(J. Draisma, oral communication, 1992). Does each of these women (re)invent the doubling method spontaneously? Or does there exist a tradition? If so, how is the method taught and learned?

Unschooled Peasants in Nigeria

Shirley (1988) and his students (e.g. Azenge, 1988) at the Ahmadu Bello University in Nigeria conducted oral interviews with unschooled, illiterate members of the students' home communities. They found that "although some of the (arithmetical) algorithms used by the informants are similar to those taught in schools, some interesting non-standard techniques were also found" (Shirley, 1988, p. 5). "Several respondents displayed a good number sense, especially in using decomposition, associativity, and distributivity to simplify into round numbers" (Shirley, 1988, p. 6). For instance, to determine 18 + 19 =..?, some used the following procedure: 1 removed from 18, leaves 17 and 1 combined with 19 gives 20, therefore 17 + 20 gives the same result as 18 + 19. Schematically:

$$\frac{18 + 19 = ..?}{18 - 1 = 17 \text{ and } 1 + 19 = 20}$$
$$17 + 20 = 37$$

Yet another approach to the same problem rounded both 18 and 19 to 20, noting that 2 and 1 had been respectively added; the two 20s give 40, and then the 2 and 1 are subtracted to yield 37. They noted that the illiterate respondents presented some creative solutions to division problems. "For 45 ÷ 3 =..?, one respondent apparently knew that 21 ÷ 3 is 7, and so decomposed 45 into 21 + 21 + 3. Dividing all the terms of this sum by 3 and adding gave 7+7+1 = 15" (Shirley, 1988, p. 7). Shirley gives the advise to assign teacher-student to find (ethno)algorithms in their communities—literate or illiterate, rural or urban, as "Too often, school lessons leave the impression that there is only one way to do a given task" (Shirley, 1988, p. 9).

Ethnomathematics as a Field of Research

In the previous sections, ethnomathematics was the mathematics of a certain sub-culture. In that sense so-called academic mathematics is also a concrete example of ethnomathematics. When all ethnomathematics is mathematics, why call it ethnomathematics? And not simply the mathematics of this and that sub-culture?

Doing so, ethnomathematics may be defined at another level, as a research field, that recognizes the existence of many mathematics, particular in a way to certain sub-cultures.

As a research field, ethnomathematics may be defined as the cultural anthropology of mathematics and mathematical education, or in the formulation of D'Ambrosio in 1977: "*Ethnoscience* as the study of scientific and, by extension, technological phenomena in direct relation to their social, economic and cultural background" (D'Ambrosio, 1987, p. 74). In this sense, it includes "the study of mathematical ideas of non-literate peoples," which is how Ascher defined ethnomathematics in 1986.[10]

Among ethnographers and anthropologists the concept of "ethnoscience" has been used since the end of the nineteenth century. Their use of the concept seems to be more restricted and to have a different ideological connotation than the concept as it is used nowadays by mathematicians and mathematics educators.

In the ethnological dictionary of Panoff and Perrin (1973), two definitions of the concept of ethnoscience are presented. In the first case, it is a "branch of ethnology that dedicates itself to the comparison between the positive knowledge of exotic societies and the knowledge that has been formalized in the established disciplines of western science" (Panoff & Perrin, 1973, p. 68) This definition raises immediately some questions. "What is *positive* knowledge?" "In what sense exotic?" "Does there exist a *Western* science?" In the second case, "each application of one of the Western scientific disciplines to natural phenomena which are understood in a different way by indigenous thinking" is called ethnoscience (Panoff & Perrin, 1973, p. 68). Both definitions belong to a tradition that traces back to the colonial time when ethnography was born in the most "developed" countries as a "colonial science," that studied almost exclusively the cultures of subjected peoples, also a "science" that opposed the so-called primitive thinking to the Western thinking as somehow absolutely different.

Among ethnographers there exists also another current, one that considers ethnoscience in a strikingly different way. For example, C. Favrod (1977) characterizes ethnolinguistics in his introduction to social and cultural anthropology as follows: "Ethnolinguistics tries to study language in its relationship to the whole of cultural and social life" (Favrod, 1977, p. 90). When we transfer this characterization of ethnolinguistics to ethnomathematics, we obtain by analogy: "*Ethnomathematics tries to study mathematics (or mathematical ideas) in its (their) relationship to the whole of cultural and social life.*" In this sense, ethnomathematics comes near to the *sociology of mathematics* (1942) of D.

Struik: "Mathematics has its roots in the outside world, or nature . . .; society is only part of this, directing what and how these roots will continue to sprout" (Struik, 1986, p. 297). H. Bos and H. Mehrtens (1977) published a literature review on the interactions between mathematics and society.

According to T. Crump, the term ethnoscience became popular among ethnographers in the 1960s: "It may be taken to refer to the 'system of knowledge and cognition typical of a given culture'" (Crump, 1990, p. 160).[11] In Crump's *"The Anthropology of Number"* (1990) there is almost no reference to the work of "ethnomathematicians." He points out that "first, few professional mathematicians have any interest in the cognitive assumptions in their work; second, few anthropologists are numerate in the sense of being able to realize how significant the numbers that occur in the course of their field work might be in the local culture" (Crump, 1990, p. viii). Still in the 1990s, anthropologists, historians of science, and mathematicians need to collaborate with each other to develop ethnomathematics as an anthropology of mathematics (D'Ambrosio uses sometimes the expression "anthropological mathematics," e.g. 1985b; cf. Gerdes, 1985c).

Ethnomathematical Movement (cf. Gerdes, 1989a)[12]

The scholars who are engaged in ethnomathematical research are normally highly motivated. In this sense we might speak of an *ethnomathematical movement*, that may be characterized as follows:

- *Ethnomathematicians* use a broad concept of mathematics, including, in particular, counting, locating, measuring, designing, playing, explaining;[13]
- *Ethnomathematicians* emphasize and analyze the influences of sociocultural factors on the teaching, learning, and development of mathematics;
- Ethnomathematicians draw attention to the fact that mathematics (its techniques and truths) is a *cultural product*; they stress that every people—every culture and every subculture—develops its own particular mathematics. Mathematics is considered to be a *universal, pan-human activity*. As a cultural product, mathematics has a history. Under certain economic, social, and cultural conditions, it emerged and developed in certain directions; under other conditions, it emerged and developed in

other directions. In other words, the development of mathematics is *not unilinear* (Cf. Ascher & Ascher, 1986, p.139, 140).

- *Ethnomathematicians* emphasize that the school mathematics of the transplanted, imported "curriculum" is *apparently* alien to the cultural traditions of Africa, Asia, and South America. Apparently this mathematics comes from the outside of the Third World. *In reality*, however, a great part of the contents of this "school mathematics" is of African and Asian origin. It became expropriated in the process of colonization that greatly destroyed the (scientific) culture of the oppressed peoples.[14] Then colonial ideologies ignored or despised the survivals of African, Asian, and American-Indian mathematics. The mathematical capacities of the peoples of the Third World became negated or reduced to rote memorization. This tendency has been reinforced by the curriculum transplantation (New Math) from the highly industrialized nations to Third World countries in the 1960s.
- *Ethnomathematicians* try to contribute to the knowledge of the mathematical realizations of the formerly colonized peoples. They look for culture elements, that survived colonialism and that reveal mathematical and other scientific thinking. They try to *reconstruct* these mathematical thoughts.
- *Ethnomathematicians* in Third World countries look for mathematical traditions that survived colonization and for mathematical activities in people's daily life and analyze ways to incorporate them into the curriculum.
- *Ethnomathematicians* also look for other *culture elements and activities* that may serve as a *starting point* for doing and elaborating mathematics in the classroom.
- In the educational context *Ethnomathematicians* generally favor a critical mathematics education that enables the students to reflect on the reality they live in and empowers them to develop and use mathematics in an emancipatory way. The influence of the well-known radical Third World pedagogue P. Freire is visible.

Paulo Freire and Ethnomathematics

A series of scholars working in the field of ethnomathematics, like P. Gerdes (1975),[15] M. Frankenstein (1981, 1983, 1989), S. Mellin-

Olsen (1986), M. Borba (1987a), E. Ferreira (1992), and M. Frankenstein and A. Powell (1994) have paid tribute to the Brazilian pedagogue P. Freire. His ideas, in particular, through his books *"Pedagogy of the Oppressed"* (1970) and *"Education for Critical Consciousness"* (1973) influenced deeply their reflection. In Mellin-Olson words: "If knowledge is related to culture by the processes which constitute knowledge—as Freire expresses it—this must have some implication for how we treat knowledge in the didactic processes of (mathematical) education" (Mellin-Olsen, 1986, p. 103). Freire (1987) himself included the paper *"Ethnomathematics: The Mathematics of a 'Favela' (slum) in a Pedagogical Proposal"* written by his and D'Ambrosio's student, M. Borba, in his book *"In The School We Make . . .: An Interdisciplinary Reflection on People's Education."* M. Frankenstein and A. Powell argue that reconsidering what educators value as mathematical knowledge, considering the effect of culture on mathematical knowledge, and uncovering the distorted and hidden history of mathematical knowledge are significant contributions of a *Freirean*, ethnomathematical perspective in reconceiving the discipline of mathematics and its pedagogical practice (Frankenstein and Powell, 1991, p. 14). The use of Freire's dialogical methodology is seen as essential in developing "the curricular praxis of ethnomathematics by investigating the ethnomathematics of a culture to construct curricula with people from that culture and by exploring the ethnomathematics of other cultures to create curricula so that people's knowledge of mathematics will be enriched" (Frankenstein and Powell, 1991, p. 32).

In the following, I will present an overview of ethnomathematical literature continent by continent, ending the chapter with a discussion of ethnomathematical research in Mozambique.

Americas

M. and R. Ascher, mathematician and anthropologist, published in 1981 *"Code Of The Quipu: A Study in Media, Mathematics and Culture,"* showing how Peruvian pieces of string served to embody a rich, logical, numerical tradition. M. Closs edited the book *"Native American Mathematics"* (1986). The editor states in his preface that ". . . . native American mathematics can best be described as a composite of separate developments in many individual cultures" (Closs, 1986, p. 2). The book analyses number systems, numerical representation in rock art, calendrial systems, tallies and ritual use of number, and some aspects of geometry. Ethnomathematical research methodology is a point of concern. Closs remarks that the "papers give some idea as to the form which the history of mathematics must take if it is to incorporate material outside of its traditional boundaries. It is a form in which an almost total reliance on the historical approach is supplemented or replaced by drawing on the resources and methodologies of other disciplines such as anthropology, archaeology and linguistics" (Closs, 1986, p. 2). R. Pinxten et al. (1983) studied the "geometrical" world view of the Navajo and formulated suggestions for mathematics education (Pinxten, 1989). C. Moore (1986) analyzed the use of string figures for "Native American mathematics education" (cf. Moore, 1987, 1994). J. Marschall (1987) elaborated an "atlas" of American Indian geometry.

E. Ferreira (1988, 1989) and his students at Campinas State University have analyzed mathematics and mathematics teaching among Indian communities in Brazil. For example, L. Paula and E. Paula (1986) studied string figures among the Tapirapé Indians.

C. Cossio and A. Jerez (1986) published a study on mathematics in the Quichua (Ecuador) and Spanish language.

A series of important studies have been realized by D'Ambrosio's students. For example, the aforementioned M. Borba (1987) analyzed the mathematics in the daily life of the population of a slum in Campinas; S. Nobre studied the mathematics of the popular animal lottery and wrote a masters thesis entitled "Social and Cultural Aspects of Mathematics Curriculum Development" (1989a) (cf. Nobre, 1989b); R. Buriasco completed a masters thesis on "Mathematics outside and inside School: From Blockage to Transition" (1989); G. Knijnik concluded a doctoral dissertation on the mathematics in the struggle for life of landless peasants in the southern Brazilian state Rio Grande do Sul (cf. Knijnik, 1993, reprinted here as chapter 18).

In Pernambuco (northeastern Brazil), important research on the borderline between ethnomathematics and cognitive psychology is done by the school of T. Carraher (Nunes), A. Schliemann, and D. Carraher, to which I referred already above.[16] For example, Schliemann (1984) analyzed mathematics among carpentry apprentices; T. Carraher (1988) compared street mathematics and school mathematics, and T. Carraher et al. (1987) analyzed the differences between written and oral mathematics; G. Saxe (1988) reported on candy selling and math learning. The first edition of the Brazilian journal *"A Educação Matemática"* published in 1993 is dedicated to ethnomathematics. It contains contributions from U. D'Ambrosio, E. Ferreira, L. Meira, G. Knijnik, and M. Borba.

There has been cooperation in ethnomathematical studies between Mozambique and Brazil. In Brazil, Mozambican researchers conducted a series of workshops on ethnomathematics: in April and May 1988, P. Gerdes (State University of São Paulo, Rio Claro), in January 1992 P. Gerdes with the assistance of M. Cherinda and D. Soares (Federal University of Santa Catarina, Florianópolis) and in January 1993, D. Soares (Federal University of Santa Catarina, Florianópolis). P. Gerdes (1989b) conducted a study on arithmetic and geometrical decoration of Indian baskets from Brazil.

In Colombia, V. Albis (1988) analyzed some aspects of ritual geometry among Indian populations. A. Cauty and his collaborators analyze possibilities of mathematics education in the context of what they call "ethnoeducation of the indigenous populations of Colombia" (cf. Cauty, 1994, 1995).

The newsletter of the International Study Group on Ethnomathematics published a series of short articles on ethnomathematical research and education in North America. D. Orey (1989) analyzed "Ethnomathematical Perspectives on the NCTM Standards"; G. Gilmer (1990) proposed an "Ethnomathematical Approach to Curriculum Development"; B. Lumpkin (1990) commented on "A Multicultural Mathematics Curriculum"; C. Zaslavsky (1989) argued for "Integrating Math With The Study Of Cultural Traditions," and "World Cultures In The Mathematics Class"; L. Shirley (1991) analyses mathematics in "kid culture" "Video Games For Math: A Case For 'Kid Culture.'" J. Stigler and R. Baranes (1988) published a review of research on culture and mathematics learning.

One of the forerunners of ethnomathematical interest in the U.S.A. seems to be H. Ginsburg and his students Petitto and Posner. In 1978, his students concluded their Ph.D. dissertations on mathematical knowledge in professional groups such as cloth merchants and tailors and two ethnic groups in Côte d'Ivoire and did comparative testing. In his paper "Poor Children, African Mathematics and the Problem of Schooling," Ginsburg draws as a lesson: "The moral for American researchers is clear. If poor children do badly on some tests, the likelihood is greater that there is a problem with the test than with the child" (Ginsburg, 1978, p. 41). Therefore ". . . teaching of basic skills could be more effective if the curricula were oriented to the particular styles of each culture." "For African children, the answers seem obvious: to be effective, curricula must be responsive to local culture." Ginsburg maintains that "the same is likely to be true for subgroups of the American poor" (Ginsburg, 1978, p. 42, 43).

In the context of the influence of P. Freire, I referred already to the work of M. Frankenstein and A. Powell. Their Criticalmathematics Educators Group is actively involved in the mathematical empowering of nontraditional students (cf. Frankenstein & Powell, 1989).

The multiculturalization of the mathematics curriculum[17] is one way to increase (cultural) self-confidence among nontraditional students. M. Ascher (1991) joined and adapted a series of her earlier papers on mathematical ideas in "non-Western" societies in the book *Ethnomathematics: A Multicultural View of Mathematical Ideas.* The book contains chapters on numbers, graphs in the sand, logic of kin relations, chance and strategy in games and puzzles, organization and modeling of space and symmetric strip decorations. G. Gilmer, M. Thompson, and C. Zaslavsky (1992) prepared multicultural mathematical activities for children from kindergarten through grade eight (cf. Zaslavsky, 1992). J. Ratteray (1992) published an African-centered approach for the multicultural curriculum, including mathematics.[18] B. Lumpkin has prepared a multicultural mathematics book for children.

Asia, Oceania, and Australia

R. Souviney (1989) described the results with the Indigenous Mathematics Project that was started in 1976 in Papua New Guinea. Earlier in 1983, D. Lancy published the book *Cross-cultural Studies In Cognition And Mathematics* where results of cognitive testing in Papua New Guinea and in the U.S.A. are compared. A. Bishop (1978, 1979) analyzed spatial abilities. G. Saxe (1981, 1982a, 1982b) conducted a series of studies of body counting and arithmetic among the Oksapmin of Papua New Guinea. Lean published in 1986 a research bibliography on counting systems on the same island.

P. Harris (1984) and K. Crawford (1984, 1989) analyzed mathematics education in Australian Aboriginal communities. The Western Australia Institute of Technology published in 1985 the book *Learning, Aboriginal World View and Ethnomathematics* by R. Hunting (cf. Hunting, 1987). Graham (1988) analyses the mathematical education of Aboriginal children.

G. Knight (1984) published two papers on the geometry of Maori art—on weaving and rafter patterns. M. Ascher studied mathematical aspects of Maori games. B. Barton wrote a paper entitled "Using The Trees to See The Wood: An Archaeology Of Mathematical Structure In New Zealand" (1990), and prepares a Ph.D. dissertation that has as a theme "A Philosophical Justification For Ethnomathematics And

Some Implications For Education" (cf. discussion paper with the same title, 1992).

In the last chapter of his book Language And Mathematics Education, R. Zepp provides a theoretical discussion of ethnomathematics.

M. Ascher (1988a) and P. Nissen (1988) analyzed mathematical aspects of sand drawings in the New Hebrides. P. Gerdes (1989c, 1994a: chapter 11) did a study of threshold designs among the Tamil of South India.

In a series of studies J. Turner discussed primary mathematics education and ethnomathematics in Bhutan in the eastern Himalayas (e.g. Turner, 1992).

M. Fasheh (1982, 1989) analyzed the cultural conflicts arising in mathematics education on the West Bank of occupied Palestine.

Europe

Mellin-Olsen and his colleagues at the Bergen Institute of Education organized in 1985 a seminar on "Mathematics and Culture" with participants from the Scandinavian countries, Great Britain, France, Mexico, and Mozambique (cf. Bonilla-Rius, 1986). It seems to have been the first meeting with cultural issues as a specific theme in mathematics education in the European context. M. Harris expressed (1987, p. 26) how ethnomathematical research in Africa stimulates critical reflection on mathematics education in the English setting, in view of women, "working class" and minority emancipation. Her "Maths in Work Project" based at the University of London Institute of Education attempts to "make a reconciliation between school ideal and work practice by bringing mathematically rich activities of daily life into school as resources to be developed mathematically by teachers" (Harris & Paechter, 1991, p. 278). Textiles and textile activities are an example of such a "very rich mathematical resource, one that is common and natural to all cultures and both sexes" (M. Harris, 1988, p. 28). If pupils of both sexes and all social and cultural backgrounds "become self confident in their recognition of mathematics as some thing they do and enjoy as part of everyday working life, then there is surely more chance of them developing the positive and confident attitudes that employers say they want" (Harris and Peachter, 1991, p. 282).[19]

Interest in multicultural issues in mathematics education is growing in France. In 1992, the French edition of the newsletter of the International Study Group on Ethnomathematics was launched by the Institute for Research in Mathematics Education. P. Damerow (1992) underlines that it is urgent in Germany to reflect on the issues raised by ethnomathematicians all over the world. S. Shan and P. Bailey (1991) and D. Nelson, G. Joseph, and J. Williams (1993) analyze the necessity and practice of multicultural mathematics education for a more just and equal society in the British context. O. Skovmose (1994) sees ethnomathematical studies as an important contribution to the realization of a "critical mathematics education."

Ongoing Ethnomathematical Research In/On Africa[20]

I have already mentioned Raum's *Arithmetic in Africa* (1938), Gay and Cole's *New Mathematics and an Old Culture* (1967), and Zaslavsky's classical *Africa Counts* (1973).

From the San hunters in Botswana, Lea (1987, 1989a, b, c, 1990a, b) and her students at the University of Botswana have collected information. Her papers describe counting, measurement, time reckoning, classification, tracking, and some mathematical ideas in San technology and craft. Educational suggestions are included in Stott and Lea (1993).

During the last years, a whole series of research projects on spoken and written numeration systems in Africa is being carried out, for example, on:

- counting in traditional Ibibio and Efik societies (I. O. Enukoha, University of Calabar, Calabar, Nigeria);
- numeration among the Fulbe (Fulani) (S. O. Ale, Ahmadu-Bello-University, Bauchi, Nigeria);
- pre-Islamic ways of counting (Y. Bello, Bayero University, Nigeria);
- counting in Nigerian languages (Ahmadu-Bello-University, Zaria);
- pre-colonial numeration systems in Burundi (J. Navez, University of Burundi, Bujumbara);
- learning of counting in Côte d'Ivoire (cf. Tro, 1980, Zepp, 1983);
- numeration systems used by the principal linguistic groups in Guinea (S. Oulare, University of Conakry);

- counting among the various ethnic groups in Kenya (J. Mutio, Kenyatta University, Nairobi);
- traditional counting in Botswana (H. Lea, University of Botswana, Gaberone);
- number and pattern in selected cultures in Uganda (E. Segujja-Munagisa);
- numeration systems and popular counting practices in Mozambique.

E. Kane's Ph.D. thesis "The Spoken Numeration Systems of West-Atlantic Groups and of the *Mandé*" (1987) analyses numeration in about twenty languages spoken in Senegal. He shows that spoken numeration systems, like the one of the *Mandé*, are susceptible to reform and evolution. Kane develops a methodology for the analysis of numeration systems that is adapted to the specificities of "oral cultures." V. Mubumbila (1988) wrote a study on numerology in Central Africa.

Zaslavsky (1973, 109–110) presents a riddle from the Kpelle (Liberia) about a man who has a leopard, a goat, and a pile of cassava leaves to be transported across a river, whereby certain conditions have to be satisfied. Ascher (1990) places this river crossing problem in a cross-cultural perspective and analyses mathematical-logical aspects of story puzzles of this type from Algeria, Cape Verde Islands, Ethiopia, Liberia, Tanzania, and Zambia. Kubik (1990) recorded "arithmetical puzzles" from the Valuchazi (eastern Angola and northwestern Zambia). S. Doumbia (1992, 1993a, 1993b) conducts pedagogical experiments with traditional verbal and cowry games (Côte d'Ivoire) (cf. Doumbia and Pil, 1992).

Continuing earlier research by Crowe (1971, 1973, 1975, 1982a, b), Washburn, and Crowe's book *Symmetries of Culture, Theory and Practice of Plane Pattern Analysis* (1988) classifies a number of patterns from African contexts on the basis of the so-called twenty-four plane groups due to Federov (1891). More recently Washburn (1990: ch. 5) showed how a symmetry analysis of the raffia patterns can differentiate patterns produced by the different Bakuba groups. Zaslavsky (1979) gives some examples of strip and plane patterns, and of bilateral and rotational symmetries, occurring in African art, architecture, and design. Why do symmetries appear in human culture in general, and in African craft work and art, in particular? This question is addressed by Gerdes in a series of studies (cf. Gerdes, 1985b, 1990a, 1991a). He analyses the origin of axial, double axial, and rotational symmetry of order four in African basketry. In "Fivefold Symmetry and (basket) Weaving in Various Cultures," it is shown how fivefold symmetry emerged quite "naturally" when artisans were solving some problems in (basket)weaving (Gerdes, 1992a). The examples chosen from Mozambican cultures range from the weaving of handbags, hats, and baskets to the fabrication of brooms.

Langdon (1989, 1990) describes the symmetries of "adinkra" cloths (Ghana) and explores possibilities for using them in the classroom. In a similar perspective, M. Harris (1988) describes and explores not only the printing designs on plain woven cloths from Ghana, but also symmetries on baskets from Botswana and "buba" blouses from the Yoruba (Nigeria).

I reported already the work of Doumbia on traditional African games (cf. also Doumbia, 1994). Vergani (Open University, Lisbon) prepares a monograph on mathematical aspects of intellectual games in Angola. Mve Ondo (1990) published a study on two "calculation games," that is, the "Mancala" games, Owani (Congo) and Songa (Cameroon, Gabon, Equatorial Guinea) (cf. A. Deledicq and A. Popova, 1977).

The Faculty of Education of the Ahmadu-Bello-University (Zaria, Nigeria) has been very dynamic in stimulating ethnomathematical research, for example, on the mathematics used by unschooled children and adults in daily life, and the possibilities to embed this knowledge in mathematics education.[21] S. Ale (1989) does research on the mathematical heritage of the Fulbe (Fulani) and the possibilities to construct a curriculum that builds upon this heritage and fits the needs of the Fulbe (Fulani) people. G. Thomas-Emeagwali included a chapter on mathematics in her recent book on science and technology in African history with case studies from Nigeria, Sierra Leone, Zimbabwe, and Zambia.

G. Aznaf prepares a Ph.D. dissertation on ethnomathematics in Ethiopia. D. Mtetwa started a research project on "Mathematical Thought in Aspects of Shona Culture."

At the regional conference "Mathematics, Philosophy, and Education" (Yamoussoukro, Côte d'Ivoire, Jan. 1993), S. Doumbia (Côte d'Ivoire) and P. Gerdes (Mozambique) conducted jointly a workshop on the didactical uses of traditional African games, drawings, and craft work. The newly created, post-Apartheid Association for Mathematics Education in South Africa (AMESA) organized at its first national congress in 1994 a round table on ethnomathematics and education. Later in the year AMESA formed a study group on ethnomathematics.

Ethnomathematical Research in Mozambique

Ethnomathematical research started in Mozambique in the late 1970s. As most "mathematical" traditions that survived colonization and most "mathematical" activities in the daily life of the Mozambican people are not explicitly mathematical, that is, the mathematics is "hidden," the first aim of this research was to "uncover" the "hidden" mathematics. The first results of this "uncovering" are included in the book *On the Awakening of Geometrical Thinking* (1985b, c) and slightly extended in *Ethnogeometry: Cultural-anthropological Contributions to the Genesis and Didactics of Geometry* (1991a).

In the papers "On Culture, Mathematics and Curriculum Development In Mozambique" (1986) and "On Culture, Geometrical Thinking and Mathematics Education" (1988), Gerdes summarized his experiments by incorporating traditional African cultural elements into mathematics education. The second paper presents alternative constructions of Euclidean geometrical ideas developed from the traditional culture of Mozambique. As well as establishing the educational power of these constructions, the paper illustrates the methodology of "cultural conscientialization" in the context of teacher training. The papers "A Widespread Decorative Motif and the Pythagorean Theorem" (1988) and "How Many Proofs of the Pythagorean Proposition do there Exist?" are more elaborated in the book African Pythagoras. In a paper entitled "A Study In Culture And Mathematics Education" (1992b, 1994c), Gerdes shows how diverse African ornaments and artifacts may be used to create a rich context for the discovery and the demonstration of the so-called Pythagorean Theorem and of related ideas and propositions. A series of ethnomathematical papers are included in the books *Etnomatemática: Cultura, Matemática, Educação,* (1991b), and Ethnomathematics And Education In Africa. (1995a).

In *SONA Geometry: Reflections on The Tradition of Sand Drawings in Africa South of the Equator* (1993a, b, 1994a, 1994d, 1995c), Gerdes tried to reconstruct mathematical components of the Tchokwe drawing-tradition (Angola)[22] and to explore their educational, artistic and scientific potential. In an earlier article "On Possible Uses of Traditional Angolan Sand Drawings in the Mathematics Classroom" (1988d, e), Gerdes analyzed already some possibilities for educational incorporation of this tradition. In the paper "Find The missing Figures" (1988f) and in the book *Lusona: Geometrical Recreations of Africa* (1991c), Gerdes presents mathematical amusements that are inspired by the geometry of the sand drawing tradition. For children (age 10–15) the booklet *Living Mathematics: Drawings of Africa* (1990b) has been elaborated. Experimentation with the use of "sona" in teacher education is described in "Exploring Angolan Sand Drawings (sona): Stimulating Cultural Awareness in Mathematics Teachers" (1993c). A overview of this research is given in "On Mathematical Elements in the Tchokwe *sona* Tradition" (1990c).

In recent years, more lecturers and in particular young lecturers, who returned home after having studied abroad have become interested in and started ethnomathematical research. At the third Pan-African Congress of Mathematicians (Nairobi, 1991) A. Ismael presented a communication on "The Origin of the Concepts of 'Even' and 'Odd' in Macua Culture (Northern Mozambique)" (cf. Ismael, 1994) and M. Cherinda delivered a paper on "Mental Arithmetic and the Tsonga Language (Southern Mozambique)." At the eighth Symposium of the Southern Africa Mathematical Sciences Association (Maputo, 1991) D. Soares presented a communication on "Popular Counting Practices in Mozambique" (cf. Soares and Ismael, 1993, 1994). At the First Conference on Mathematics Education in Africa (Cairo, 1992) M. Cherinda presented a paper on "A Children's 'Circle of Interest' in Ethnomathematics" (cf. Cherinda, 1994a, b) and J. Draisma a communication on "Mental Addition and Subtraction in Mozambique" (cf. Draisma, 1994). Also a series of students at Mozambique's Pedagogical University (Maputo and Beira) became interested in ethnomathematical research. Two students completed in 1991 master's theses in the field of ethnomathematics: "Symmetries of Ornaments On Baskets of the 'khuama' type" (E. Uaila), and "Symmetries of Ornaments On Metallic Window Gratings In the City of Maputo" (A. Mapapá) (cf. Mapapá, 1994).

In *Numeration in Mozambique—Contribution to a Reflection on Culture, Language and Mathematical Education* (Gerdes, 1993d), papers by M. Cherinda, A. Ismael, D. Soares, A. Mapapá, E. Uaila, and J. Draisma are included. Another collective study, *Explorations in Ethnomathematics and Ethnoscience in Mozambique* (Gerdes, 1994e), contains ethnomathematical contributions from A. Ismael, M. Cherinda, D. Soares, and A. Mapapá. Gerdes and Bulafo (1994) published a book on the geometrical knowledge of the mostly female weavers of the *sipatsi* hand bags. This investigation of mathematical knowledge by females has been continued in the study by Gerdes (1995b) on women and geometry in Southern Africa, where he suggests further research.

Notes

1. Very informative is the newsletter of the International Study Group on Ethnomathematics (ISGEm Newsletter), that may be obtained from the editor P. Scott, c/o College of Education, University of New Mexico, Albuquerque, NM 87131, U.S.A.

2. For the list of his publications, see: K. Reich, M. Folkerts, and C. Scriba, Schriftenverzerchnis von Ewald Pettweis (1881–1967) samt einer Würdigung von Olindo Falsirol, *Historia Mathematica* Vol. 16 (1989): 360–372.

3. Cf. U. D'Ambrosio's analysis, 1987, p. 80

4. Cf. also the bibliographic guide by B. Wilson (1981).

5. Cf. e.g. the studies of G. Joseph (1987, 1989, 1991) and S. Anderson (1990).

6. D'Ambrosio used in 1976 the same term in the context of Brazil. See U. D'Ambrosio (1976).

7. Students and colleagues of D'Ambrosio, like Carraher, Schliemann, Ferreira and Borba published many interesting examples of this spontaneous mathematics.

8. In the context of his historical research on Ancient Mesopotamia, J. Høyrup introduced the concept of "subscientific mathematics" (cf. Høyrup, 1994).

9. Interesting examples are given in the book by S. Doumbia and J. Pil (1992), and in Doumbia (1993a).

10. At that time the Aschers were not aware of the work of D'Ambrosio. Cf. M. Ascher and R. Ascher (1986), and M. Ascher (1984).

11. Cf. also W. Sturtevant (1972).

12. The international journal *For the Learning of Mathematics* published in 1994 a special issue on ethnomathematics—edited by U. D'Ambrosio and M. Ascher—with contributions from C. Zaslavsky, C. Moore, A. Bishop, P. Gerdes, R. Pinxten, V. Katz, R. Bassanezi, M. Ascher and U. D'Ambrosio.

13. See the chapter "Environmental activities and mathematical culture," in A. Bishop (1988a); important is his forthcoming book *Mathematical Acculturation—Cultural Conflicts in Mathematics Education*, where it is assumed that all mathematics education is in a process of cultural interaction, and that every child experiences some degree of cultural conflict in the process (cf. Bishop, 1994).

14. Cf. e.g. Bishop (1989):

"One of the greatest ironies is that several different cultures and societies contributed to the development of [the so-called] Western Mathematics—the Egyptians, the Chinese, the Indians, the Moslems, the Greeks as well as the Western Europeans. Yet when Western cultural imperialism imposed its version of mathematics on the colonized societies, it was scarcely recognizable as anything to which these societies might have contributed . . ."; cf. A. Bishop (1990).

15. Cf. the sections "The theory of Paulo Freire" and "The problemizing method, a problem?" in Gerdes (1975); cf. Gerdes (1985a).

16. See "oral mathematics" and "sugar cane farmers in Brazil."

17. P. Wilson (Mathematics Education Department, University of Georgia, 105 Aderhold Hall, Athens, GA 30602) prepared an annotated bibliography of multicultural issues in mathematics, 1992.

18. See also the journal "Africa and the World" edited by Ratteray's Institute for Independent Education.

19. Cf. also T. Smart & Z. Isaacson (1989) and J. Evans (1989).

20. For a bibliography on ethnomathematics and the history of mathematics in Africa south of the Sahara, see Gerdes (1994b). Cf. the information on the history of mathematics in Africa in the *AMUCHMA Newsletter* (AMU = African Mathematical Union; AMUCHMA = AMU Commission on the History of Mathematics in Africa). In order to obtain the English version of the version, write to the editor, P. O. Box 915, Maputo, Mozambique.

21. For a survey of these research projects, see: L. Shirley (1986a, b, 1988).

22. Cf. also M. Ascher (1988b) and chapter 2 in M. Ascher (1991).

References

G. Abreu and D. Carraher. (1989). The mathematics of Brazilian sugar cane growers. In C. Keitel et al. (eds.). *Mathematics, education, and society*. Paris: UNESCO, 68–70.

V. Albis. (1988). "La division ritual de la circunferencia: Una hipótesis fascinante." Bogotá, Colombia: (mimeo).

S. Ale. (1989). Mathematics in rural societies. In C. Keitel, P. Damerow, A. Bishop, and P. Gerdes (eds.). *Mathematics, Education, and Society*. Paris: UNESCO, 35–38.

U. D'Ambrosio. (1976). Matemática e sociedade. *Ciência e Cultura*. Vol. 28. São Paulo: 1418–1422.

———. (1982). *Mathematics for rich and for poor countries*. Paramaribo, Brazil: CARIMATH.

———. (1985a). *Socio-cultural bases for mathematics education*, Campinas, Brazil: UNICAMP.

———. (1985b). Ethnomathematics and its place in the history and pedagogy of mathematics. *For the Learning of Mathematics*. Vol. 5 (no.1): 44–48.

———. (1987). *Etnomatemática: raízes socio-culturais da arte ou técnica de explicar e conhecer*, Campinas, Brazil: UNICAMP.

———. (1990). *Etnomatemática: arte ou técnica de explicar e conhecer*. São Paulo: Editora Ática.

U. D'Ambrosio and P. Gerdes. (1994). Ethnomathematics, ethnoscience, and the recovery of world history of science—Zaragoza Symposium report. *Physis, Rivista Internazionale di Storia della Scienza*. Firenze, Vol. 31, (fasc. 2): 570–573.

S. Anderson. (1990). Worldmath curriculum: fighting eurocentrism in mathematics. *Journal of Negro Education*. Vol. 59 (no.3): 348–359.

H. Anzenge et al. (1988). *Indigenous mathematical algorithms*. B. Ed. project. Zaria, Nigeria: Ahmadu Bello University.

M. Ascher. (1984). Mathematical ideas in non-western cultures. Historia Mathematica. Vol. 11: 76–80.

———. (1988a). Graphs in culture: A study in ethnomathematics I. *Historia Mathematica*. Vol. 15, 201–227.

———. (1988b). Graphs in cultures (II): A study in ethno-mathematics. *Archive for History of Exact Sciences*. Vol. 39, no.1: 75–95.

———. (1990). A River-crossing problem in cross-cultural perspective. *Mathematics Magazine*. Vol. 63, no.1: 26–29.

———. (1991). *Ethnomathematics: A multicultural view of mathematical ideas*. Pacific Grove, CA: Brooks.

M. Ascher and R. Ascher. (1981). *Code of the Quipu: A study in media, mathematics and culture*. Ann Arbor: The University of Michigan Press.

———. (1986). Ethnomathematics. *History of Science*, Vol. 24. London: 125–144.

———. (1994). Ethnomathematics. In I. Grattan-Guinness, *Companion Encyclopedia of the History and Philosophy of the Mathematical Sciences*. London: Routledge and Kegan Paul, 1545–1554.

M. Ascher and U. D'Ambrosio. (1994). Ethnomathematics: A dialogue. *For the Learning of Mathematics*. Vol. 14, no. 2: 36–43.

B. Barton. (1990). "Using the trees to see the wood: An archaeology of mathematical structure in New Zealand." Auckland: (mimeo).

———. (1992). "A philosophical justification for ethnomathematics and some implications for education." Auckland: (mimeo).

R. Bassanezi. (1994). Modelling as a teaching-learning strategy. *For the Learning of Mathematics*. Vol. 14, no. 2: 31–35.

A. Bishop. (1978). *Spatial abilities in a papua New Guinea context*. Lae: Papua New Guinea: University of Technology.

———. (1979). Visualising and mathematics in a pre-technological culture. *Educational Studies in Mathematics*. Vol. 10: 135–146.

———. (1988a). *Mathematical enculturation, a cultural perspective on mathematics education*. Boston: Kluwer.

———. (Ed.) (1988b). *Culture and Mathematics Education*. Boston: Kluwer, Dordrecht.

———. (1988c). Mathematics education in its cultural context. In A. Bishop (ed.). *Culture and Mathematics Education*. Boston: Kluwer, Dordrecht, 179–192.

———. (1989). *Cultural imperialism and western mathematics: The hidden persuader*. Cambridge, MA (preprint).

———. (1990). Western mathematics: The secret weapon of cultural imperialism. *Race & Class*. Vol. 32: no. 2. 51–65.

———. (1994). Cultural conflicts in mathematics education: Developing a research agenda. *For the Learning of Mathematics*. Vol. 14, no. 2: 15–18.

E. Bonilla-Rius. (1986). Seminar on Mathematics and Culture. A viewpoint of the meeting. *Zentralblatt für Didaktik der Mathematik*. No. 2: 72–76.

B. Booss and M. Niss. (1979). *Mathematics and the real world*. Basel: Birkhäuser.

M. Borba. (1987a). Um estudo em etnomatemática: Sua incorporação na elaboração de uma proposta pedagógica para o Núcleo-Escola da Favela de Vila Nogueira e São Quirino. M. Ed. thesis, São Paulo State University, Rio Claro.

———. (1987b). Etnomatemática: A matemática da favela em uma proposta pedagógica. In P. Freire et al. (eds.). *Na escola que fazemos . . . Uma reflexão interdisciplinar em educação popular*, Petrópolis: Editora Vozes, 71–77.

H. Bos and H. Mehrtens. (1977). The interactions of mathematics and society: Some explanatory remarks. *Historia Mathematica*. Vol. 4: 7–30.

R. Buriasco. (1989). *A matemática de fora e de dentro da escola: Do bloqueio à transição*. Rio Claro, Brazil: UNESP.

T. Carraher, D. Carraher, and A. Schliemann. (1982). Na vida, dez; na escola, zero: Os contextos culturais da aprendizagem de matemática. *Cadernos de Pesquisa*. Vol. 42: 79–86.

———. (1987). Written and oral mathematics. *Journal of Research in Mathematics Education*. Vol. 8: 83–97.

T. Carraher. (1988). Street mathematics and school mathematics. *Proceedings of the 12th International Conference on Psychology of Mathematics Education*. Veszprem, Hungary: 1–23.

A. Cauty. (1994). What sort of mathematics for Amerindians. *Abstracts of the ORSTOM-UNESCO Conference '20th Century Science beyond the Metropolis'*. Paris: ORSTOM-UNESCO, 243–245.

———. (1995). What sort of mathematics for Amerindians. *Proceedings of the ORSTOM-UNESCO Conference '20th Century Science beyond the Metropolis'*. Paris: ORSTOM-UNESCO, (in press).

M. Cherinda. (1994a). Mathematical-educational exploration of traditional basket weaving techniques in a children's 'Circle of Interest.' In P. Gerdes (ed.). *Explorations in Ethnomathematics and Ethnoscience in Mozambique*. Maputo, Mozambique: ISP, 16–23.

———. (1994b). Children's mathematical activities stimulated by an analysis of African cultural elements. In C. Julie et al. (Eds.). *Proceedings of the 2nd International Conference on the Political Dimensions of Mathematics Education: Curriculum reconstruction for society in transition*. Cape Town: Maskew Miller Longman, 142–148.

M. Closs. (Ed.) (1986). *Native American Mathematics*. Austin: University of Texas.

C. Cossio and A. Jerez. (1986). *Elementos de analisis en matematicas Quichua y Castellano*. Quito, Ecuador: Pontifica Universidad Catolica.

K. Crawford. (1984). Bicultural teacher training in mathematics education for aboriginal trainees from traditional communities. In P. Damerow et al. (eds.). *Mathematics for all*. Paris: UNESCO, 101–107.

———. (1989). Knowing what versus knowing how: the need for a change in emphasis for minority group education in mathematics. In C. Keitel et al. (eds.). *Mathematics, Education, and Society*. Paris: UNESCO, 22–24.

D. Crowe. (1971). The geometry of African art I. Bakuba art. *Journal of Geometry*. Vol. 1: 169–182.

———. (1973). Geometric symmetries in African art. In C. Zaslavsky, *Africa counts*. 190–196.

———. (1975). The geometry of African art II. A catalog of Benin patterns. *Historia Mathematica*. Vol. 2: 253–271.

———. (1982a). The geometry of African art III. The smoking pipes of Begho. In C. Davis, B. Grunbaum, F. Sherk (ed.). *The geometric vein, the Coxeter Festschrift*. New York: Springer Verlag, 177–189.

———. (1982b). Symmetry in African art. *Ba Shiru, Journal of African Languages and Literature*. Vol. 3, no.1: 57–71.

T. Crump. (1990). *The anthropology of number*. Cambridge, England: Cambridge University Press.

P. Damerow. (1991). Ethnomathematik und Curriculumexport. Preface to P. Gerdes, *Ethnogeometrie*. Badsalzdetfurth / Hildesheim: Franzbecker Verlag, ix–xviii.

A. Deledicq and A. Popova. (1977). *"Wari et Solo," le jeu de calculs africain*. Paris: Cedic.

S. Doumbia. (1989). Mathematics in traditional African games. In C. Keitel, P. Damerow, A. Bishop, and P. Gerdes (eds.). *Mathematics, education, and society*. Paris: UNESCO, 174–175.

———. (1992). "Les jeux verbaux au Côte d'Ivoire." Paper presented at ICME7, Montreal.

———. (1993a). Jeux verbaux et enseignement traditionnel en Afrique. *Actes du Séminaire Interdisciplinaire Mathématiques-Philosophie et Enseignement*. Ministère de l'Éducation Yamoussoukro, Côte d'Ivoire: Nationale, 92–96.

———. (1993b). Les jeux de cauris. *Actes du Séminaire Interdisciplinaire Mathématiques-Philosophie et Enseignement*. Yamoussoukro: Ministère de l'Éducation Nationale, 97–101.

———. (1994). Les mathématiques dans l'environnement socioculturel africain, et l'exposition "jeux africains, mathématiques et société." *Plot*. No. 69: 1–31.

S. Doumbia and J. Pil. (1992). *Les jeux de cauris*. Abidjan: IRMA.

J. Draisma. (1992). Arithmetic and its didactics. In *Report of the first year of the Master's Degree Program in Mathematics Education for Primary Schools* (August 1991—July 1992). Beira, 1992: Instituto Superior Pedagógica, 89–129.

———. (1993). Numeração falada como recurso na aprendizagem da Aritmética. In P. Gerdes (ed.). *A numeração em Moçambique*. Maputo, Mozambique: ISP, 134–150.

———. (1994). How to handle the theorem 8 + 5 = 13 in (teacher) education? In P. Gerdes (ed.). *Explorations in ethnomathematics and ethnoscience in Mozambique*. Maputo: ISP, 30–48; reproduced in C. Julie et al. (eds.). *Proceedings of the 2nd International Conference on the Political Dimensions of Mathematics Education: Curriculum reconstruction for society in transition*. Cape Town: Maskew Miller Longman, 196–207.

M. El Tom. (Ed.). (1979). *Developing mathematics in Third World countries*. Amsterdam: North-Holland.

J. Evans. (1989). Mathematics for adults: community research and the 'barefoot' statistician. In C. Keitel et al. (eds.). *Mathematics, education, and society*. Paris: UNESCO, 65–68.

M. Fasheh. (1982). Mathematics, culture and authority. *For the Learning of Mathematics*. Vol. 3, no.2: 2–8.

———. (1989). Mathematics in a social context: math within education as praxis versus within education as hegemony. C. Keitel et al. (eds.). *Mathematics, education, and society*. Paris: UNESCO, 84–86.

C. Favrod. (1977). *A antropologia*. Lisbon: Publicações Dom Quixote.

E. Ferreira. (1988). *The teaching of mathematics in Brazilian nature communities*. UNICAMP.

———. (1989). The genetic principle and ethnomathematics. In C. Keitel et al. (eds.). *Mathematics, education, and society*. Paris: UNESCO, 110–111.

———. (1991). Por uma teoria da etnomatemática. *BOLEMA*, Vol. 7: 30–35.

———. (1992). A matemática no pensamento de Paulo Freire. Campinas: UNICAMP, (mimeo).

M. Frankenstein. (1981). A different third R: Radical maths. *Radical teacher* No. 20: 14–18.

———. (1983). Critical mathematics education: an application of Paulo Freire's epistomology. *Journal of Education*. Vol. 165: 315–339.

———. (1989). *Relearning mathematics: A different third R—Radical maths*. London: Free Association Books.

M. Frankenstein and A. Powell. (1989). Mathematics education and society: Empowering non-traditional students. In C. Keitel, P. Damerow, A. Bishop, and P. Gerdes (eds.). *Mathematics, education, and society*. Paris: UNESCO, 157–159.

———. (1994). Towards liberatory mathematics: Paulo Freire's epistemology and ethnomathematics. In P. McLaren & C. Lankshear (eds.). *The politics of Liberation: Paths from Freire*. London: Routledge and Kegan Paul, 74–99 (preprint 1991).

P. Freire. (1970). *Pedagogogy of the oppressed*. New York: Publisher.

———. (1973). *Education for critical consciousness*. New York: Publisher.

P. Freire et al. (Eds.). (1987). *Na escola que fazemos . . . Uma reflexão interdisciplinar em educação popular*. Petrópolis, Brazil: Editora Vozes.

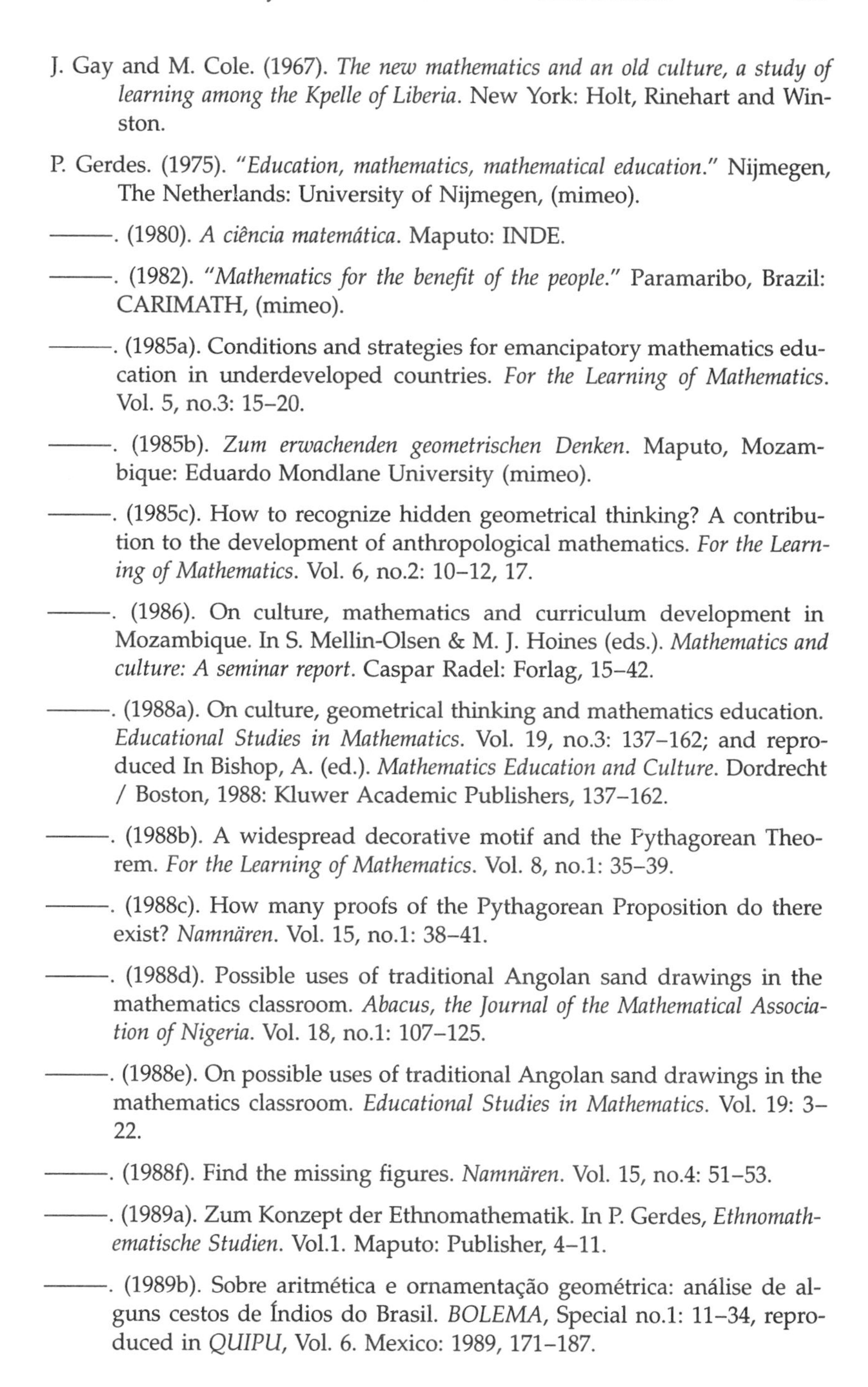

J. Gay and M. Cole. (1967). *The new mathematics and an old culture, a study of learning among the Kpelle of Liberia.* New York: Holt, Rinehart and Winston.

P. Gerdes. (1975). *"Education, mathematics, mathematical education."* Nijmegen, The Netherlands: University of Nijmegen, (mimeo).

———. (1980). *A ciência matemática.* Maputo: INDE.

———. (1982). *"Mathematics for the benefit of the people."* Paramaribo, Brazil: CARIMATH, (mimeo).

———. (1985a). Conditions and strategies for emancipatory mathematics education in underdeveloped countries. *For the Learning of Mathematics.* Vol. 5, no.3: 15–20.

———. (1985b). *Zum erwachenden geometrischen Denken.* Maputo, Mozambique: Eduardo Mondlane University (mimeo).

———. (1985c). How to recognize hidden geometrical thinking? A contribution to the development of anthropological mathematics. *For the Learning of Mathematics.* Vol. 6, no.2: 10–12, 17.

———. (1986). On culture, mathematics and curriculum development in Mozambique. In S. Mellin-Olsen & M. J. Hoines (eds.). *Mathematics and culture: A seminar report.* Caspar Radel: Forlag, 15–42.

———. (1988a). On culture, geometrical thinking and mathematics education. *Educational Studies in Mathematics.* Vol. 19, no.3: 137–162; and reproduced In Bishop, A. (ed.). *Mathematics Education and Culture.* Dordrecht / Boston, 1988: Kluwer Academic Publishers, 137–162.

———. (1988b). A widespread decorative motif and the Pythagorean Theorem. *For the Learning of Mathematics.* Vol. 8, no.1: 35–39.

———. (1988c). How many proofs of the Pythagorean Proposition do there exist? *Namnären.* Vol. 15, no.1: 38–41.

———. (1988d). Possible uses of traditional Angolan sand drawings in the mathematics classroom. *Abacus, the Journal of the Mathematical Association of Nigeria.* Vol. 18, no.1: 107–125.

———. (1988e). On possible uses of traditional Angolan sand drawings in the mathematics classroom. *Educational Studies in Mathematics.* Vol. 19: 3–22.

———. (1988f). Find the missing figures. *Namnären.* Vol. 15, no.4: 51–53.

———. (1989a). Zum Konzept der Ethnomathematik. In P. Gerdes, *Ethnomathematische Studien.* Vol.1. Maputo: Publisher, 4–11.

———. (1989b). Sobre aritmética e ornamentação geométrica: análise de alguns cestos de Índios do Brasil. *BOLEMA,* Special no.1: 11–34, reproduced in *QUIPU,* Vol. 6. Mexico: 1989, 171–187.

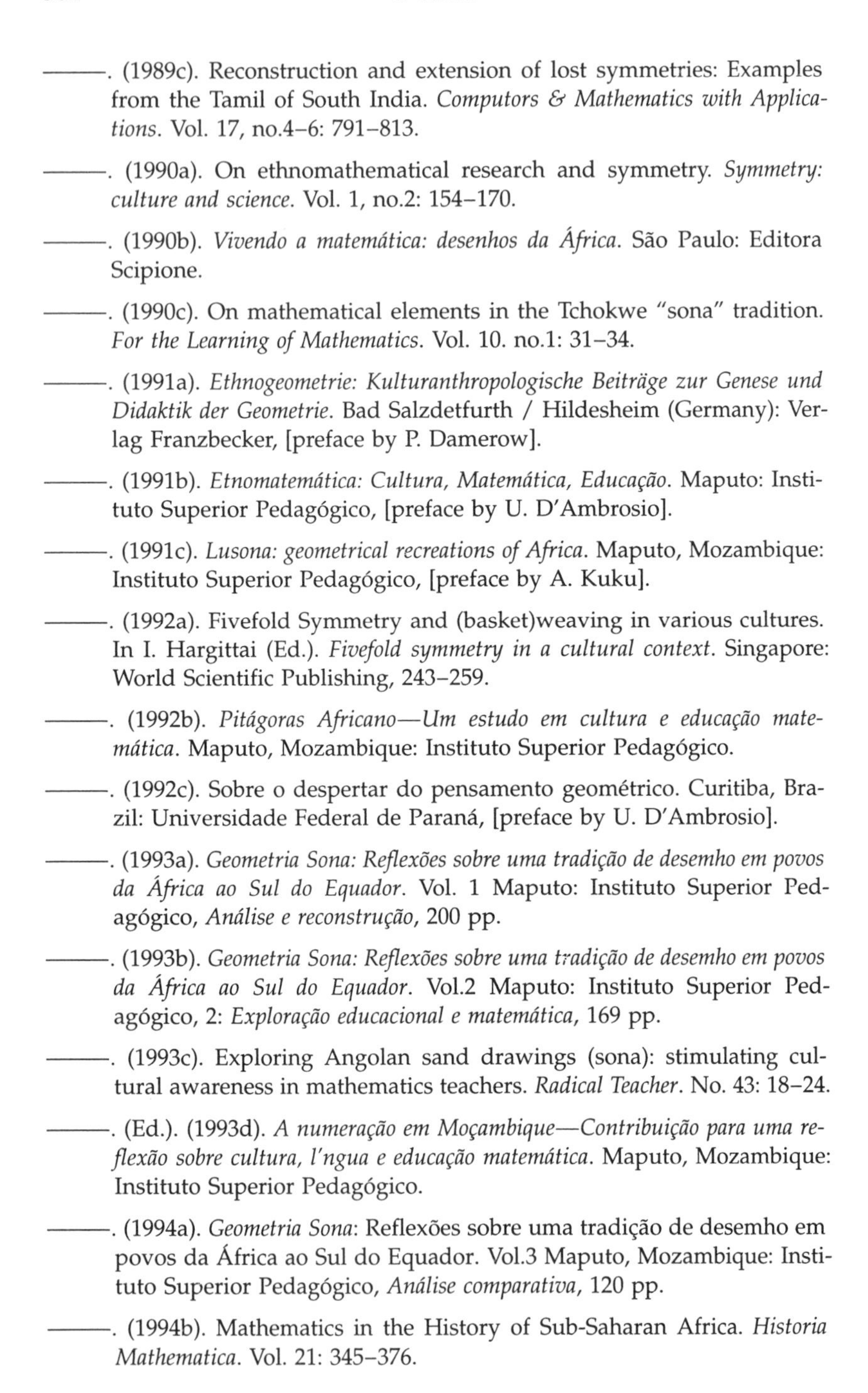

———. (1989c). Reconstruction and extension of lost symmetries: Examples from the Tamil of South India. *Computors & Mathematics with Applications*. Vol. 17, no.4–6: 791–813.

———. (1990a). On ethnomathematical research and symmetry. *Symmetry: culture and science*. Vol. 1, no.2: 154–170.

———. (1990b). *Vivendo a matemática: desenhos da África*. São Paulo: Editora Scipione.

———. (1990c). On mathematical elements in the Tchokwe "sona" tradition. *For the Learning of Mathematics*. Vol. 10. no.1: 31–34.

———. (1991a). *Ethnogeometrie: Kulturanthropologische Beiträge zur Genese und Didaktik der Geometrie*. Bad Salzdetfurth / Hildesheim (Germany): Verlag Franzbecker, [preface by P. Damerow].

———. (1991b). *Etnomatemática: Cultura, Matemática, Educação*. Maputo: Instituto Superior Pedagógico, [preface by U. D'Ambrosio].

———. (1991c). *Lusona: geometrical recreations of Africa*. Maputo, Mozambique: Instituto Superior Pedagógico, [preface by A. Kuku].

———. (1992a). Fivefold Symmetry and (basket)weaving in various cultures. In I. Hargittai (Ed.). *Fivefold symmetry in a cultural context*. Singapore: World Scientific Publishing, 243–259.

———. (1992b). *Pitágoras Africano—Um estudo em cultura e educação matemática*. Maputo, Mozambique: Instituto Superior Pedagógico.

———. (1992c). Sobre o despertar do pensamento geométrico. Curitiba, Brazil: Universidade Federal de Paraná, [preface by U. D'Ambrosio].

———. (1993a). *Geometria Sona: Reflexões sobre uma tradição de desemho em povos da África ao Sul do Equador*. Vol. 1 Maputo: Instituto Superior Pedagógico, *Análise e reconstrução*, 200 pp.

———. (1993b). *Geometria Sona: Reflexões sobre uma tradição de desemho em povos da África ao Sul do Equador*. Vol.2 Maputo: Instituto Superior Pedagógico, 2: *Exploração educacional e matemática*, 169 pp.

———. (1993c). Exploring Angolan sand drawings (sona): stimulating cultural awareness in mathematics teachers. *Radical Teacher*. No. 43: 18–24.

———. (Ed.). (1993d). *A numeração em Moçambique—Contribuição para uma reflexão sobre cultura, l'ngua e educação matemática*. Maputo, Mozambique: Instituto Superior Pedagógico.

———. (1994a). *Geometria Sona*: Reflexões sobre uma tradição de desemho em povos da África ao Sul do Equador. Vol.3 Maputo, Mozambique: Instituto Superior Pedagógico, *Análise comparativa*, 120 pp.

———. (1994b). Mathematics in the History of Sub-Saharan Africa. *Historia Mathematica*. Vol. 21: 345–376.

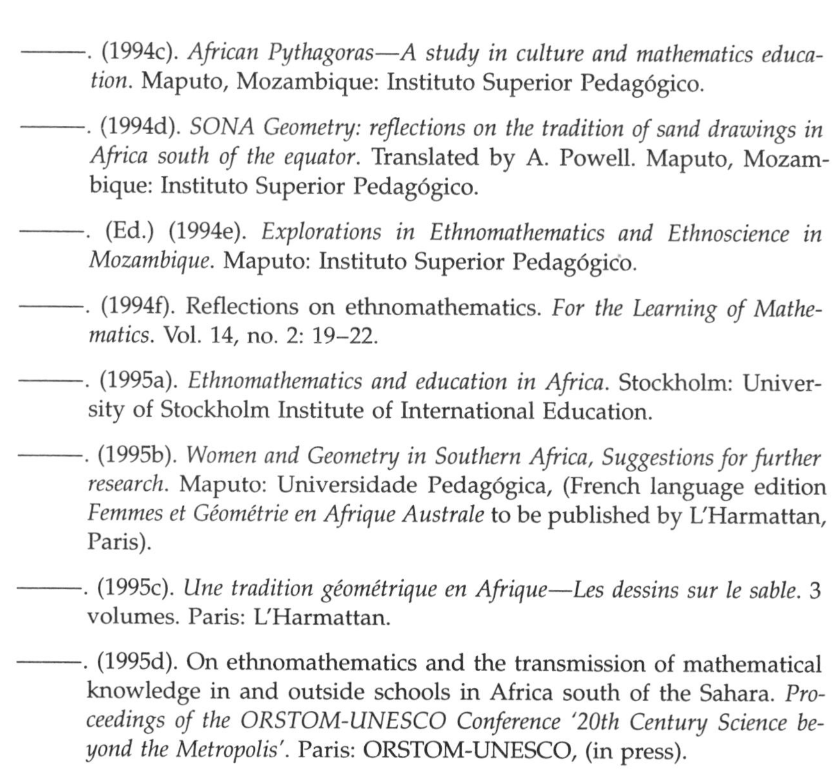

———. (1994c). *African Pythagoras—A study in culture and mathematics education.* Maputo, Mozambique: Instituto Superior Pedagógico.

———. (1994d). *SONA Geometry: reflections on the tradition of sand drawings in Africa south of the equator.* Translated by A. Powell. Maputo, Mozambique: Instituto Superior Pedagógico.

———. (Ed.) (1994e). *Explorations in Ethnomathematics and Ethnoscience in Mozambique.* Maputo: Instituto Superior Pedagógico.

———. (1994f). Reflections on ethnomathematics. *For the Learning of Mathematics.* Vol. 14, no. 2: 19–22.

———. (1995a). *Ethnomathematics and education in Africa.* Stockholm: University of Stockholm Institute of International Education.

———. (1995b). *Women and Geometry in Southern Africa, Suggestions for further research.* Maputo: Universidade Pedagógica, (French language edition *Femmes et Géométrie en Afrique Australe* to be published by L'Harmattan, Paris).

———. (1995c). *Une tradition géométrique en Afrique—Les dessins sur le sable.* 3 volumes. Paris: L'Harmattan.

———. (1995d). On ethnomathematics and the transmission of mathematical knowledge in and outside schools in Africa south of the Sahara. *Proceedings of the ORSTOM-UNESCO Conference '20th Century Science beyond the Metropolis'.* Paris: ORSTOM-UNESCO, (in press).

P. Gerdes and G. Bulafo. (1994). *Sipatsi: Technology, art and geometry in Inhambane.* Translated by A. Powell. Maputo, Mozambique: Instituto Superior Pedagógico, 102 pp. (Also published in Portuguese and French).

G. Gilmer. (1990). Ethnomathematical approach to curriculum development. *ISGEm-Newsletter* Vol. 5, no.2: 4–5.

G. Gilmer, M. Thompson, and C. Zaslavsky. (1992). *Building bridges to mathematics: Cultural connections*, San Francisco: Addison-Wesley.

H. Ginsburg. (1978). Poor children, African mathematics and the problem of schooling. *Educational Research Quarterly.* Special Edition, 1978, Vol. 2, no.4: 26–43.

B. Graham. (1988). Mathematical education and Aboriginal children. In A. Bishop (ed.). *Culture and mathematics education.* Dordrecht/Boston: Kluwer, 119–136.

M. Harris. (1987). An example of traditional women's work as a mathematics resource. *For the Learning of Mathematics.* Vol. 7, no.3: 26–28.

———. (1988). Common threads, mathematics and textiles. *Mathematics in school.* 24–28.

M. Harris et al. (1988). *Cabbage, mathematics in work*. London: Institute of Education, University of London.

M. Harris and C. Paechter (1991). Work reclaimed: Status mathematics in non-élitist contexts. In M. Harris (ed.). *Schools, mathematics and work*. London: Falmer, 277–283.

P. Harris. (1984). The relevance of primary school mathematics in tribal Aboriginal communities. In P. Damerow et al. (eds.). *Mathematics for all*. Paris: UNESCO, 96–100.

J. Høyrup (1994). *In measure, number, and weight—Studies in mathematics and culture*. Albany: State University of New York Press.

R. Hunting. (1985). *Learning, aboriginal worldview and ethnomathematics*. Western Australia: Institute of Technology.

———. (1987). Mathematics and Australian aboriginal culture. *For the Learning of Mathematics*. Vol. 7, no. 2: 5–10.

A. Ismael and D. Soares. (1993). Métodos populares de contagem em Moçambique. In P. Gerdes (ed.). *A numeração em Moçambique*. Maputo, Mozambique: ISP, 114–120.

A. Ismael. (1994a). On the origin of the concepts of "even" and "odd" in Makhuwa culture. In P. Gerdes (ed.). *Explorations in Ethnomathematics and Ethnoscience in Mozambique*. Maputo, Mozambique: ISP, 9–15.

———. (1994b). Motivations for the learning of mathematics—in view of Mozambique's historical, social and cultural development. In C. Julie et al. (eds.). *Proceedings of the 2nd International Conference on the Political Dimensions of Mathematics Education. Curriculum reconstruction for society in transition*. Cape Town: Maskew Miller Longman, 53–56.

G. Joseph. (1987). Foundations of Eurocentrism in mathematics. *Race & Class*. Vol. 28, no.3: 13–28.

———. (1989). Eurocentrism in mathematics: The historical dimensions. In C. Keitel et al. (eds.). *Mathematics, Education, and Society*. Paris: UNESCO, 32–35.

———. (1991). *Crest of the peacock: Non-European roots of mathematics*. London: Tauris.

E. Kane. (1987). *Les systèmes de numération parlée des groupes ouest-atlantiques et Mande. Contribution à la recherche sur les fondaments et l'histoire de la pensée logique et mathématique en Afrique de l'Ouest*. Lille.

V. Katz. (1994). Ethnomathematics in the classroom. *For the Learning of Mathematics*. Vol. 14, no. 2, 26–30.

C. Keitel, P. Damerow, A. Bishop, and P. Gerdes. (Eds). (1989). *Mathematics, Education, and Society*. Paris: UNESCO.

C. Keyser. (1932). Mathematics as a Culture Clue. *Scripta Mathematica*. Vol. 1: 185–203.

G. Knight. (1982a). The geometry of maori art-rafter patterns. *The New Zealand Mathematics Magazine*. Vol. 21, no.2: 36–40.

———. (1982b). The geometry of Maori art-weaving patterns. *The New Zealand Mathematics Magazine*. Vol. 21, no.3: 80–87.

G. Knijnik. (1993). An ethnomathematical approach in mathematics education: A matter of political power. *For the Learning of Mathematics*. 13(3): 23–26.

G. Kubik. (1990). Visimu vya mukatikati—dilemma tales and "arithmetical puzzles" collected among the Valuchazi. *South African Journal of African Languages*. Vol. 10, no.2: 59–68.

D. Lancy. (Ed.) (1978). The indigenous mathematics project. *Papua New Guinea Journal of Education*. Vol. 14: 1–217.

———. (1983). *Cross-cultural studies in cognition and mathematics*. New York: Academic Press.

N. Langdon. (1989). Cultural starting points for mathematics: A view from Ghana. *Science Education Newsletter*. No.87: 1–3.

———. (1990). *Cultural starting points*. Paper presented at SEACME 5, Brunei, (mimeo).

H. Lea. (1987). Traditional mathematics in Botswana. *Mathematics Teaching*. Vol. 119.

———. (1989a). Informal mathematics in Botswana. *Proceedings of the 41st CIEAEM Meeting of the International Commission for the Study and Improvement of Mathematics Teaching*. Brussels: 43–53.

———. (1989b). *Mathematics in a cultural setting*. Gaberone: University of Botswana.

———. (1989c). *Informal Mathematics in Botswana: Spatial concepts in the Kalahari*. Gaberone: University of Botswana, 1989.

———. (1990a). *Informal Mathematics in Botswana: Mathematics in the Central Kalahari*. Gaberone: University of Botswana.

———. (1990b). Spatial concepts in the Kalahari. *Proceedings of the 14th International Conference on Psychology of Mathematics Education*. Vol. 2, Oaxtepec, Mexico: 259–266.

G. Lean. (1986). *Counting systems of Papua New Guinea*. Lae: Papua New Guinea University of Technology.

R. Linton. (1936). *The Study of man*. New York: Appleton.

B. Lumpkin. (1990). A multicultural mathematics curriculum. *ISGEm-Newsletter*. Vol. 6, no.1, 2.

G. Luquet. (1929). Sur l'origine des notions mathématiques: Remarques psychologiques et ethnographiques. *Journal de Psychologie*, 733–761.

A. Mapapá and E. Uaila. (1993). Tabelas e mapas comparativos relativos à numeração falada em Moçambique. In P. Gerdes (ed.). *A numeração em Moçambique*. Maputo, Mozambique: ISP, 121–132.

A. Mapapá. (1994). Symmetries and metal grates in Maputo—Didactic experimentation. In P. Gerdes (ed.). *Explorations in Ethnomathematics and Ethnoscience in Mozambique*. Maputo, Mozambique: ISP, 49–55.

J. Marschall. (1987). An atlas of American Indian geometry. *Ohio Archaeologist*. Vol. 37, no.2: 36–49.

S. Mellin-Olsen. (1986). Culture as a key theme for mathematics education. Postseminar reflections. In *Mathematics and Culture, a seminar report*. Radal: Caspar Forlag, 99–121.

———. (1987). *The Politics of Mathematics Education*. Dordrecht: Kluwer.

S. Mellin-Olsen, and M. J. Hoines. (Eds.). *Mathematics and Culture: A seminar report*. Radel: Caspar Forlag.

C. Moore. (1986). *The implication of string figures for Native American mathematics education*. Flagstaff, AZ: Publisher (mimeo).

———. (1987). Ethnomathematics. *Science*. Vol. 236, Washington, DC: 1006–1007.

———. (1994). Research in Native American mathematics education. *For the Learning of Mathematics*. Vol. 14, no. 2: 9–14.

V. Mubumbila. (1988). *Sur le sentier mystérieux des nombres noirs*. Paris: L'Harmattan.

B. Mve Ondo. (1990). *L'Owani et le Songa. Deuz jeux de calculus africains*. CCF St Exepéry/Sépia, Libreville (Gabon): Collection Découvertes du Gabon.

D. Nelson, G. Joseph, and J. Williams. (1993). *Multicultural mathematics: Teaching mathematics from a global perspective*. Oxford: Oxford University Press.

P. Nissen. (1988). Sand drawings of Vanuatu. *Mathematics in School*: 10–11.

G. Njock. (1985). Mathématiques et environnement socio-culturel en Afrique Noire. *Présence Africaine*. no.135, 3rd Quarterly. Paris: New Bilingual Series, 3–21.

S. Nobre. (1989a). *Aspectos sociais e culturais no desenho curricular da matemática*. Rio Claro, Brazil: UNESP.

———. (1989b). The ethnomathematics of the most popular lottery in Brazil: the "animal lottery." In C. Keitel et al. (eds.). *Mathematics, Education, and Society*. Paris: UNESCO, 175–177.

D. Orey. (1989). Ethnomathematical perspectives on the NCTM standards. *ISGEm-Newsletter*. Vol. 5, no.1: 5–7.

M. Panoff and M. Perrin. (1973). *Dicionário de Etnologia.* São Paulo: Lexis.

L. Paula, and E. Paula. (1986). *XEMA'EAWA, jogos de barbante entre os Índios Tapirapé,* Campinas, Brazil: UNICAMP, (mimeo).

R. Pinxten, I. v. Dooren, and F. Harvey. (1983). *The anthropology of space: Exploration into the natural philosophy and semantics of the Navajo.* Philadelphia: University of Pennsylvania.

R. Pinxten. (1989). World view and mathematics teaching. In C. Keitel et al. (eds.). *Mathematics, Education, and Society.* Paris: UNESCO, 28–29.

———. (1994). Ethnomathematics and its practice. *For the Learning of Mathematics.* Vol. 14, no. 2: 23–25.

J. Posner. (1982). The development of mathematical knowledge in two west African societies. *Child Development.* Vol. 53: 200–208 [based on her Ph.D. diss., 1978].

J. Ratteray. (1992). *Center shift, an African-centered approach for the multicultural curriculum.* Washington, DC: Institute for Independent Education.

O. Raum. (1938). *Arithmetic in Africa.* London: Evans Brothers.

K. Reich, M. Folkerts, and C. Scriba. (1989). Das Schriftenverzeichnis von Ewald Fettweis (1881–1967) samt einer Würdigung von Olindo Falsirol. *Historia Mathematica.* Vol. 16: 360–372.

G. Saxe. (1981). Body parts as numerals: A developmental analysis of numeration among remote Oksapmin village populations in Papua New Guinea. *Child Development.* Vol. 52: 306–316.

———. (1982a). Developing forms of arithmetical thought among the Oksapmin of Papua New Guinea. *Developmental Psychology.* Vol. 18: 583–594.

———. (1982b). Culture and development of numerical cognition: Studies among the Oksapmin of Papua New Guinea. *Children's logical and mathematical cognition.* New York: Publisher, 157–176.

———. (1988). Candy selling and math learning. *Educational Researcher.* Vol. 17, no.6: 14–21.

———. (1991). *Culture and cognitive development: Studies in mathematical understanding.* Hillsdale: Erlbaum.

A. Schliemann. (1984). Mathematics among carpentry apprentices: Implications for school teaching. In P. Damerow et al. (eds.). *Mathematics for all.* Paris: UNESCO, 92–95.

S. Shan, and P. Bailey. (1991). *Multiple factors: Classroom mathematics for equality and justice.* Staffordshire: Trentham Books.

L. Shirley. (1986a). History of mathematics in Nigerian mathematics classrooms: Values and problems. *Abacus, Journal of the Mathematical Association of Nigeria*. Vol. 12: 123–133.

———. (1986b). *Ethnomathematics and the history of African mathematics*. Paper presented at the Second Pan-African Congress of Mathematicians, Jos, Nigeria: (mimeo).

———. (1988). *Historical and ethnomathematical algorithms for classroom use*. Budapest: ICME VI.

———. (1991). Video games for math: A case for 'kid culture'. *ISGEm-Newsletter*. Vol. 6, no.2: 2–3.

O. Skovsmose. (1994). *Towards a Philosophy of Critical Mathematics Education*. Dordrecht: Kluwer.

T. Smart, and Z. Isaacson. (1989). "It was nice being able to share ideas," women learning mathematics. In C. Keitel et al. (eds.). *Mathematics, Education, and Society*. Paris: UNESCO, 116–118.

C. Smorynski. (1983). Mathematics as a cultural system. *Mathematical Intelligencer*. Vol. 5, no.1: 9–15.

D. Soares, and A. Ismael. (1994). Popular counting methods in Mozambique. In P. Gerdes (ed.). *Explorations in Ethnomathematics and Ethnoscience in Mozambique*. Maputo, Mozambique: ISP, 24–29.

R. Souviney. (1989). The indigenous mathematics project: Mathematics instruction in Papua New Guinea. In C. Keitel et al. (eds.). *Mathematics, Education, and Society*. Paris: UNESCO, 106–110.

J. Stigler, and R. Baranes. (1988). Culture and mathematics learning. *Review of Research in Education*. Vol. 15: 253–306.

L. Stott, and H. Lea. (1993). *Common threads in Botswana*. Gaberone: British Council.

D. Struik. (1942). On the sociology of mathematics. *Science and Society*. Vol. 6: 58–70.

———. (1986). The sociology of mathematics revisited: A personal note. *Science & Society* . Vol. 50, no.3: 280–299.

W. Sturtevant. (1972). Studies in ethnoscience. In J. Spradley (ed.). *Culture and Cognition*. San Francisco: Chandler Publ., 129–167.

G. Thomas-Emeagwali. (1992). *Science and technology in African history wirh case studies from Nigeria, Sierra Leone, Zimbabwe and Zambia*. New York: Edwin Mellen Press.

S. Touré. (1984). Preface. In *Mathématiques dans l'environnement socio-culturel Africain*. Abidjan, Côte d'Ivoire: Institut de Recherches Mathématiques d'Abidjan, 1–2.

G. Tro. (1980). *Étude de quelques systèmes de numération en Côte d'Ivoire*. Abidjan, Côte d'Ivoire: (mimeo).

J. Turner. (1992). Complementarity, ethnomathematics, and primary education in Bhutan. *Canadian and International Education*. Vol. 21, no.1: 20–43.

D. Washburn, and D. Crowe. (1988). *Symmetries of culture, theory and practice of plane pattern analysis*. Washington, DC: University of Washington Press.

D. Washburn. (1990). *Style, classification and ethnicity: Design categories on Bakuba raffia cloth*. Philadelphia: American Philosophical Society.

L. White. (1947). The locus of mathematical reality: an anthropological footnote, reproduced in J. Newman (ed.). *The World of Mathematics*. Vol. 4. New York: Tempus, 1956, 2348–2364.

R. Wilder. (1950). The cultural basis of mathematics. *Proceedings International Congress of Mathematicians*. 1950, Vol. 1: 258–271.

———. (1968). *Evolution of mathematical concepts*. New York: John Wiley, 1968; London: Transworld Publ., 1974.

———. (1981). *Mathematics as a cultural system*. Oxford: Pergamon Press.

B. Wilson. (1981). *Cultural contexts of science and mathematics education*. Leeds, England: University of Leeds.

C. Zaslavsky. (1973). *Africa counts: Number and pattern in African culture*. Boston: Prindle, Weber and Schmidt, 328 pp.; paperback edition: Brooklyn, NY: Lawrence Hill Books, 1979.

———. (1979). Symmetry along with other mathematical concepts and applications in African life. *Applications in school mathematics*. Reston VA: National Council of Teachers of Mathematics, 82–97.

———. (1989). Integrating math with the study of cultural traditions. *ISGEm-Newsletter*. Vol. 4 no.2: 5–7.

———. (1992). *Multicultural mathematics: Interdisciplinary cooperative-learning activities*. Portland, ME: Weston Walch.

———. (1994). "Africa Counts" and ethnomathematics. *For the Learning of Mathematics*. Vol. 14, no. 2: 3–8.

R. Zepp. (1983). *L'apprentissage du calcul dans les langues de Côte d'Ivoire*. Place: Institut de Linguistique Appliqueé, Université d'Abidjan. Vol. 99.

———. (1989). *Language and mathematics education*. Hong Kong: API Press, 211–236.

Chapter 17

Applications in the Teaching of Mathematics and the Sciences

Rik Pinxten

To the memory of H. Freudenthal

Editors's comment: Rik Pinxten, an anthropologist, uses his research on the Navajo conception of space to propose a geometry curriculum for Navajo children that starts with their culture's conception of space. In our analysis, this chapter presents a challenge to the notion of the universality of mathematical ideas. In contrast to Gerdes (chapter 11) who argues that without colonialism it is possible that Mozambicans would have been credited, for example, with the discovery of the so-called Pythagorean Theorem, Pinxten posits that Navajo geometry is incommensurable with static conceptions of space such as the so-called Pythagorean Theorem. This chapter is a revised and expanded text, based largely on chapter 5 of R. Pinxten, I. van Dooren, and F. Harvey. (1983). *Anthropology of space: Explorations into the natural philosophy and semantics of the Navajo*. Philadelphia, PA: University of Pennsylvania.

From the outset of my study on Navajo space I continuously searched for possible applications of its results in a useful and beneficial way for the Navajo people. The first and foremost domain of application was found in the education of young Navajo Indians (and maybe other groups of people of non-Western origin), particularly regarding instruction in disciplines that involve spatial conceptualization and representation.

Statement of the Problem

When reading the works of Piaget, Bruner, and others on genetic psychology and ensuing educational procedures, one is compelled to reconsider the premises of these authors. In general terms it can be stated that all the authors agree on the genetic primacy of some kind of part/whole distinction, of some kind of "qualitative geometry" (or, more precisely, topology), and of some definite notions of order (conservation principles in Piaget, for example). With this preschool knowledge in a more or less developed form, the child enters the formal instruction in mathematics, geography, history (as well as a more systematic study of the environment). The fact that Piagetian strategies and insights are generally applied in Western European schools now, and that the so-called new mathematics (set theory mathematics) has conquered the West by now, is evidence in favor of the applicability—even successful applicability—of the theory.[1]

When looking for possible applications of similar insights in the same disciplinary instruction on the Navajo reservation in the southwest on U.S.A. (even in the best schools that are to be found there), one is struck by the utter ineffectualness, not to say the total inappropriateness of the same material, the same psychological convictions and findings in this different cultural setting. What has happened? Why is there this difference between the two types of education? In my view these questions will be answered if we look at the presupposed conditions of knowledge ability. The fact must be mentioned that my questioning of instructors/teachers, both Navajos and whites, on these matters and on their individual understandings of the problems yielded very little useful information. Very different outlooks of the educational task and setting are realized. The school at Rock Point is quite free or "progressive" in educational and social matters, while the school at, for example, Round Rock is rather classical, a Bureau of Indian Affairs (BIA)-dominated school with low aspirations and rather poor results. Still, teachers from both schools, particularly the Navajo teachers I talked to, showed a very poor understanding of the difference in outlook, system, preconceptions, and actual notions between the so-called Western (e.g., Piagetian) and the Navajo way of talking about, manipulating, acting upon, and actual formalization building of spatial differentiations in both cultural knowledge systems. Only one Navajo consultant was really able to shift consciously and swiftly between both "systems" (BB).[2] He started a deep discussion on the differences he saw between both spatial systems and the way "intranslatability" is felt when one looks at one from the perspective of the other, and vice versa. He encouraged me to go about the present study, and I can only hope that I have reached a sensible result, one that will bridge the gap of intranslatability and promote mutual understanding. Most of the teachers (Navajos as well as white teachers, eventually) occasionally told me about concrete instances of "misunderstand"; the spatial interpretation of time in English ("before/after in time") was generally considered to be a very difficult distinction for Navajo teachers and children alike, as well as the varying interpretations of the presumably "evident" part/whole distinctions in English.[3] With the gradual rediscovery of Vygotsky's work (e.g., Wassman, 1995) and the detailed psycho-ethnography of M. Cole's group (e.g. Cole and Scribner, 1974) such distractions are better understood in the 1990s.

My experiences convince me that the widespread feeling of ineffectiveness of schooling may be due—among other, more social and cultural factors—to the fundamental and poorly understood lack of commensurability of Western and Navajo knowledge systems, since they already exist and play a role in further, formal education within the school settings. The success of Piagetian approaches in education must at least partly be explained by the fact that, for example in geometry, they are founded on and built up from logically satisfying and closely "similar" or "naturally presupposed" preschool notions. In other words, the preschool knowledge is highly appropriate to and compatible with the contents of instruction in school. Further, our culture has built on a spatial systematization that has clearly continuous and smoothly developing stages that stretch from the natural environment (indeed the "carpentered world" Campbell, 1964, is talking about) and its conceptualization, all the way to more abstract and complex geometries. This seeming relativization of Piagetian notions is not new: both at the center at Geneva (e.g. Bovet, 1974) and elsewhere (e.g. Pinxten, 1976) this critique on the Western biases in Piaget's theory of intellectual development has been heard. The problem with Navajo education (and with any education in non-Western cultures) can then be summarized as follows: apart from other factors (social, emotional, etc.), the impact and precise content of preschool native knowledge is different in the Navajo culture than it is in ours, and consequently the curriculum and its actual implementation in the Navajo situation should be adjusted in order to reach a smooth, understandable, really integrated development in the knowledge acquired at school. Most certainly, in order to prevent a merely "schizoid" situation (like the one that exists currently, where Navajos obtain some knowledge from their tradition, some woefully incomplete knowledge from the Western tradition, and few and badly worked out means of comparing, transforming, shifting, or translating from one to the other), it is very necessary to gain a better understanding of the differences between both cultural systems. The possibility, then, of building formal schooling on the Navajo natural knowledge background acquired in the preschool age, can be pursued. It is this task that I confront when I consider the value of the present study for education.[4]

The Problem with Spatial Representations

I shall present a sketch of the problem of spatial and mathematical preschool conceptualization conveyed to Navajo and Western children to indicate the range and depth of the problem under discussion, for which I propose a solution. The presentation is a "sketch" in the sense that only some basic notions will be highlighted, whereas the more sophisticated or derived ones will be omitted from consideration. An elaborate curriculum for intuitive geometry teaching to Navajo children can be found elsewhere (Pinxten et al., 1987).

The Western Case

I see three main points in the description of Western psychological development of formal and mathematical thought that are developing in a mutually exclusive way in Navajo thought:

1) The Specific Hierarchy

In his explanation on the child's construction of space, Piaget (with others like Bruner, Furth, Goodnow, etc.) gives a detailed analysis of the way more sophisticated notions are linearly deduced or construed in a systematic one-to-one progression from notions acquired earlier. (Piaget and Inhelder, 1947; cf. also in schematic form: Pinxten 1976). The notion of "distinctness" is built upon (or deduced from) that of "neighborhood." The notion of "order" is built upon those of "neighborhood" and "distinctness." The notion of "border" is built upon that of "order," and so on until all more sophisticated notions (the projective geometric and the Euclidean notions) are integrated in the total conception by similar, quite linear, and systematically progressing procedures. It is pertinent to emphasize that these notions and the system of progressive acquisition is not all meant as a mere external or formal interpretation, but aims to describe the genetic progression that is actually taking place in the child's mind over the years. In my semantic terms of meaningful spatial representation, it can be said that the progression of semanitic constitution follows a quite systematic and nearly necessary progression of the following type:

Near: has no constituents; Distinct: has a constituent: near; Order: has as constituents: Near, distance; and so on.[5]

One fundamental feature of the model of the Western development can thus be summarized as follows: it is a hierarchical progression in the sense that each "higher" (more complex) notion necessarily and exclusively implies "lower" (less sophisticated and earlier acquired) notions as constituents. In practice, the Euclidean notions are clearly constructed from topological notions, while all topological notions in turn have one basic or "primitive" notion in common, that is, "neighborhood."

The educational consequences of this description are straightforward: since children start with topological spatial preschool notions, the curriculum can take these for granted (in a specific order of acquisition) and construe the logically and genetically "subsequent" notions quite naturally on top of and on the basis of these topological notions. The success of this procedure in actual schooling has been accepted as corroboration of the theory. The hierarchy has validated itself in practice.

2) The Part/Whole Distinction is Omnipresent

A certain atomism is obviously characteristic of the current Western style of thought. The "objectification" of the environment is taken for granted in the sciences (situations, static entities can be abstracted from their environment for a certain time) and in school instruction (we study particular animals, places, objects in themselves). A particularly powerful aspect of this approach, it seems, is the segmentability of objects (and of space and time as aspects of objects). The physical and, to some degree, the psychic and cultural world is considered to be sliced, segmented into smaller parts, which can be studied in themselves in greater detail.[6] It is not surprising that genetic psychologists and cognitive psychologists have in general been concerned most of all with this perspective when they wanted to study thought and the development of thought. Neither is it surprising that modern mathematics, which is meant to have more natural links with the Western child's preschool knowledge, is essentially set theory, that is, a formal system that basically deals with and operates in wholes of which the initial entity is but a part.[7] In all disciplines taught at primary and other school levels, the types and characteristics of segmentation of part/whole systems are of central importance, since indeed this atomistic perspective is basic to our cultural knowledge. Consequently, one should emphasize that in order to successfully approach or learn the more systematic presentation of Western knowledge (as is attempted in the school setting), a clear and common understanding of part/whole notions is absolutely required.

3) The Static World

A very general and possibly somewhat metaphysical sounding characteristic of Western knowledge, as it is practiced and taught, is the static interpretation: The outside world is primarily interpreted as a composition of situations, objects, transitions between situations, and so forth, and not as a composite of processes and actions. Physics, biology, and the social sciences decide on constants and thus turn to the most static characterizations of static phenomena in the first place (e.g., taxonomies in biology: see Atran (1990), only to introduce dynamic phenomena in the range of study at a later stage of development (a most typical and well-known case occurs again in biology: Darwinism comes up in the middle of the nineteenth century). The perspective is most clearly detectable again in discussions on the philosophy of science: after numerous generations sought for sense data, atoms, minimal characteristics, and so on, it is only in very recent years that attempts at praxiology and theories of event and change are discussed fully.[8]

Space and time are construed orthogonally that is, aspects of the world (objects, situations) are characterized in a three dimensional space upon which, as it were, time is defined. Time is, in its turn, most often represented as segmentably spatialized, cut up in identical units (atoms so to speak), which is similarly an attempt to make it manageable in a static framework (cf. Grunbaum, 1964; Einstein, 1949, particularly emphasizes this problem as one of his main points; Smart, 1964). The present outline may appear farfetched in regard to the problems at hand; this may be because of the conciseness of the discussion offered here.[9] However, the relevance of the point will become clear when I deal with the "Navajo correlates" to these fundamental aspects of knowledge.

Some more down-to-earth arguments may illustrate the point as well. Geography, from the primary school level and on, deals with segmented and segmentable units of land ("continent" is a constant, "state" is a static unit, mountains, prairies, and even agricultural land and cities are conceived of as surfaces, lines, and points, that is as static aspects). Natural sciences speak about the different species, families, and individuals as static units that are distinguishable in constant interrelationships; teaching about the human body emphasizes the part/whole relationships, together with the constancy of the body and its organs (it is not the continuous and never-ceasing processes of metabolism, of destruction or reconstruction that are emphasized, but rather the constant forms, the morphology of it all).

In a general way, I would like to emphasise that Western schooling (and Western knowledge for that matter) looks at the world, first and foremost in terms of objects, states, situations, constant aspects of phenomena, and that dynamic aspects are introduced in a lesser and at least secondary way. The discipline of geometry is evidence of this point: topology, and most certainly Euclidean geometry as it is taught in Western schools, concern themselves with points, lines, volumes, sets, and the like as static units. Dynamic aspects of spatial phenomena are seldom, if at all, introduced in these curricula. The appearance of this emphasis in geometry and other disciplines in considered to be compatible, indeed quite consistent, with the view of the general emphasis on static aspects of phenomena, on form, on structure, and on segmentability in both the Westerner's natural and his learned knowledge.[10]

The Navajo Case

The Navajo conceptualization of the external world differs from the Western one on fundamental points. I shall take up the three correlated characteristics in Navajo to bring out the difference, on the basis of which the educational implications can be outlined:

1. The Specific Hierarchy

In contrast to the neatly regular hierarchical structure and development that were shown in the Western space conceptualization, Navajo space (according to our semantic analysis, in Pinxten et al., 1983) appears to be founded on at least three equally important "basic" notions (movement, volumeness/planeness, dimensions). All three are topological in character. Moreover, none of them are really "primitive" in the sense that the (Piagetian) Western notions are: They are clearly composites themselves, and have spatial notions as their constituents; they codetermine themselves. Hence, they exhibit a certain circularity. Finally, they cannot, in any strict sense, be said to be the sole "basic constituents" since they have this status by virtue of the fact that they were selected as the notions involved in the constitution of most other notions. If this is a criterion of "basicality" (and no other criterion could be defended by such strong arguments), it is certainly weaker than the deducibility criterion Piaget forwarded.[11]

Hence, the hierarchical structure or, more generally, the organizational structure in Navajo spatial knowledge is not of the type that was found by Piaget in Western children. Consequently, the educational procedures that would be appropriate and satisfactory for the teaching of spatial notions in Navajo cultural settings cannot take the Western hierarchical spatial structure for granted.

2. The Part/Whole Distinction Is Secondary

The part/whole distinctions that proved so important in Western thought and knowledge systems play a minor role in Navajo knowledge. Navajos tend to speak of the world in terms of process, event, and fluxes, rather than parts and wholes or clearly distinguishable static entities. The emphasis is on continuous changes rather than on atomistic structure. This becomes more understandable when one envisions the world as constituted of dynamic "things," forces, changes, actions, and interactions: it is then not quite so obvious (as it is in a static world, a world of objects and constant forms) to try to segment perceptual and actional aspects, to distinguish parts in the continuous flux as a first and self-evident characterization. The difference is crucial, I feel, and it must be made totally explicit: "objects" cannot be defined in the same way, "form" cannot be understood in quite the same way as in the Western outlook, since all aspects of reality of Navajo knowledge are process-like and not thing-like. In the first place, processes or fluxes should not be "cut up" and considered as (statically defined) combinations of (statically defined) chunks of processes or fluxes. This point is important: even in Thom's "dynamic topology" (1977, p. 1–10), the dynamic phenomena are rendered understandable and manageable by petrifying them, that is, by defining them as changes between initial and final states. In other words, a dynamic phenomenon can be grasped in the Western outlook primarily by defining it as a transformation of a static phenomenon (state, situation) since it can then be segmented. This is the primary (though not exclusive) way of approaching reality in the Western atomistic outlook. This approach is generally alien, inappropriate, and unnatural to the Navajo eye, both at the level of practical manipulation (you cannot divide land and fence it off because by doing so you would then miss the essential character of land as a living, changing aspect), and at the theoretical level (you cannot speak sensibly of the parts of the human body; the complete phenomenon of a human being is a highly solid system).

The foregoing analysis does not mean that Navajos do not distinguish obviously distinct forms, bodies, objects, and the like in their environment. First, the foregoing is an analysis of the interpretation of (partially) similar precepts in Navajo and Western world knowledge, not of the precepts themselves. Second, and this is the most important point, Navajos react to the outside world as if it were primarily constituted of regular phenomena, of presumably patterned or "constant instances" (objects) of processes or fluxes. Consequently this means that if a notion of "object" is introduced to describe these constancies of processes,[12] this notion has but a minor kinship with the Western notion of "object." The Navajo emphasis is clearly on process in form instead of form as such.

The relevance of all this to educational strategies must be demonstrated: reference to point, line, atom, and other part/whole instances is obvious for instance, in teaching geometry and the sciences in the Western cultural context. Even when starting out with presumably the most general spatial notions of topology in education, the central importance of the notions of point, line, surface, and volume in a clearly segmentational (indeed set theoretic) sense is noncontroversial (cf. e.g., Thom, 1977, starting with "form," chap. 2). After that, distinctly Euclidean notions can be taught and actually construed on their basis, since they consist of similar (though more specific) notions of point, and so forth. This entry into the construction of geometric notions and models is inapplicable in an arithmetic sense to the Navajo students. Rather, a genuinely dynamic topological entry—as yet I have failed to find such a thing worked out in the literature—is required: clearly a formal system of "dynamic volumes," "dynamic dimensions," and "dynamic interpretations of movement" (cf. the results of Pinxten et al., 1983), is needed, which will translate the Navajo natural knowledge notions of space efficiently. From there, further metric notions should be worked out as refinements of this genuine basis of Navajo spatial knowledge. I will suggest a detailed solution to this problem in the following paragraphs. In any event, work should be started, beginning with a topology and following with a system of geometry that emphasizes segmentability and atomism to a lesser degree than contemporary approaches do. (see, again: Pinxten et al., 1987 for curriculum material along these lines).

3. *The Dynamic World*

In earlier sections, it was pointed out that the Navajo world view stresses the dynamic rather than the static aspects of reality. Although it cannot be said that these two terms are used in any exclusive sense in either the Navajo or the Western representation system, the primacy of one is obvious for each.

Space and time orthogonality are present to a certain degree in Navajo knowledge, although the clear "dynamization" of spatial phenomena obscures this fact. Indeed, some aspects of time in Western thought (notably the time aspect of change, "existence in time," and the like) are not distinguished in Navajo: rather, spatial phenomena are considered to possess these qualities inherently (that is, in a Westerner's terms). Other aspects of time (e.g., the notions of "generation" and "linear progression of time") are held as separate and "genuinely temporal," features (cf. chap. 1). Consequently, to a certain extent one can speak of "timespace" in the Navajo world view (cf. Whorf, 1956), referring to the relatively "time-laden" identity of space. In other words, spatial phenomena always have this spatial dynamic aspect, which—from a Western viewpoint—makes them different, less strictly, or purely "spatial" than, for example, Euclidean geometrical notions.[13] Again, the instruction in a system of purely static spatial representations as is currently practiced in Navajo schools easily leads to miscomprehension.

How Can the Problem of Cultural Alienation Be Met within Education?

I distinguish among three possible approaches:

1. Teach the Western System

In the present school situation, even where the Navajo view on reality receives due attention, I have seen the question resolved in the wrong way. The Navajo knowledge input was tolerated or even encouraged in matters of traditional moral and social concern, but was considered completely irrelevant or, even worse, nonexistent in other matters. To the teacher trained in Western schools, it appears obvious that the Navajos (or, for that matter, any other non-Western people) have nothing really to say or to think, within the confines of their cultural knowledge, on subject matter that is considered in the realm of the sciences. The argument goes as follows: Sometime during the classical age a few people started a completely new way of thinking and nothing of the other peoples' knowledge has any real value as compared to that. This attitude is understandable when one reckons the technological explosion that is linked with this type of thinking, the "scientific thought" that conquers huge domains of nature and society wherever it is implanted. However, as I explained above, the problem of the input of other peoples is more delicate and makes itself felt at another level. The two following alternative approaches will make this clear. Meanwhile, the effect of the present approach is, as illustrated in previous paragraphs, that the "victims" of Westernized education, in this case the Navajo Indians, gain a hybrid and badly integrated knowledge. That is, they end up being not really knowledgeable in their own tradition and having another distinct, separately organized, and definitely incomplete second system of knowledge, the Western system. They are often unable to give good and valuable information on their traditional system, and also unable to discuss and question critically some of the aspects of the schooled knowledge. Moreover, the integration and the degree of translatability or even substitutability of elements of one system in the other, and vice versa, is most often alarmingly poor. Such were my experiences, at least, when I talked to and questioned children and schooled adults in this situation. In other words, the net result is the gain of a partial and relatively unintegrated knowledge of both systems (a bad example of "biculturalism" according to Fishman, 1979), and an alienation from the relevance and mutual relationships of both systems.

2. Elaborate the Navajo System

The second possibility I envision would take the Navajo spatial distinctions as genuinely qualitative material for the teaching of geometry and would work toward the usual Western geometric notions from its Euclidean foundation onward. In practice, I found, this approach would amount to the translation, translating the basic Navajo notions and the relationships between all notions distinguished in the course of the present semantic analysis into existing formal notations and an existing formal theory.

To realize this strategy, the different Navajo spatial notions require a notation system that (*a*) deals with or is appropriate for representing the more or less qualitative (rather than strictly metric) interpretation of spatial distinctions in Navajo, and that (*b*) has the capacity to emphasize the dynamic aspect (rather than the static character that is conveyed by most geometric systems). The only formal approach available that to some extent honors this type of interpretation of space is the dynamic topology or catastrophe theory as worked out mainly by Thom (e.g., 1977).

In general terms, Thom set out to construct a formal theory that would account for several dynamic aspects like change, alternation, cessation, wavelike movement, and so on. He sought applications of a model in several sciences, including social sciences (mainly linguistics). Waiving details of his theory, it can be stated that he presents an approach to dynamic aspects according to which they are changes between two states (initial and final state), and a formal characterization of types, features, regularity, and so forth of such changes. His theory is a topological theory, dealing with presumable the most general characteristics of space (or those that are common to different sophisticated spaces like Euclidean space, non-Euclidean spaces, and so on). Topology, as is well known, offers at least the advantage of speaking about the foundations of spatial models (e.g., the Euclidean model), and therefore dynamic topology could be tried out as a candidate for a general and foundational approach to geometries, specifically starting out with dynamic spatial aspects. It is this type of program, I feel, that should be tried out with Navajo spatial differentiation.

The problem with the Thom approach, though, is that there is no easy and thoroughgoing introductory text of the theory, as we have for general (static) topology (e.g., Sauvy and Sauvy, 1972, is a guide for application in the classroom). It may be hoped that mathematicians may find something valuable in the present explanation and thus be encouraged to work out a teacher's guide.

The procedure would run as follows: Western geometries can be reduced to their foundations and expressed in topology. Navajo spatial conceptualisation could be formulated in a like manner (emphasizing dynamic topology) and could thus be reduced to a particular system of topological propositions.[14] In a second stage, the "Western" and the "Navajo" system of topological premises, axioms, and statements should then be compared. Finally, exact adaptation procedures should be devised in order to ease the transition from the "Navajo" statements to the "Western" system and the related "Western" geometry.

This strategy will have the tremendous advantage that the transition into "Western" biased geometrical notions is understood and explicitly taken into account, while in the former strategy it was bluntly imposed without consideration for the Navajo native categories. Indeed, starting from the presumably more basic formulations and notions of topological aspects in both cases would offer the opportunity to compare both systems of spatial structuring (the Western and the Navajo system) as alternatives to each other. Further, it would make an adaptation or "guided transition" possible for the Navajo child. I would start with the differences between the child's notion and the Western ones and train the transition between the two explicitly and purposefully, in order to carefully convey the detailed and difficult "pitfall" (from the point of view of the "Westernizing" teaching) that confront the child because of the difference in outlook of the systems. In further development of geometric thinking, the "Western" basis could then safely be taken as a point of reference since it was instructed in the proper way.

Since this strategy has the tremendous advantage in that at least the gap of translatability has been taken into account, it alone could improve schooling. However, it has two serious drawbacks. In the first place, it unquestionably endorses a reductionist frame. Statements and outlooks described in the representation are always reduced to topological phrasings within a formal system. Since, as I mentioned, existing formats (as far as I know) would cause a very serious reduction of the Navajo conceptual content (with no alternative for "Navajo volume," that is, without the notion of "point," and numerous difficulties with the hierarchical and not strictly interrelated notions), the drawback may outweigh the advantage in the case of Navajo space. A decisive balance can only be made when the detailed mathematical "translation" in a dynamic topological model of the semiformal spatial representation model (of chapter 4 in Pinxten et al., 1983) is completed. A second drawback, and one that is plain without complex argument, has to do with the employment of topological phrasings in grade school processes. The procedure, as outlined, aims at integrating the Navajo conceptualization into a general set of spatial differentiations. The inevitable upshot would be that Navajo children and teachers would have a better grasp and in fact a better entry into the field of geometry since they would be able to understand the basic, foundational link it has with topology and thus with their specific "topological system." However, all this is but an avenue to Western knowledge and thus a "Westernization" of communication and interaction in schools.

So, although the second alternative would offer a better understanding of the link between Navajo thought and Western thought (at least in some fields), it rests on the same implicit conviction that all is well as long as the West expands its hegemony. Not only is it morally and politically a very questionable position to hold, but moreover it could lead to an ensuing disastrous uniformity in the world. Indeed, it does not take too vast an imagination to understand that the investment of all survival dependence on one, presumable "true," outlook on the world in every detail is in fact an incredible risk; the risk is so much greater because only very limited areas of the world, in which we have to survive, can be said to be reasonably well understood by this outlook. Nevertheless, that is the gamble we are actually engaging in by the present attempts at colonial and semicolonial institutional "development" of all non-Western peoples, be they Navajo Indians or others. Good intentions are certainly not lacking, but consciousness of the payoff in risks is.

3. Integrate the Western System

An alternative consists of the integration of the Western outlook within the Navajo model and in terms of the Navajo spatial model.

The net result of this strategy would lead to the integration of Navajo and Western thought in a Navajo-biased frame, with both politically and educationally justified ends.[15]

I intend to use the model of spatial notions (as presented in Pinxten et al., 1983, IV) in two ways: in the first place, the model should facilitate an unambiguous and clearly explicit education in the Navajo knowledge; in the second place, it should facilitate the introduction of Navajos to a cognitive context of foreign notions (i.e. Western notions or any other) which they learn in school. Further, these notions should be translated into the native system of knowledge as much as possible. In a way, then, this strategy is comparable to that of the foregoing section, the result being exactly the inverse of the previous one: Instead of reducing Navajo knowledge to Western knowledge, I rather defend the opposite. The general political or sociocultural reason for this maneuvre is quite straightforward. In view of the impossibility (for the time being) of any clearly englobing meta-system approach (there is no adequate meta-language and no rules of correspondence and translation that I know of), I try to safeguard as much as possible the typically native frame at the cost of the imported one, because its survival will enrich us and actual communication between cultural knowledge systems is certainly at the cost of the native one at the present time.

Education in the Native (Navajo) Spatial Knowledge

I conjecture that in education the explicit treatment of the Navajo spatial knowledge in schools, in geometry courses, and also in other parts of the curriculum (cf. below) would change the present unhappy situation dramatically. Consonant with the results of Cole and his coworkers on schooling and literacy, I argue that the explicit treatment of this material in a classroom setting would improve teacher's and pupil's understandings of spatial notions (Scribner and Cole, 1981). I propose the following scheme, based on the results of the semantic analysis:

POSTULATES

1. Space (in Navajo knowledge) is finite, bounded, and absolute. This can be demonstrated through the model of the world that is held: every existing thing is contained in a saucer-like grand

structure, which assigns a definite place to any imaginable entity. (One need not identify this structure with the Earth between the four Mountains, as long as the idea of an absolute, finite, and bounded space is rendered).
2. Space, like the entities or objects within it, is dynamic. That is, all "entities," "objects," or similar units of action and perception must be considered as units that are engaged in continuous processes. In the same way, spatial units and spatial relationships are "qualitative" in this same sense and cannot be considered to be clearly defined, readily quantifiable and static in essence.

Elaborations: while the preceding "postulates" should be explored and demonstrated through numerous, indirect examples and can draw directly on the meanings in Navajo terminology and phrases, the elaborations of all specific notions should be more rigidly governed in the classroom.

In practice, all notions that were isolated from the Navajo vocabulary (in chapter 1 and 3 of Pinxten et al., 1983) can be taught systematically, starting at whatever point one chooses. Perhaps the most convenient entry would be the more basic notions like "volumeness," dimension," and "movement." In any event, a practical and pedagogical approach seems fairly easy.

The following procedure can be followed step by step in order to convey the abstract notion of "volumeness" in Navajo:

1. The teacher has his pupils play with different types of objects, sets of objects, and the like, so as to cover all instances of volumeness encountered in the Navajo lexicon. In the first place, all different classificatory verbs in a way define different types of nonrandom objects, and other terms and expressions complete this picture.
2. Throughout the process, and at several different stages of this conscious and explicit exploration of the Navajo world of objects, grows the more abstract and englobing notion that all these phenomena are but instances of "volumeness" (a Navajo correlate to this noun is needed, as is the case for all technical and noncommon notions). Particular exercises to train the acquisition of the notion of "volumeness" can be introduced. The difference between this notion and notions like "movement" can be trained by a thorough and explicit comparison of instances of both notions.

Subsequently, all other notions of the Navajo spatial knowledge system can be taught in similar ways. Once the more basic (and less constituted) notions are acquired, one can introduce more sophisticated notions, that is, those that show several "constituents" in their semantic paraphrases. For example, let me explain:

"Distance," a more or less metric notion that is acquired by Western children in a full and systematic way as late as the fourth or fifth year of schooling (cf. ten years of age Piaget and Inhelder, 1947), can be introduced through the following processes of learning:

1. The teacher has the children play with rods, displacements of volumes, and so on, and tries to make the notion of "spatial distance" clear. It is treated as a more sophisticated and highly abstract form of "separateness" (which was acquired earlier because it is less sophisticated).
2. In a subsequent phase it is made clear in the same way (by trial and search with volumes, objects, sticks, or steps between them, etc.) that at least "something or someone" must be at a distance from "something or someone else," introducing in this way the constituent character of "volumeness" (object or ego) in the notion of distance; distance exists between instances of volumeness.
3. In similar ways the other constituents are introduced as necessary (but insufficient) components of "distance." In the end, at least the semantic characterization of "distance" (given in the semantic analysis of chap. 4 in Pinxten et al., 1983) is acquired. It should be born in mind, at all times, that the schooling process aims at acquiring the abstractions. Concrete instances as well as names for these instances can only serve as examples at the beginning of the procedure (of each attempt to get at each specific notion), but cannot hold as the definitions or abstractions themselves.[16]

This procedure can be followed until all notions of the semantic model (given at the end of chapter 4) are taught in their specific characterization, that is, in their double role of having constituents and being constituents (for other notions). In this way, the Navajo notions of space and the total structure of Navajo spatial knowledge can be taught explicitly and with proper pedagogical means.

I scarcely elaborate on the actual and detailed educational means and devices that can be used in this process. I am confident that they would be somewhat similar to those used in the modern play-and-learn schools of the West, although proper and quite different strategies may be defensible. However, this is outside the domain of the present study. These questions are technical in nature and are treated extensively in Pinxten et al., 1987: we develop educational contexts here which are common to Navajo children's experiential world: the hooghan, the rodeo, sheepherding, weaving, and so on. I am satisfied here to have provided a general outline and hopefully to have demonstrated that a proper teaching of the Navajo spatial knowledge in (modern) schools is sensible, desirable, and indeed possible. It is up to the specialists in the curriculum services to try to test this information in the proper pedagogical ways.

Education in the Western Spatial System

Once a fully developed and well-trained knowledge of the Navajo spatial system is acquired, the (more or less corresponding) Western spatial knowledge can be instructed. Again, in order to avoid a blind and misunderstood training in a completely "foreign" knowledge system and, on the other hand, to restrict the reduction ultimately of the native categories to a Western-biased education, I propose to teach the Western notions against the fully conscious and explicit background of Navajo knowledge. If reduction need be, it will then be on the part of Western notions. I thus propose to teach the Western alternative at a later age, as a specific and differential elaboration on the same basic problems of interaction with the outside world, somewhat in the way the non-Western approaches to problems are, to a certain degree at least, taught to Western children once they have reached an age of understanding in a Western cultural context. Again, for concrete examples, I refer to Pinxten et al., 1987.

CONSIDER THE FOLLOWING POSTULATES:

1. Space (in the Western knowledge system) is infinite, bounded, and relative. This perspective, that so drastically differs from the Navajo view, can be taught in similar ways through reference to the basic scientific model of the world and can be visualized through substractions from the Navajo view: imagine a never ending line, imagine a circle as a continuous movement along its contour, imagine the characteristics of a ball (in fact very much the same way we were taught about the characteristics of these phenomena); imagine a relativistic view-

point working with mere relations between things and nothing more—the Leibniz description, almost to the word, in his discussions against Newtonian absolute space (cf. Smart, 1964).

2. Space is static, just as the objects and entities within it are static. This may be teachable through the idea that Western space takes account of "snapshots," momentary pictures of Navajo spatial phenomena and acts upon them "as if" they were the things or phenomena (in a Navajo sense) themselves.

Elaborations: some training is, of course, required in order to instruct the above postulates, so that reasoning within such a world view becomes feasible to some degree. (I experienced difficulty in switching frames during fieldwork and I do not intend to minimize the difficulty of such an acquisition).

From this point on, then, several Western notions of space, in fact geometrical systems can be instructed, always securing the delicate "translations" that were shown to be presupposed.[17]

For "volume," the following procedure can be followed:

1. In rehearsing the Navajo notion of volumeness, the ways in which (Western) geometric notions of volume are similar to and at variance with the former notion can easily be pointed out. A major difference that is easily clarified concerns the static-dynamic dichotomy. Again, the volume notion is special in the sense that it results in a "snapshot" view or a momentary slice of the dynamic volumeness instances known in Navajo. Besides such features, the strictly metric nature can be explained in a similar way (for instance, segmentability and respecting of constancy of form is basic and easily instructed by drawing on the work of Piaget). Instances of volumeness can be reduced or manipulated, in practice as well as theoretically, to result in the type of regular and highly symmetrical forms that Westerners deal with as volumes. a high degree of similarity, if not identity, between Western and Navajo volumeness lies in the fact that volumes can be manipulated, perceived, classified, reconstructed, and so on in the same way as instances of volumeness can. The difference between the notions is a set of supplementary, and therefore highly refined, features.

   Similar explorations of differences and likenesses between Navajo and Western concepts can be performed, both at a practical (active, concrete) level and at a theoretical level. A supplementary and more difficult task is the teaching and systematically explicit analysis of the difference in relational definitions (or semantic characterizations) between pairs of notions. This theoretical work, that is, construing the appropriate sets of relations and definitions, is most often done through contrasts with the corresponding structural information in the Navajo spatial knowledge system. I propose that this level of reasoning, which informs the actual axiomatic way of relating and structuring that is typical of all sophisticated Western knowledge, be introduced through drills or through a detailed analysis of the use of logical rules in the Western system. The latter are easily linked with native Navajo logical rules that are, no doubt, much the same as the Western ones (cf. examples in analysis and field material): consistency, coherence, and logical operators are stressed in explaining and teaching the Western (absolutely crucial) dependence on them in the elaborated form of geometrical reasoning.

But let me, for the sake of clarity and strength of argument, point out in very general terms what procedure can be followed to introduce the basic notions of Western geometry starting from "volume."

2. In a second step, the notion of "line" can be introduced: whenever two concrete volumes (in the above sense) touch each other at several points in a row, the structure that can be seen, drawn, or "walked" (eventually) through all these points, by passing and touching each point once, constitutes a line. Visually, the connection between sky and heaven is such a line. As well as the connection between two "volumes," a concatenation of places where both touch each other, like a mountain and a river, is seen as an ever progressing concatenation.
3. Whenever two volumes touch or overlap with one "side," the notion of "surface" can be reached.
4. Whenever two volumes touch or overlap in such a way that their place of being together is but one and cannot be seen or visualized or made to progress in any direction, but to be always the same place, we have a point. And so on.

The descriptive sophistication of these steps can be improved greatly. Again, I refer those interested to the brilliant work of Whitehead on these topics to get an idea of a much more appropriate (i.e., less intuitive and "logically" messy) version of this line of thought. What I intend to do here is merely to point out a possible way of communication between the two systems, without worrying about the appropriateness of the steps and notions implied. It is up to a well-trained pedagogue or psychologist to "translate," as it were, Whitehead's suggestions into the proper and correct form that would suit the instruction of Navajos. The general nature of such translations should have become clear from this exposition.

Elaboration of Other Materials along These Lines

The preceding type of instruction is not restricted, of course, to mathematical (or, more strictly still, geometrical) reasoning alone. Similar "programs" can, and should, be worked out for other domains of the curriculum where space is implied.

Geography obviously employs plenty of spatial representation. It would be quite interesting to develop a geographical segment of a social studies curriculum that would take into account the dynamic and holistic perspective of Navajo space. For example:

(*a*) A mountain or a valley should not necessarily be introduced the way "objects" are in the Western context, but should rather be represented as a complex unit of mutual interaction, where valley and mountain aspects change status over the years. The dynamic interactive relationship between the two is what counts most in the teaching. A beautiful example is the way orientation works as an exploring device for Navajo children when herding sheep in a canyon (Pinxten et al., 1987).

(*b*) The earth should not be subdivided easily into states, but should primarily be seen as a relatively constant body surrounded by water, wherein different life-circles or territories are identified, that is, the cities, the spaces of peoples, the communities, and so on. These life-circles exhibit a relatively dense pattern of interaction with each other (some more, some less) and have a dense inner scheme of interaction; they have a certain amount of growth, decay, change, and so on.

This approach to geography can, of course, be worked out in full detail and may yield very interesting results. The approach is not at all outlandish, since the so-called New Geographers have tried to work upon Western geographical notions in somewhat similar (but less radically elaborated) ways: see Gould and White (1974) and the pioneering work by Lynch (1960). On the other hand, the beautiful book on Ifugao geography by Conklin (1980) could be used as an example of this approach in other cultures.

Natural sciences can, obviously, be recast in a similar way. The organismic view of persons and animals as "systems that were placed to work in specific ways" can easily be emphasized in education, instead of the objective and segmentability-centered approach of contemporary Western teaching. Again, this alternative is not so alien to modern perspectives in the life sciences: Waddington (1977, chaps. 1, 2) stresses the importance of an alternative, dynamic, and even organismic approach in these disciplines today. Nature seen as a symbolic system with ecosystems and relatively autonomous biological systems (man, animals, plants) can be a rewarding outlook.

Even in disciplines such as history, social science, and linguistics, a similar approach can yield important results. In each of these, the systemic and diachronic outlook of Navajo knowledge could be worked into the syllabus.

What Is the Use of All This?

It will have become clear that I push toward a genuinely native approach to aspects of knowledge that seem to be rather "established" in the Western context. The general question that arises against such a laborious and painstaking reorientation of parts of knowledge and schooling will be: what is the use of all this? or, why engage in such a hazardous and unwarranted enterprise instead of sticking to established data and disciplines?

I have a number of different and important (not to say universally relevant) reasons for proposing such a complicated approach:

First, *the Navajo reason*: in the very first place, there is today the phenomenon of the totally inappropriate education, leading to misunderstanding and sociocultural and psychological alienation of Navajo children and adults. With its almost complete lack of consideration for the authentic Navajo world view, the school curriculum is scarcely integrated into the native context. The result is that people at some point have to choose between their native system of knowledge and the Western alternative. Most people never choose and come to live and think in a "divided world," partly Navajo and partly Western. Nobody deserves this second-rate treatment in a democratic society. The alternative, as I emphasized earlier in this chapter, is a redesign of the curriculum in terms of the native frame. This is precisely what I have tried to accomplish in this third alternative. I know of some institutions that try to work in this fashion, notably the Division for

Curriculum Services at Albuquerque, the community schools at Rock Point and Rough Rock, and some individual workers. It is hoped that the present work can be helpful for them in the domains that have been treated here.

Second, *the epistemological reason from a Western point of view*: it is fascinating and theoretically rewarding to try to work out alternatives to the historically and culturally specific outlook that predominates nowadays. The challenge of bridging the tremendous gap of quasi-intranslatability and incommensurability between Western and non-Western outlooks is a most rewarding and important task for any (Western) scientist who claims to work towards universal truths. The formal and interpretational intricacies of such a program codetermine the adequacy of a particular proposition. So, apart from the political and social dimensions of this and similar researches, I want to emphasize the epistemological relevance.

Third *the evolutionary reason*: finally, and from the most general point of view, an evolutionary argument appears compelling. Through a systematic superimposition of the world view and thought system of the West on traditional non-Western systems of thought and action all over the world, a tremendous uniformization is taking hold. In the past, inputs from other cultures into the Western pool of thought have proved significant and at some points even crucial (e.g., the impact of some Chinese concepts via Leibniz's transformation of modern mathematics and physics), while the orderly communication and interaction with Western ideas proves beneficial to other cultures at some points. However, through the systematic extinction of other systems of knowledge that is now taking place, the original "pool of responses" (a metaphor, of course) is dramatically reduced and eventually ends up being identical with the Western pool. The risks we take on a worldwide scale, and the impoverishment we witness is—evolutionary speaking—quite frightening. As long as science cannot pretend to have valid answers to all basic questions (as is the case in our contemporary situation, because only some aspects of nature and social life can be treated with satisfactory validity and confidence in the still-incomplete scientific system), it is foolish to exterminate all other, so-called primitive, prescientific, or otherwise foreign approaches to world questions.

One does not have to believe with J. Needham (1956) that Chinese models are appropriate for the (Western) life sciences where the native (Western) outlook fails to apply in a satisfactory way. However, mutual inspiration and interaction at a fundamental level of epistemology and ontology (natural philosophy, strictly speaking) will probably always be rewarding and illuminating in all instances. In view of this general appreciation of the possible relevance for survival and the value of a multitude of perspectives (rather than the unique approach at the cost of all others or whatever culturally biased perspectives one may cherish), I mean to defend the sensibility and the possible power of the conscious and full-fledged elaboration of native knowledge (e.g., the Navajo system). This should be maintained to the extent where comparison and mutual influences can be mapped, or even planned, in an egalitarian way. As a concretization of this general strategy I propose, for example, that it may be very inspiring for a mathematician to try to construe an axiomatization of dynamic topology, in a broader and less confined way than Thom has tried to do, starting with the insights of Navajo spatial knowledge. The elementary and intuitive inspiration that springs from the confrontation with a system of thought that differs on such basic levels cannot be measured beforehand, that is, without full consciousness and primary consideration of the spatial knowledge system as such.

The foregoing arguments unite in a general appreciation of the strength and validity of particular theories and models of science, converging to the position held by Don Campbell (1973, 1989). All human beings confront the world in much the same way, notwithstanding their particular cultures, and deal with almost identical questions, even if they receive somewhat different elaborations. If an evolutionary model of science and truth has validity, as I agree with Campbell that it does, then the variability and the multitude of perspectives is valuable, certainly in areas where the validity and strength of presumed scientific models and theories are so weak as to prohibit valid prediction. I share the conviction that inability in this aspect is characteristic of a lot of disciplinary knowledge today, and I therefore defend the thesis that as many rational approaches as possible should be supported and elaborated to serve as a comprehensive analysis of life questions at some later point. The Navajo thought system certainly is rational in several aspects and can therefore, when it is fully elaborated, illuminate the Western system at some points, and vice versa. I hold that it is the task of philosophers of science and anthropologists to try to elaborate more adequate criteria of rationality (a task currently being undertaken by several scholars) and to organize an effective and nonalienating communication with, and interaction between, non-Western cultural knowledge systems that are not extinguished, through a confrontation with foundational propositions of the modern sciences.

Notes

1. It is impossible at this point to go into this matter. However, numerous publications on the topic exist. The publications by the Genevan school of thought offer ample examples.

2. BB, a brilliant young Navajo man, reacted to the idea of studying Navajo spatial knowledge by saying: "The moment I heard of your study, I thought you would run into great and highly complicated problems. Space is a deep notion and poorly understood in the Navajo studies."

3. One consultant, LB, told me about a test on taxonomic classification that was carried out for some time. Although consultants followed the instructions and filled in the test, they confessed not to have understood what was "known" from them in this way: they simple would not work in this manner on their own. The principle did not really apply.

4. The lack of understanding of the incommensurability between the two cultural knowledge systems (Western and Navajo) that is so apparent in discussions with teachers and students on this matter is not difficult to grasp. It is a well-known fact that Kant considered Euclidean geometry to be innate/evident, the only possible geometry of the outside world, and that thinkers until the first half of this century (including Frege and Husserl) could not really cope with the non-Euclidean geometries worked out by Riemann and Lobachevsky in the nineteenth century. The latter seemed mere "faits divers," counterintuitive products that could not really claim compatibility with anything in the physical world, until they did show it (Einstein, 1949, discusses this point). The matter is not yet fully ingrained in our education, let alone the more basic "shifts" through topology (cf. Reichenbach, 1958; Thom, 1977).

5. The progression does not always present itself in that rigid form, but the idea of a systematic and clearly hierarchical progression certainly holds for the total system of spatial representation.

6. See the discussions in philosophy of science: Russell (1921) sought sense data, Goodman (1977) seeks for "qualia as atoms". Cf. also Quine (1969), "Speaking of Objects."

7. See the discussion on these matters in Ernest (1991), Pinxten (1994).

8. I am referring to the fundamental work on catastrophe theory (e.g., Thom, o.c.) and dissipative structure theory/chaos theory (e.g., Prigogine and Stengers, 1990).

9. The interested reader is referred to the works cited (especially Atran). I have made a detailed analysis on these points elsewhere (Pinxten, forthcoming, Pinxten and Farrer, 1994).

10. I am quite aware of the fact that this characterization is an oversimplification, and that it tends to draw a somewhat one-sided picture of the topic. Still, my general conviction seems to be justified (cf. references to philosophers of science), while the one-sidedness is meant to give the point extra emphasis.

11. Hence the point is not that Navajos are shown to think more primitively or "prelogically," but rather that the same type of notions and the same operations are used in different ways (cf. the concurring position on the use of logical operators by Kpelle in Cole et al., 1971).

12. Thom (1977, pp. 10–11) tries to develop a more or less dynamic notion of "objects," which contrasts only with the percept-dominated notion. His notion is not applicable here since it starts off from states and construes "object" as the result of transitions between stable states.

13. Throughout, as will be clear, "dynamic aspect" cannot be identified with "movement" or displacement through space. The latter category has relevance both in Navajo and Western systems, but is alien to the "static/dynamic" dichotomy.

14. I am aware that there would be considerable technical problems involved, for example, the emphasis on "volumeness" in the Navajo system is difficult to render in any existing topological notation system or formal theory, since most (if not all) topologies work with "point" as the basic unit.

15. At first it was my intention to try to build a model of parallel development of Western and Navajo spatial organization. However, I would need a meta-language that would at least be able to describe simultaneously and adequately both the static and the dynamic outlook. No formal sophistication whatever could be reached in this outlook, landing us with mere vague and analogous comparisons. The actual search for "correlations" between both systems of spatial representation thus became highly hazardous, leaving no possibility of describing the correlations in any significantey clear and unambiguous way. In view of all these drawbacks, I decided to abandon this program and concentrate on a pedagogically feasible alternative. I acknowledge Dirk Batens' insights and advice on these points.

16. English, like most European languages, is ambiguous in this sense: the same word is used for the abstract notion and the geometric notion (most of the time) that is used to denote the concrete instance. The noun plays a double role, at two levels of thought.

17. I propose to start with notions that are generally closer to the Navajo native system than, for example, point (which was shown to be lacking completely). Whitehead (1953), in a brilliant little book, demonstrates that geometry and all geometrical theses can be built with equal validity and equal strength, starting alternatively from "point," or "volume," or "surface" as primitive notions. Starting from the primitive "volume," one can subse-

quently satisfactorily define "surface" and any other geometric notion, ending with "point" as the most detailed and complex notion of the whole system. In other words, the approach is possible with equal validity and strength from a multitude of perspectives. It is this idea that I would like to refer to here (in a less strict and less sophisticated way, to be sure). I consequently propose to introduce volume first in the instruction of Navajos, eventually ending up with other notions of more "primitive" status, but at a much later stage.

References

Atran, S. (1990). Cognitive foundations of natural history. Towards an anthropology of science. Cambridge: Cambridge University Press.

Bovet, M. (1974). Cognitive processes among illiterate children and adults. In *Culture and cognition*. Edited by J. Berry and P. Dasen. London: Methuen.

Campbell, D. T. (1964). Distinguishing differences of perception from failures to communicate in cross-cultural studies. In *Crosscultural understanding*. Edited by F. S. C. Northrop and H. Livingstone. New York: Harper and Row.

———. (1973). Ostensive instances and entitativity in language learning. In *Unity through diversity*. A Festschrift for Ludwigh von Bertalanffy. Edited by William Gray and Nicholas D. Rizzo. New York: Gordon and Breach.

———. (1989). Descriptive epistemology. In Campbell D. T. *Methodology and epistomology for social sciences*. Chicago: Chicago University Press.

Cole, M., J. Gay, J. Glick, and D. Sharp. (1971). *The cultural context of learning and thinking*. London: Methuen.

Cole, M., and S. Scribner. (1974). *Culture and psychology*. New York: Methuen.

Conklin, H. C. (1980). *An ethnographic atlas of ifugao*. New Haven: Yale University Press.

Einstein, A. (1949). Einstein's autobiography. In *Albert Einstein: Philosopher-scientist*. Edited by Paul A. Schilpp. Evanston, Ill.: La Salle.

Ernest, P. (1991). *The philosophy of mathematics education*. London: Falmer Press

Fishman, J. (1977). In *On going beyond kinship, sex and the tribe: Interviews with contemporary anthropologists in the U.S.A.* By Rik Pinxten. Gent: Story.

Goodman, N. (1977). *The structure of appearance*. Boston Studies in Philosophy of Science. Dordrecht: Reidel.

Gould, P., and R. White. (1974). *Mental maps*. London: Penguin.

Grunbaum, A. (1964). *Philosophical problems of space and time*. London: Routledge and Kegan Paul.

Lynch, K. (1960). *The image of the city*. Cambridge, Mass.: MIT Press.

Needham, J. (1956). *Science and civilization in China*. London: Cambridge University Press.

Piaget, J., and B. Inhelder. (1647). *La construction de l'espace chez l'enfant*. Paris: PUF.

Pinxten, R. (1976). Epistemic universals: A contribution to cognitive anthropology. In *Universalism versus relativism in language and thought*. Edited by R. Pinxten. The Hague: Mouton.

———. (1994). An anthropologist in the mathematics classroom? In S. Lerman (ed.). *Cultural perspectives on the mathematics classroom*. Dordrecht: Kluwer Academic Publishers, 85–98. Forthcoming Philosophy and Anthropology. M.S.

Pinxten, R., I. van Dooren, and F. Harvey. (1983). *Anthropology of space: Explorations into the natural philosophy and semantics of the Navajo*. Philadelphia: University of Pennsylvania Press.

Pinxten, R., I. van Dooren, and E. Soberon. (1987/1994). *Towards a Navajo indian geometry*. Gent: KKI Books.

Pinxten, R., and C. Farrer. (1994). On learning and tradition. In W. De Graaf and R. Maier (eds.) *Sociogenesis revisited*. New York: Springer, 169–184.

Prigogine, I. and I. Stengers. (1990). *Order out of chaos*. New York: Wiley Interscience.

Quine, W. V. O. (1969). *Ontological relativity and other essays*. New York: Columbia University Press.

Reichenbach, H. (1958). *The philosophy of space and time*. New York: Dover.

Russell, B. (1921). *The analysis of mind*. London: Allen and Unwin.

Sauvy, J., and Sauvy, S. (1972). *L'enfant à la découverte de l'espace*. Paris: Casterman.

Scribner, S., and M. Cole (1981). *The psychology of literacy*. Cambridge, MA.: Harvard University Press.

Smart, J. J. (1964). *Problems in space and time*. New York: Macmillan.

Thom, R. (1977). *Stabilité structurelle et morphogénèse*. Paris: Interéditions.

Waddington, C. H. (1977). *Tools for thought*. St. Albans, England: Paladin.

Wassman, J. (1995). The final requiem for the omniscient informant? In *Culture & Psychology* I: 167–202.

Whitehead, A. N. (1953). On mathematical concepts of the material world. In Alfred N. *Whitehead: An anthology*. Edited by F. S. C. Northrop and H. Gross. Cambridge, England: Cambridge University Press.

Whorf, B. L. (1956). *Language, thought, and reality: Selected writings of Benjamin Lee Whorf*. Edited by John B. Carroll. Cambridge, MA: MIT Press.

Chapter 18

An Ethnomathematical Approach in Mathematical Education: A Matter of Political Power

Gelsa Knijnik

Editors's comment: Gelsa Knijnik, a Brazilian mathematics educator, reports on her research with an organized movement of landless Brazilian rural workers. She grapples with the interrelations between academic and popular knowledge and with the contribution that an ethnomathematical approach in education can make to the process of social change. This chapter, which has been slightly modified, first appeared in *For the Learning of Mathematics* 13(3): 23–26, in 1992.

The research took place in the rural area of the state of Rio Grande do Sul, in Brazil. The country's total area is approximately 8.5 million square kilometers, corresponding to almost 50 percent of South America's total area and more than 20 percent of the American continent. Nowadays, Brazil has about 146 million inhabitants, with a demographic density of sixteen people per square kilometers.

One of the biggest economic problems in Brazil, which is also a social and political problem, is its land structure with its very dense

---

This article explains and discusses the author's work in a rural school in the southernmost Brazilian state. This school is linked with organized rural movements, mainly the MST (Movimento dos Sem-Terra: landless people's movement). Two specific practices of the people are presented using an ethnomathematical approach, the author developed some educational work which deals with the interrelations between erudite and popular mathematical knowledge in the context of the struggle for the land. Without glorifying popular knowledge, the paper discusses the contributions this kind of pedagogical work can give to the process of social change.

concentration. Rural properties of less than twenty-five acres—which represent 53 percent of all Brazilian rural properties—occupy only 3 percent of the country's total area. However, rural properties with more than 2,500 acres—which represent only 1 percent of all Brazilian rural properties—occupy 44 percent of the country's total area.

The author's fieldwork was developed in the rural area of Braga, a small district of Rio Grande do Sul, about 370 miles away from the State capital, with about 5,000 inhabitants (15 percent less than its population in 1980). With 73 percent of its population living in the rural areas, it is essentially an agricultural district.

The DER/FUNDEP rural school,[1] located in Braga, where the author developed her research, was founded in 1989, as a response to organized rural movements, mainly the MST. The school aims at implementing a form of popular education, understood as a methodological approach which should contribute to social changes. Today, the school has to work, as a priority, with the organized rural workers themselves because, as they say, "there is a historic urgency in the need to educate the main agents involved in the social change process."

The school has different types of formal and nonformal courses. The author's research was more directly linked with one of these courses, which aims at preparing and giving a certificate to elementary school teachers. Almost all the students were, in fact, teaching, even before getting the required degree. They are called "lay teachers." In Brazilian rural areas where teachers are scarce and, sometimes, there are no schools, a community member who has a more advanced level of study holds the teacher's position, helping children and also adults in the learning process. In this situation, it is absolutely necessary to provide better conditions for those people involved with education. Usually the schools in those very distant communities have only one class for children in all the grades. The teachers teach part time and farm part time.

These "lay teachers" go to the DER/FUNDEP rural school, as students, for four periods during their school vacations. This represents six months of all-day study at the school and "long-distance study" while they are teaching in their communities. After these four periods there is a probation semester, when they teach in the communities, supervised by DER/FUNDEP teachers.

Today, the MST is one of the most important rural movements in Brazil. The slogan of the Movement—"occupying, resisting, producing"—describes the struggle for land reform in order to avoid a rural exodus, giving inland people better conditions of life through the redistribution of the country's wealth.

When MST members occupy a piece of land, they live on precarious campsites, in very rudimentary plastic tents. This is the time when they organize themselves into groups and start discussions about political, economic, and social issues. They learn how to resist police violence. They also plan how to produce and market goods in the settlements as well as how to manage the land they are supposed to receive from the government. At present (October 1996), there are about 145,000 people in the settlements and more than 24,000 in the camps linked to MST, spread over nineteen out of the twenty-six states of Brazil.

It is clear to the Landless People's Movement that education is a strategic issue for the land reform; the Brazilian peasants have always been deprived of academic knowledge and in this situation it is hard to live and produce goods successfully in the rural areas. The Movement is aware that it is absolutely necessary to have academic and therefore technological, knowledge when organizing, administering, and planning production.

An Ethnomathematical Approach

The author's research—taking for granted that pedagogical practice in mathematical education is fundamentally a political issue—tries to establish concrete links between broad questions of emancipatory popular education in the Third World and the processes of learning and teaching mathematics. It deals with the interrelations between academic and popular mathematical knowledge in the context of the struggle for land; it is also inserted in the broader educational movement called Ethnomathematics. The expression *ethnomathematics* was coined by the Brazilian Professor Ubiratan D'Ambrosio in the mid-1970s. Since then he has made important theoretical contributions as well as laid down outstanding research guidelines in ethnomathematics.

I use the expression *ethnomathematical approach* to designate

> the investigation of the traditions, practices, and mathematical concepts of a subordinated social group and the pedagogical work which was developed in order for the group to be able to interpret and decode its knowledge; to acquire the knowledge produced by academic mathematicians; and to establish comparisons between its knowledge and academic knowledge, thus being able to analyze the power relations involved in the use of both these kinds of knowledge.

This approach assumes that mathematics is cultural knowledge and that its birth and development are linked to human needs. It implies placing the ethnomathematical approach, from an epistemological viewpoint, in the confluence of mathematics and cultural anthropology.[2] The author would like to introduce into this confluence pedagogical and sociological knowledge as well.

A theoretical analysis of the present fieldwork emphasizes the sociological dimension of the ethnomathematical approach. The object of the research refers to two social practices involved in the productive activities of the inland people where mathematical knowledge is absolutely necessary. The investigation discusses the interrelations between popular and academic knowledge in the context of two specific practices of the group concerning mathematical knowledge: *cubação de terra* (estimating the area of a piece of land) and *cubagem da madeira* (estimating the volume of a tree trunk).

These two rural social practices are not exclusive to the southernmost countryside of Brazil. In the northeast region of Brazil, the researcher Guida Abreu has also looked into them. Her study investigated the mathematics used by sugar cane farmers in activities related to farming.[3] What is relevant, from the author's viewpoint, is that the students who, before coming to DER/FUNDEP, knew the popular methods were those who taught them to the group, during the math classes. They were "teachers" as well as "students." In one of the student's own words: "Before our math class I knew the counts, now I know the mathematics." This sentence can be understood as saying that the student was aware that in the math classes he was able to decode and understand the mathematics involved in his social practices. Furthermore, with the given pedagogical approach, he also gained academic knowledge—"book mathematics," as the group called it. It is important to emphasize that popular knowledge was taken into account in the pedagogical work without any intention of "exalting" it.

Two Social Practices

Here I present the mathematical practices which were the object of the investigation, starting with those referring to land area measurement. They consist of two different methods, which were called by the group "Jorge's Method" and "Adão's Method." The students Jorge and Adão themselves taught the group the specific way of "measuring the land" in their communities. In estimating the volume of a tree trunk, "Roseli's Method" will be shown below; she also taught it herself to the group. Other students were familiar with other methods of figuring out the volume of a tree trunk, but Roseli's was the method most used in the southern Brazilian countryside.

Table 18.1
Jorge's Method of Estimating Area

PEASANT'S WORDS	ACADEMIC'S WORDS
Here is a land with four walls.	This is a convex quadrilateral.
First, we add all the walls.	First, we find the perimeter of this convex quadrilateral.
Second, we divide the sum by four.	Second, we divide the perimeter by four.
Third, we multiply the obtained number by itself.	Third, we find the area of the square whose side was determined after dividing the perimeter by four.
This is the "cubação" of this land.	This is the area of the square obtained from the perimeter of the convex quadrilateral.

Table 18.2
Adão's Method of Estimating Area

PEASANT'S WORDS	ACADEMIC'S WORDS
This is a land with four walls.	This is a convex quadrilateral.
First, we add two of the opposite walls and divide them by two.	First, we find the average between two opposite sides.
Second, we add the other two walls and also divide them by two.	Second, we find the average between the other two opposite sides.
Third, we multiply the first obtained number by the second one.	Third, we find the area of the rectangle formed by the two averages reached before.
This is the "cubação" of this land.	This is the area of the rectangle whose sides were determined by the average of the two pairs of the opposite sides of the convex quadrilateral.

Table 18.3
Roseli's Method of Estimating Volume

PEASANT'S WORDS	ACADEMIC'S WORDS
This is a trunk of a tree.	This is a frustrum of a cone.
First, we select the medium section of the trunk of a tree.	First we transform the frustrum of a cone into a cylinder.
Second, we take a rope. Then, put it around the middle section. Then, we find the rope length and divide it by four.	Second, we find the cylinder base perimeter. Then, we find its fourth part.
Third, we multiply the obtained number by itself.	Third, we find the area of a square whose side was obtained after finding the fourth part of the perimeter of the base of a cylinder.
Fourth, we multiply the obtained number by the length of the stem of a tree.	Fourth, we multiply the square area by the cylinder's height.
This is the "cubagem da madeira."	This is the volume of a quadrangular prism, whose base side was obtained through one fourth of the circumference length. This circumference is, in fact, the circumference of the cylinder base; the cylinder had been previously obtained from the transformation of the frustrum of a cone.

It is not difficult to show, using elementary mathematics, that in the case of a convex quadrilateral area, Jorge's result always exceeds Adão's, which itself overestimates the measure of the "land" obtained by academic calculations. When Jorge and Adão were explaining their methods to the group, the students decodified and understood what the methods meant in terms of academic mathematics. This represented the passage from the first to the second column in the above tables.

Final Comments

The final part of this chapter emphasizes some theoretical aspects of the current research. It is important to note that the work tries to avoid "glorifying" popular knowledge; on the contrary, it aims at understanding how popular mathematical practices—the product of a relation of social inequality—represent a limitation, a disadvantage. It is necessary to understand that the mere perpetuation of this popular mathematical knowledge involved mechanisms which reinforce social subordination. It is clear to the MST that grasping and appropriating academic and technological knowledge is a strategic issue in the concretization of land reform. Without this specific knowledge it would be hard to organize, to administer, to plan, and to commercialize production. Taking into account these arguments, it is important to reaffirm that merely glorifying popular knowledge does not contribute to the process of social change. When a specific subordinate group becomes conscious of the economic, social, and political disadvantages which its scarce knowledge brings about, and tries to learn erudite knowledge, this type of consciousness may contribute to the process of social change.

Pierre Bourdieu[4] gives support to this view, when he says: "If in order to resist I have no better argument than to claim the right to hold to precisely that which makes me dominated, is this resistance? When, on the contrary, dominated people endeavor to lose what makes them "vulgar" and endeavor to appropriate those things in relation to which they seem "vulgar," is this submission?" Bourdieu finishes the paragraph saying: "Resistance may become alienation and submission may become liberation."

The second point is the pedagogical work which is part of what the author has called the "*ethnomathematical approach.*" To start with, it tries to probe into and rescue popular mathematics. This mathematics is not legitimated by the dominant culture and has survived only through a process of oral transmission; it tends to disappear, without any "synthesis-knowledge" occupying its place. Nevertheless this pedagogical work does not simply try to rescue popular knowledge. It also tries to decodify and understand it, giving the students the opportunity to become aware of the limitations of their methods and the reason why these methods, even without being exact, are utilized by subordinate rural groups. Furthermore, the process of *cultural awareness*, in Paulus Gerdes' words,[5] allows the birth of a "synthesis-knowledge," which is constructed taking popular knowledge as its starting point but which, however, transcends it.

Acknowledgments

I would like to acknowledge Professor Tomaz Tadeu da Silva's help and acute critical sense which guided me through the theoretical interpretation of this research. I also wish to thank Dick Tahta and Ubiratan D'Ambrosio for comments made on the first version of this paper.

Notes

1. DER/FUNDEP: Departamento de Educação Rural da Fundação para o Desenvolvimento, Educação e Pesquisa da Região Celeiro.

2. ISGEm Newsletter: *International Study Group on Ethnomathematics Newsletter* 1.1 (1985).

3. ABREU, Guida M. C. P., Disscertação de Mestrado, Universidade Federal de Pernambuco (Brasil: Recife, 1988).

4. Pierre, BOURDIEU, "Os usos do povo," in *Coisas Ditas* São Paulo: Brasiliense, 1987, 181–187. Translated from a lecture given by the author in Lausanne, Switzerland, 1982. (Author's own translation into English)

5. Paulus, GERDES, *Ethnomatematica: cultura, matematica, educacao*. Maputo, Mozambique: Instituto Superior Pedagógico, 1991, 62–79.

Afterword

Gloria F. Gilmer

Editors' note: Gloria F. Gilmer, mathematician, mathematics educator, and consultant, is the founding president of the International Study Group on Ethnomathematics. Along with others, she has been instrumental in gaining international recognition for the group and, thereby, the scientific field of ethnomathematics.

Introduction

From 1985 to 1996, I have had the responsibility of presiding over the International Study Group on Ethnomathematics (ISGEm), whose goals are twofold: (1) to promote the teaching and learning of mathematics in cultural contexts and (2) to encourage further research in ethnomathematics. In this afterword, I have the challenge of exploring the tasks ahead for this organization. Our vice president, Ubiratan D'Ambrosio, has done much to shape the field of ethnomathematics. Our newsletter editor, Rick Scott, together with our country distributors, has done much to disseminate information from our membership to a broader readership. I have worked to build and retain relationships with other organizations in the mathematical community, especially in the United States, by arranging for our inclusion on their programs at national meetings and, each year, planning an ISGEm business and program meeting. ISGEm has four special interest groups (SIGs) through which research activities are encouraged among our members: (1) Conceptual and Theoretical Perspectives, (2) Research in Culturally Diverse Environments, (3) Out-of-School Applications, and (4) Curriculum Development and Classroom Applications.

Background

My knowledge of ethnomathematics and subsequent presidency of the ISGEm followed closely on the heels of my tenure on the faculty of Atlanta University (now Clark-Atlanta University), my at-large membership on the Board of Governors of the Mathematical Association of America (MAA), and my membership on the Mathematics team of the National Institute of Education (NIE, formerly the research arm of the United States Department of Education).

At Clark-Atlanta University, an historically African American university in the state of Georgia, I had the opportunity to return to my roots, recover some of what I had lost, and guide research and practices of future African American mathematicians and teachers of mathematics. During this period, I also succeeded Etta Falconer in directing MAA's Blacks and Mathematics project (BAM)—a high school lectureship program designed to encourage Black students to learn more about uses of and careers in mathematics. Through BAM coordinators such as Della Bell, Henry Gore, Jim Donaldson, and Jack Alexander, new ideas, new faces, and new income came to MAA. As a member of MAA's Board of Governors, I had the opportunity to visit many of our nation's classrooms in the United States and experience the culture of the mathematical community by working with mathematicians in leadership positions like Dorothy Bernstein, Henry Alder, Gerald Alexanderson, Richard Anderson, Don Bushaw, Bill Chinn, Ed Dubinsky, R. Craighton Buck, Debra Haimo, Donald Hill, Richard Griego, John Kennelly, Pat Kenschaft, Annelli Lax, Joanne Leitzel, Lee Lorch, Henry Pollak, Gerald Porter, G. Baley Price, David Roselle, Ken Ross, Alan Schoenfeld, Martha Siegel, Donald Small, David Smith, Lynn Arthur Steen, Marcia Sward, Alvin White Willcox, and many others.

From 1981 to 1984, while at the National Institute of Education, I worked with project officers from the National Science Foundation, to manage the first government-funded studies on the use of computers in the teaching and learning of mathematics and the Second International Study of Mathematics Achievement, which involved twenty-four countries. In addition, I gathered information for the now-famous report of the National Commission on Excellence in Education, released in 1983–*A Nation at Risk: The Imperative for Educational Re-*

form. While at NIE, I also joined the 1983 American Delegation of approximately fifty mathematicians to the People's Republic of China and prepared an extensive report for NIE on mathematics education in China that could inform practices in the United States (Gilmer, 1983). At NIE, I viewed up close and in a global context some problems and pressing issues in mathematics education today. For example, in the U. S. A., mathematics teachers were so preoccupied with lecturing that mathematical ideas and techniques that students acquire without without formal schooling are virtually unknown to teachers as are the processes by which such ideas and techniques are transmitted from one generation to the next. Realizing this, I spent several years studying, writing, and speaking about the mathematics learning of central-city residents in Milwaukee, Wisconsin and developing awareness of issues in mathematics education through such civic organizations as the National Council of Negro Women, the National Urban League, and the National Urban Coalition, all of whom have programs in mathematics and science education.

In 1984 at the Fifth International Congress on Mathematics Education (ICME-5) in Adelaide, Australia, I first learned of ethnomathematics from D'Ambrosio's plenary address and wrote of my perceptions of this concept of mathematics (Gilmer, 1985). Before then, like Bertrand Russell, I too had experienced the cold and austere beauty of mathematics. Early in my career as a student and, later, colleague of Clarence F. Stephens,[1] I learned to value the rigor, precision, and resilience of mathematicians and to appreciate social and humanitarian values implicit in this scholarly community such as respect, solidarity, and cooperation (Gilmer, 1990). Then as now, my modes of understanding, learning styles, intuition, emotions, and use of mathematics are all closely bound to my cultural heritage both as an African American Christian and as an active member of the mathematical community. Now, however, I am more aware of the immense potential in the development and acquisition of mathematical knowledge by the inclusion of the concept of ethnomathematics. Ethnomathematics as a field of study connects mathematical concepts, their acquisition and application through cultural origins. In this way, ethnomathematics paves the way for reform in mathematics education and new horizons in mathematical research.

History of the Organization

In 1985 at the Annual Meeting in San Antonio, Texas, of the National Council of Teachers of Mathematics (NCTM), Ubiratan D'Ambrosio, Rick Scott, Gilbert Cuevas, and I formed the International Study Group on Ethnomathematics. In March 1986, Rick Scott published the first ISGEm newsletter. The first business meeting was held in April 1986 in Washington, DC, as part of the NCTM Annual Meeting. In September 1987, Claudia Zaslavsky became our secretary and Elisa Bonilla joined the board to translate the newsletter into Spanish. Each year we were able to hold Research Presessions in connection with the NCTM annual meetings. In 1990, the NCTM Directors approved ISGEm for affiliation.

Also in 1985, Julia Robinson, then president of the American Mathematical Society (AMS), appointed me chair of what became the Joint MAA, AMS, AAAS (American Association for the Advancement of Science) Committee on Opportunities in Mathematics for Underrepresented Minorities (COMUM). I used this committee to connect the minority community to the broader professional community by creating "Friends of COMUM." ISGEm members also became "Friends of COMUM." They attended meetings of the COMUM and shared ideas about ethnomathematics with mathematicians in NAM (National Association of Mathematicians), AMS, and MAA. ISGEm members were also invited to participate in other activities of the COMUM such as panel presentations and contributed paper sessions at annual meetings. When COMUM initiated the effort which resulted in the Mathematical Sciences Education Board's project "Making Mathematics Work for Minorities," many ISGEm members participated. In this way, COMUM provided a voice for ideas of the ISGEm at MAA, AMS, and NAM meetings.

Beside activities and recognition in the United States, ISGEm has been active internationally. We extended our newsletter distribution to twenty countries. In 1988, we held our first International meeting at ICME-6 in Budapest, Hungary. Our second international meeting was held in 1992 at ICME-7 in Quebec and our third meeting was held in 1996 at ICME-8 in Seville, Spain.

The Tasks Ahead

Into the next century, the ISGEm and generally researchers in the field of ethnomathematics will be faced with the continuing tasks of mobilizing human energies to generate and disseminate more ethnomathematical knowledge and to use ethnomathematical approaches for promoting human development. For ISGEm, these tasks may be achieved through SIGs, which must be strengthened. To generate knowledge, we must design strategies for retaining our openness to new ideas. One such strategy is to continue to develop meaningful relationships with the MAA, AMS, and NCTM and with more diverse groups, such as the History and Pedagogy of Mathematics, the Criticalmathematics Educators Group, the Humanistic Mathematics Network, the American Education Research Association, the American Association for the Advancement of Science, the American Indian Science and Engineering Society, the National Technical Association, the National Association of Mathematicians, the Bannaker Association, the Interamerican Conference on Mathematics Education, and the African Mathematical Union. Along with professional groups, the ISGEm must target schools and universities to include ethnomathematical thinking in their course offerings, staff development, and research programs. In particular, ISGEm must promote the inclusion of individuals from social, racial, class, or gender groups, who are presently on the margins of professional mathematical communities, into academic mathematics and mathematics-related careers and involve them as serious researchers in mathematics education studies. In this way, instinctive kinds of mathematical knowledge that exists among adults and children in their groups will become widely known. In addition, more studies in the United States and elsewhere are needed to reveal the nature of informal and formal processes used to transmit mathematical ideas from one generation to another within groups. Strategies needed to accomplish these tasks should be part of a long range strategic plan for the development of ISGEm. Such a plan should include the following components: (1) building meaningful relationships that encourage members to know each other and their work; (2) developing leadership and sharing responsibilities among the membership; and (3) developing and implementing strategies for acting effectively in the public arena to further ISGEm's mathematics education agenda.

Changing the Paradigm

The need for the above strategies is well-documented by recent events at the University of Rochester. In November 1995, members of the mathematics department at the University of Rochester learned that the university administration intended to implement a Renaissance Plan which would eliminate their department's doctoral program, reduce the size of the department by half, and employ part-time and full-time members of other departments to teach entry level mathematics courses. These cuts were part of a general downsizing of the university's operations. While the administration also intended to eliminate graduate programs in chemical engineering, comparative literature, and linguistics, the faculties in these departments had established interdepartmental doctoral programs in which they could continue to participate. Twenty-seven departments claimed the doctoral program of the mathematics department was not essential to their own programs. In addition, the administration noted that there were few linkages between the mathematics faculty and the science and engineering faculties. Fearing a possible ripple effect of such action at universities across the United States, the mathematics community rallied to support the Rochester mathematics faculty, including the creation of a special AMS committee. This led to unprecedented conversations between the mathematics faculty, their colleagues in other departments, and the administration. Subsequently, with the aid of the Department of Physics and Astronomy, the decision to close the Ph.D. programs in mathematics was reversed. As part of the rescue package, the department also agreed to review courses it offers undergraduates who are not majors in mathematics and to forge linkages with the research specialties of faculty in other departments. A trustee gave the university several million dollars to set up an award system for good undergraduate teaching at the university (*Focus*, February 1996, p. 1, and June 1996, p. 1).

The experience of the Mathematics Department at the University of Rochester illustrates an important principle from the science of chaos which is that "relationships are fundamental." Students of chaos are discovering that nothing can be isolated and studied as an individual. Everything that exists is free and yet simultaneously bound by its relationship to everything else that exists. This interconnectedness is called "sensitive dependence on initial conditions." The interconnectedness of all things is a way of thinking about mathematics that parallels the thinking of ethnomathematicians. Thus, an important task ahead for the ISGEm's SIG on "Conceptual and Theoretical Foundations" is to continue and expand its use of an interconnectedness paradigm in developing the foundations of this discipline.

A 1994 AMS policy statement (American Mathematical Society, 1994) characterizes mathematics as the study of measurement, forms, patterns, and change which evolved from efforts to describe and understand the natural world. Further, mathematics developed a rich and sophisticated culture that feeds back into the natural sciences and technology. Now mathematics reaches beyond the physical sciences and engineering into medicine, business, the life sciences, and the social sciences. Therefore, when D'Ambrosio characterizes ethnomathematics as a comparative study of techniques used by diverse cultures to explain and cope with reality in different natural and cultural environments, the Eurocentric notion of mathematics is subsumed and respected. Importantly, however, ethnomathematicians respect the natural talents, knowledge, skills, techniques, tools, and power of persons in every identifiable cultural group in such a way that they too feel challenged to advance the mathematical knowledge of the group. This leads to a deepened view of other people's understanding of what mathematics is, what it does, and who creates it. Clearly, the scope and breath of research represented in this volume attests to important and critical efforts toward achieving a theoretical paradigm for ethnomathematics that expands and connects the foundations of the field.

Notes

1. It is important to note that, for eighteen years, until the summer of 1987, Clarence F. Stephens, an African American mathematician, chaired the mathematics department of the liberal arts college, The State University of New York at Potsdam. In 1985, The State University of New York at Potsdam graduated 184 undergraduate mathematics majors. This was the third largest number in the nation behind only the University of California at Berkeley and UCLA. Over the three year period from 1985 to 1987, an average of twenty-four percent of the entire Potsdam college graduating class were mathematics majors whereas this is only true of one percent of baccalaureate degrees granted in North America. In 1981, the Potsdam State mathematics department was cited by a panel of the Committee on the Undergraduate Program in Mathematics of MAA as a successful mathematics program. In addition, it was noted that annually several dozen Potsdam mathematics graduates were hired by leading technological companies such as Bell Labs, IBM, and General Dynamics. After a visit to the Potsdam campus, John Poland of Ottawa's Carleton University described the undergraduate mathematics program at Potsdam with its extraordinary record of success as "a modern fairy tale" (Poland, 1987). Stephens had similar but unheralded accomplishments decades before at Morgan State University, a historically black institution which is part of the Maryland State system of higher education.

References

American Mathematical Society. (1994). *AMS National Policy Statement 94–95.* American Mathematical Society, Providence, RI.

Gilmer, G. F. (1983). *A Report on the 1983 Mathematics Delegation to the People's Republic of China,* NIE Unpublished Report.

———. (1985). Sociocultural Influences on Learning. In W. Page (ed.), *American Perspectives on the Fifth International Congress on Mathematical Education, MAA Notes Number 5,* The Mathematical Association of America, pp. 95–96.

———. (1990). An interview with Clarence Stephens, *UME Trends,* p. 1.

Poland, J. (1987). A modern fairy tale? *American Mathematical Monthly,* 94 (3): 291–295.

Devlin, K. (1996). University of Rochester Eliminates Ph.D. Program, *FOCUS,* 16 (1), p. 1.

———. Rochester Reinstates Ph.D. Program, *FOCUS,* Vol. 16 (3), June 1996, p. 1.

Contributors

S. E. Anderson: I was born in New York City and live in Harlem. I am a veteran activist/educator and have been involved in the Black Liberation Movement on many levels. As a young activist, I was a member of the Student Nonviolent Coordinating Committee (SNCC); since 1964, I have been active in the African Liberation Movement; in 1966, I helped found the Black Panther Party in Harlem; and, in 1968, I participated in the historic Black student/community struggle against Columbia University's encroachment into Harlem. Ironically, from 1986–1987, I was a Columbia University Revson Fellow. In 1969, I created a department at Sarah Lawrence College that included mathematics and the natural sciences as part of a Black Studies curriculum. I have been a mathematics professor at Rutgers University-Newark and have taught mathematics, science, and Black Studies at Queens College of the City University of New York. I am also a Senior Editor of *NOBO: Journal of African American Dialogue,* and a founding member of the Network of Black Organizers. In 1976, I co-edited *The Third World Confronts Science and Technology* (Livros Horizonte, Portugal) along with French physicist Maurice Bazan. Most recently, I wrote *The Black Holocaust for Beginners* (New York: Writers and Readers, 1995) and co-edited *In Defense of Mumia* (New York: Writers and Readers, 1996). Currently, I am working on two books: *Race For Beginners* and *Slavery For Beginners* and am collaborating with Sri Lankan scientist Susantha Goonitilake on a *World Science For Beginners* book.

Maria Ascher and Robert Ascher: I (Marcia) am a recently retired Professor of Mathematics at Ithaca College and I (Robert) am a Professor of Anthropology at Cornell University. Our interest in the mathematical ideas of traditional peoples began when we collaborated in a study of an Inca artifact that was said to be somehow numerical. The study was inspired simply by the desire to work together on a project. The study became more interesting and extensive than we anticipated; in 1981 it resulted in our book *Code of the Quipu: A Study in Media, Mathematics and Culture.* The project required rethinking many previously learned ideas about mathematics and about the relationship of mathematics and culture. I (Marcia) also became increasingly conscious of the omission or misrepresentation of traditional peoples in the mathematics literature. My central research focus then became ethnomathematics. Because of his contemporary anthropological knowledge, I requested that Robert join me in the preliminary exposition of ethnomathematics put forth in the paper reprinted here. Our mailing address is 524 Highland Road, Ithaca, New York 14850.

Martin Bernal: I was born in London in 1937 and educated at Kings College Cambridge, Peking University, the University of California, and Harvard University. I earned a doctorate in Chinese studies from Cambridge in 1966 and was a Fellow of Kings College from 1964 to 1972. In 1972, I became a professor of government at Cornell University and an adjunct professor of Near Eastern Studies in 1986. My major publications are *Black Athena: The Afroasiatic Roots of Classical Civilization,* Volumes I (1987) and II (1991), Rutgers University Press and *Cadmean Letters,* 1990, Eisenbrauns. Both of these works have been widely reviewed and criticized and two films have been made about the academic and political controversies around my work.

Marcelo C. Borba: I was born in Rio de Janeiro and graduated from the Federal University of Rio de Janeiro in 1983 with a major in mathematics. In 1984, I entered a master's program in mathematics education at the State University of Rio Claro (UNESP-Rio Claro). In this program, I attempted to combine my political involvement with academic work and education. Consequently, I developed a participatory research in a *favela* ("slum"), with a multi-disciplinary team that included some of Paulo Freire's students. This was the first research project based on D'Ambrosio's ethnomathematics research program. The chapter included in this book represents my theoretical reflections on this experience.

In 1992, I earned a doctorate in mathematics education from Cornell University based on research concerning computers in mathematics education. While in the United States, through my involvement with colleagues in the International Study Group on Ethnomathematics and the Criticalmathematics Educators Group, I further developed an international perspective on ethnomathematics. Since 1993, I have been an assistant professor of mathematics education at UNESP-Rio Claro, where my research centers on the relationship between different media (orality, computers, and so on) and ethnomathematics.

Ubiratan D'Ambrosio: I was born in São Paulo, Brazil, on December 8, 1932 and, in 1963, earned a doctorate in mathematics from the Universidade de São Paulo. I am Professor Emeritus at the Universidade Estadual de Campinas (UNICAMP), Brazil, where I was professor of mathematics until I retired in 1993 and where I had been the Director of the Institute of Mathematics, Statistics and Computer Science from 1972 to 1980 and Pro-Rector for University Development from 1982 to 1990. Earlier, from 1970 to 1980, I was also a visiting professor in the UNESCO/UNDP program at the Centre Pédagogique Supérieur de Bamako, a graduate program in the Republic of Mali. From 1980–1982, I was head of the unit for the Improvement of Educational Systems of the Organization of American States in Washington, D. C. From 1990 to 1994, I was coordinator of the Research Institutes of the Secretary of Health for the State of São Paulo. In the United States, I have served as the graduate chairman of mathematics at the State University of New York at Buffalo, on the mathematics faculty at the University of Rhode Island, Brown University, the University of Illinois in Chicago, and as Ida Beam Distinguished Visiting Professor at the University of Iowa. In Switzerland, I was visiting lecturer at the Biozenturm of the University of Basel. From 1982–1986, I was vice president of the International Commission for Mathematics Education and, from 1979 to 1987, president of the Interamerican Committee of Mathematics Education. I am a Fellow of the American Association for the Advancement of Science and was cited "For imaginative and effective leadership in Latin American mathematics education and in efforts toward international cooperation." I am a founding member of several institutions, among them the Sociedad Latinoamericana de Historia da las Ciencias y la Tecnologia, the Academia de Ciências do Estado de São Paulo, the Centre International de Recherches Transdisciplinaires (CIRET), and the International Study Group on Ethnomathematics (ISGEm).

Munir Fasheh: I don't know which started earlier: my rebellion against the fragmentation of Palestinians or my rebellion against the fragmentation of knowledge. What I increasingly felt over the years, however, was that the two are strongly linked and that they were imposed from outside.

Fragmentation became the "virus" I started seeing in almost all aspects of my life, shaping my thinking, my perceptions, and my work. My chapter in this book resulted from my reflection on my work with math education during the 1970s. It was one of my first attempts to articulate the interconnectedness I was increasingly seeing in life and, in particular, in my teaching of math.

Prior to 1971, I treated math, and taught it for eight years, as if it were the "queen of the sciences." However, in spite of the image of its aloofness and the claims of its universality and neutrality, in my little home and as a little boy I was intuitively and secretly relating math to parts of my experience as well as to other subjects in school. I say "secretly" because I didn't feel free or confident, at the time, to express openly my feelings and what was going on in my mind. It was like committing "adultery" against math!

When I started mixing math with the messiness of life, however, I started seeing new forms and roles of math which I couldn't see through the "pure" math perspective. I could see, for example, much clearer my illiterate mother's math, the role of math in creating dangerous trends and monopolies around the world, and how math can be helpful in human liberation.

Rather than an accumulation of facts and skills, knowledge started, more and more, to mean to me a way of seeing the "wholeness" in life; sort of a "world map" that has to be constantly redrawn in my head, in light of whatever new experience or piece of information I acquired. It was also becoming clearer to me the role of math in this "redrawing" process. In order to play this role, however, math has to be studied and taught within context: the political, social, and cultural context as well as the context of one's personal experiences.

The personal experience and expression are crucial in making sense of the world and, in particular, of seeing meaning in math. They are, however, only a beginning; it is also to seek collective meanings and to construct common "maps" which are more relevant and accurate than the ones reached at individually.

In terms of my work, what I mentioned above led me "naturally" to move from teaching math in a "pure and refined" way at both the school and college levels (both in Palestine and in the US in the 1960s) to encourage both students and teachers to connect their study or teaching of math to the context in which they lived during the 1970s. Later, during the 1980s, my interest and work were related to wider issues in education and their relationship to various aspects in life.

And, since 1989, through my work at the Tamer Institute for Community Education, I have been moving "back to square one" and stressing the process of learning as a "natural" process that (just like digesting food or growing plants) cannot be programmed. The best we can do is build environments which enhance and help this natural process of learning; for instance, help students express their experiences, work in teams, have access to information, build common maps of the world which they inhabit, deal with personal problems, and work within a vision that is beyond the interests of any specific individual and any specific moment. This struggle with teaching of math, with education and with learning was partially translated into five books (in Arabic) and into many articles (both in English and in Arabic).

Marilyn Frankenstein: I first became aware of racism, when as a White ten-year old, working-class, atheist, Brooklyn girl, I overheard a neighbor refer to my family by saying "Leave it to the kikes to bring the niggers around." At that time, my mother, who is neither a leftist nor an activist, explained to me the insanity of racism and the logic of racism under capitalism. That began my climb up the slippery slope of anti-capitalist political awareness and commitment.

In a socialist world, I would have been an artist. As it is, since 1978, I have taught math to adults at the College of Public and Community Service (CPCS), University of Massachusetts/Boston (Center for Applied Language and Mathematics, Boston, MA 02125). For the previous decade, I taught math in various public schools in New York City and New Jersey. I have written a criticalmathematics textbook for adults, *Relearning Mathematics* (London: Free Association Books, 1989) and numerous articles about criticalmathematics curriculum and teaching. I am a member of the *Radical Teacher* editorial collective, and co-founder with Arthur Powell and John Volmink of the international Criticalmathematics Educator Group (CmEG).

The goals of my mathematics teaching are to both demystify the structure of mathematics and to use mathematics to demystify the institutional structures of our society. Underlying all my teaching is an attention to questions concerning the politics of knowledge, such as "Who gets to define what counts as legitimate knowledge?"

And I love glitter, (stage) magic, dancing, and listening to and playing American classical music, usually referred to as jazz, with my friends.

Paulus Gerdes: I am a Mozambican mathematician who was born in the Netherlands. As a boy I came to like the beauty of mathematics but to dislike the way in which my best friends at high school and at university were 'eliminated' by 'mathematics.' When I became aware in the early seventies of the existence of the Army Mathematics Research Centers in the U. S. A. and their supporting role in the United States' war against Vietnam and saw that some of my own professors were attending mathematical conferences financed by NATO, I withdrew almost completely from studying mathematics. Instead, I turned to the study of cultural anthropology to understand better the world we are living in, but was not finding the answers to my questions: What is life all about?, What is mathematics all about?, Mathematics by whom, for whom, and what for?

In the mid-seventies, while struggling with these questions, I came to contribute to newly independent Mozambique because of its shortage of mathematics teachers. What mathematics and mathematical curricula did the country need? Colonial ideology and education had presented mathematics as an exclusively white man's product, difficult to be mastered by the oppressed. Inspired both by Freire's *Pedagogy of the Oppressed* and by the reflections of Eduardo Mondlane, a cultural anthropologist and first President of Mozambique's liberation movement Frelimo, and Samora Machel, first President of independent Mozambique, on the nature of colonial education and of liberatory education (in particular, the importance of cultural awareness and cultural-social self-confidence), I and my colleagues were stimulated to think about mathematical ideas in Mozambican culture(s) and how to build mathematics curricula on the base of them. Based on these efforts, in 1978, we initiated the research project "Empirical knowledge of the Mozambican population and the teaching of mathematics," a forerunner of our current project: "Ethnomathematics/ethnoscience in Mozambique." During the initial project, Ubiratan D'Ambrosio and Beatrice Lumpkin visited us. In 1980, at the Fourth International Congress on Mathematics Education, at the University of California in Berkeley, California, we met Arthur B. Powell and Claudia Zaslavsky. Also, since our initial project, we have shared our reflections in Mozambique with colleagues throughout Africa and with colleagues in other parts of the world. This has been extremely stimulating, and we can now see that the ethnomathematical movement is becoming ever stronger.

Gloria Gilmer: I am president of Math-Tech, Inc., an educational research and development corporation located in Milwaukee, Wisconsin. Math-Tech, Inc. specializes in translating research results into effective programs in mathematics teaching and learning—especially for minorities and females. I am a former research associate with the Office of Educational Research and Improvement of the United States Department of Education and a former mathematician in exterior ballistics with the United States Army at Aberdeen Proving Grounds, Maryland. I have extensive experience in mathematics teaching and learning at the university, college, school, and pre-school levels. I have been on the mathematics faculties of six major historically Black institutions of higher education in the United States—Hampton University, Virginia State University, Morgan State University, Atlanta University, Morehouse College, and Coppin State College in the Maryland State system. Also, I have lectured in mathematics on three campuses of The University of Wisconsin and taught at Milwaukee Area Technical College and in public schools in Milwaukee, WI and Lockport, NY. I directed and am an author of the Addison-Wesley project "Building Bridges to Mathematics: Cultural Connections." I have extensive involvement in professional organizations both nationally and internationally. I formerly chaired the Committee on Opportunities in Mathematics for Underrepresented Minorities (COMUM)—a joint committee of The American Mathematical Society, The Mathematical Association of America, and The American Association for the Advancement of Science which initiated the project "Making Mathematics Work for Minorities" with the Mathematical Sciences Education Board of the National Research Council. I am also a former national director of the Blacks and Mathematics project of the Mathematical Association of America. In addition, for eleven years, I was president of The International Study Group on Ethnomathematics. I hold a Bachelor of Science degree in mathematics from Morgan State University, a Master of Arts degree in mathematics from the University of Pennsylvania, and a Ph.D. from Marquette University in Curriculum and Instruction.

Herbert P. Ginsburg: I am a Professor of Psychology and Mathematics Education at Teachers College, Columbia University. For the past thirty years, I have been conducting research on the development of children's mathematical thinking, focusing on such questions as the nature of the mathematical thinking of individuals in other cultures, the difficulties children experience with school mathematics, and features of appropriate methods, particularly the clinical interview method, useful for assessing children's knowledge. My interest in these questions developed from experience with poor children in less than hospitable American schools. My goal is to achieve a deeper understanding of poor children's intellectual development and education and to apply this knowledge to the reform of American education.

Mary Harris: I had a war-torn childhood that took me right round the world by the time I was nine-years old, making me a genuine citizen of the global village. My education in New Zealand, England, and Nigeria included studying chemistry at the university, after which I became interested in the business of learning mathematics while working with one of my children who has learning difficulties. Later, while analyzing research data on the mathematics people do at work, I noted that people often deny that they do any mathematics when asked specifically about it, but reveal that they do the very mathematics they deny, when asked about some other workplace skill. The pursuit of what 'mathematics' means in these circumstances led me into ethnomathematics and the question of how and why workplace mathematics came to be ignored by formal, European-founded systems of mathematics education.

I am currently a Visiting Fellow at the Department of Mathematical Sciences, University of London Institute of Education, 20 Bedford Way, London WC1H OAL. My email address is teuemha@ioe.ac.uk.

George Gheverghese Joseph: I was born in Kerala, southern India, and lived in India for nine years. My family then moved to Mombassa, Kenya, where I started my formal schooling. I later studied at the University of Leicester and then worked for six years as an education officer in Kenya before returning to England to do my post-graduate work at Manchester. I am currently a reader in economic and social statistics at the University of Manchester in England. My teaching and research have ranged over a broad spectrum of subjects in applied mathematics and statistics, including multivariate analysis, mathematical programming, demography, and econometrics. However, in recent years, my research has been mainly in the social and historical aspects of mathematics with particular emphasis on the non-European contributions to the subject. I have travelled widely, holding visiting appointments in Tanzania and Papua New Guinea and a Royal Society Visiting Fellowship in India. I have also lectured in several universities in the United States, Canada, South Africa, and Mexico. My publications include *Women at Work* (Philip Allan, Oxford, 1983), *The Crest of the Peacock: Non-European Roots of Mathematics*

(Penguin, London, 1992), and *Multicultural Mathematics* (Oxford University, Oxford, 1993).

Gelsa Knijnik: I was born and live in southern Brazil, in the region where, fifteen years ago, burgeoned a most important social movement: the Movement of Landless Rural Workers (MST). Being a woman from the city, who began her professional career as a teacher of primary school, one of the fundamental marks of my life's trajectory concerns the work I realized, starting in 1991, in mathematics education together with MST. I completed my master's in mathematics, specializing in algebra, and wrote my doctoral dissertation in education, theorizing what I became aware of working with the rural women and men of MST.

Before my work with MST, starting around 1986, I had already been involved with ethnomathematics, conducting studies in poor communities located at the margins of the city in which I live. Effectively, however, my work with MST has provided me with the possibility of establishing a closer connection between my academic scholarship and my political involvement, seeking to contribute to the diminution of social injustices.

Beatrice Lumpkin: I am a retired associate professor of mathematics at Malcolm X College of the Chicago Community Colleges. I became a teacher after twenty years of work in industrial production jobs and technical writing (electronics). With a degree in history and a firm belief in the need for socialism, teaching history would have been a natural choice. But history teachers with my background of labor activism were being fired. So I decided to teach in my other field of interest, mathematics and physics.

At Malcolm X College in the late 1960s, because students were demanding "Black Studies" courses, I began to study the history of mathematics in Africa. Although I had learned that much of African history had been distorted or hidden from reading W. E. B. DuBois' *The World and Africa*, I was still surprised to learn how much of the mathematics taught today had its start in Africa. Although I was teaching what I had thought was a politically neutral subject, my research forced me, once again, to go against the system. I began to teach what I had newly learned, that mathematics began in Africa, not Greece. The great contributions of the Hellenistic mathematicians were based on the foundation of thousands of years of mathematics in Africa and Asia.

I am now actively developing materials for multicultural mathematics and science education. These materials include *Senefer, Young Genius in Old Egypt,* and *Senefer and Hatshepsut* for children, *Multicultural Science and Math Connections* (classroom activities), *African Cultural Materials for Elementary Mathematics* for the Illinois State Board of Education, and a "Baseline Essay" for the Portland, Oregon Public Schools on "African and African American Contributions to Mathematics." My mailing address is 7123 S. Crandon, Chicago, IL 60649; Telephone: 773-684-4553.

Brian Martin: I received my doctorate in theoretical physics from the University of Sydney and then worked for ten years as an applied mathematician at the Australian National University. I now work in the Department of Science and Technology Studies, University of Wollongong, NSW 2522, Australia; b.martin@uow.edu.au. I was first involved with radical science groups in the early 1970s, and this led me to the critique of mathematics. I have published on scientific controversies, suppression of dissent, nonviolent defense, critique of expertise, social movement strategies, information in a free society, and other topics.

Rik Pinxten: In 1975, I earned a Doctor in Philosophy and, in 1980, a Special Doctor in Philosophy from the University of Gent in Belgium. Since 1980, I have been a professor of anthropology and religious studies at the University of Gent. Before then, I was researcher at the Belgian National Science Foundation. My books include *Universalism versus Relativism* (Mouton, The Hague, 1972), *Atheistic Religiosity?* (Story, Gent, 1977), *Anthropology of Space: Navajo Natural Philosophy and Semantics* (with I. Van Doren and F. Harvey; Philadelphia, University of Pennsylvania, 1983), *Navajo Geometry* (Gent, KKI, 1987), *Evolutionary Epistemology* (with W. Callebaut; Reidel, Dordrecht, 1987), and *Culturen sterven langzaam* (*Cultures Die Hard,* Antwerp, 1994). I am the editor-in-chief of the journal *Cultural Dynamics* (Sage Publications).

I am a devoted democrat who is critically interested in cultural and religious roots of people. The power involved in using and producing knowledge, be it with native peoples or with scientists (and mathematics teachers) has intrigued me throughout my career. Power can be and often is a noble thing, I claim, but is more and more turned into domination and violence in our culture. I feel driven to study this as closely as possible and to mend or influence the impact of input on others (cultural groups or social entities) where possible.

Arthur B. Powell: Since 1981, I have been a faculty member of the Academic Foundations Department on the Newark Campus of Rutgers University, teaching courses in developmental mathematics and methods of teaching mathematics. My political activism, research, and scholarly interests are in mathematics education, particularly concerning epistemological and philosophical questions of mathematics learning as well as on sociological and political implications of mathematics education. Besides work in ethnomathematics, I have been developing and exploring the efficacy of a classroom-based, participatory research model. Through my teaching of mathematics and research in mathematics education, I labor so that underrepresented racial, ethnic, and gender groups can restructure intellectual, personal, psychological, economic, cultural, and political power relations to improve their self-esteem and to build more just communities. Over the years, in pursuit of these goals, aside from working in various communities in the United States, I have lectured and taught in China, Brazil, South Africa, and Mozambique.

Dirk Jan Struik: I am an emeritus professor of mathematics at the Massachusetts Institute of Technology. As an historian of this field, I have long been interested in the origin of mathematical concepts. I have followed with great interest the work of Zaslavsky, Gerdes, and others on ethnomathematics, seeing in it not only a fascinating academic subject, but also an aspect of the struggle against the damage colonialism has done to the culture of indigenous peoples. It also has turned out to be one of the ways to improve mathematical education in our own "Western" culture.

Valerie Walkerdine: I am a professor in the psychology of communication in Goldsmiths College, University of London, SE14 6NW, England. I have researched aspects of children's cognitive development and mathematics, and gender and mathematics for the past twenty years. My books on the subject include *The Mastery of Reason* (Routledge, 1988/90) and *Counting Girls Out* (Virago, 1989, 2nd edn. Falmer, 1996). As a result of growing up working class, I have always had keen interest in the relationship between exploitation, oppression, and psychological and cultural aspects of subjectivity.

Claudia Zaslavsky: I have taught mathematics at many levels, from middle grades to graduate courses for teachers. My years of experience in the Greenburgh Central Seven school district in New York State, where busing for integration was established in 1951, led to an investigation into the evolution of mathematical ideas in Africa, which lead to my book *Africa counts: Number and pattern in African culture,* first published in 1973 and now a classic in paperback. Subsequently, my field of interest extended to other cultures and to the development of curriculum materials that would bring a multicultural perspective into the mathematics classroom at all grade levels. Such a perspective can enrich the learning of all students, whatever their gender, ethnic, and racial heritage, or socioeconomic status. Currently, I work as a mathematics education consultant, author, and curriculum developer and may be contacted at 45 Fairview Avenue, #13-I, New York, NY 10040, 212/569-4115.

Index